Cultural Sustainability

If the political and social benchmarks of sustainability and sustainable development are to be met, ignoring the role of the humanities and social, cultural and ethical values is highly problematic. People's world views, beliefs and principles have an immediate impact on how they act and should be studied as cultural dimensions of sustainability.

Collating contributions from internationally renowned theoreticians of culture and leading researchers working in the humanities and social sciences, this volume presents an in-depth, interdisciplinary discussion of the concept of cultural sustainability and the public visibility of such research. Beginning with a discussion of the concept of cultural sustainability, it goes on to explore its interaction with philosophy, theology, sociology, economics, arts and literature. In doing so, the book develops a much needed concept of 'culture' that can be adapted to various disciplines and applied to research on sustainability.

Addressing an important gap in sustainability research, this book will be of great interest to academics and students of sustainability and sustainable development, as well as those studying sustainability within the humanities and social sciences, such as cultural studies, ethics, theology, sociology, literature and history.

Torsten Meireis is Professor of Theology, Ethics and Hermeneutics at Humboldt University, Berlin, Germany.

Gabriele Rippl is Professor and Chair of Literatures in English at the University of Bern, Switzerland.

Routledge Environmental Humanities

Series editors: Iain McCalman and Libby Robin

The *Routledge Environmental Humanities* series is an original and inspiring venture recognising that today's world agricultural and water crises, ocean pollution and resource depletion, global warming from greenhouse gases, urban sprawl, overpopulation, food insecurity and environmental justice are all *crises of culture*.

The reality of understanding and finding adaptive solutions to our present and future environmental challenges has shifted the epicentre of environmental studies away from an exclusively scientific and technological framework to one that depends on the human-focused disciplines and ideas of the humanities and allied social sciences.

We thus welcome book proposals from all humanities and social sciences disciplines for an inclusive and interdisciplinary series. We favour manuscripts aimed at an international readership and written in a lively and accessible style. The readership comprises scholars and students from the humanities and social sciences and thoughtful readers concerned about the human dimensions of environmental change.

Cultural Sustainability

Perspectives from the Humanities and Social Sciences

Edited by Torsten Meireis and Gabriele Rippl

Routledge
Taylor & Francis Group
LONDON AND NEW YORK

earthscan
from Routledge

First published 2019
by Routledge

2 Park Square, Milton Park, Abingdon, Oxfordshire OX14 4RN
52 Vanderbilt Avenue, New York, NY 10017

Routledge is an imprint of the Taylor & Francis Group, an informa business

First issued in paperback 2020

British Library Cataloguing-in-Publication Data
A catalogue record for this book is available from the British Library

Library of Congress Cataloging-in-Publication Data
A catalog record has been requested for this book

ISBN: 978-0-8153-5754-4 (hbk)
ISBN: 978-0-367-50042-9 (pbk)

Typeset in Goudy
by Apex CoVantage, LLC

Contents

List of figures viii
List of tables ix
List of contributors x
Acknowledgements xiv

PART I
Concepts of culture and cultural sustainability 1

1 **Introduction** 3
TORSTEN MEIREIS AND GABRIELE RIPPL

2 **Culture and sustainable development from a UN perspective** 12
MICHAEL GERBER

3 **The future of cultural heritage and its challenges** 25
ALEIDA ASSMANN

4 **Cultural sustainability in times of cultural genocide** 36
WOLFGANG HUBER

5 **Sustainable development and the concept of culture – an ethical view** 47
TORSTEN MEIREIS

PART II
Philosophy, sociology, economics and cultural sustainability 61

6 **Sustainability and modernisation in conflict: on Roger Scruton's conservative green philosophy** 63
ANTON LEIST

7 **Culture, consumption and sustainability in a sociological perspective** 75
ULF LIEBE

8 **Social happiness as a cultural value: an analysis of shared values for ecosystem assessment** 87
GALIT P. WELLNER

9 **The challenges of cultural sustainability on an (inter)planetary scale** 98
ANDREAS LOSCH

10 **Sustainability's promise of salvation?: A Kuhnian reconstruction of sustainability from resource management to contingency management** 109
MARIUS CHRISTEN, PETER SEELE, LUCAS ZAPF

PART III
The arts, literature and cultural sustainability: theoretical, historical and contemporary perspectives 125

11 **Culture and the arts in sustainable development: rethinking sustainability research** 127
SACHA KAGAN

12 **Cultural ecology and the sustainability of literature** 140
HUBERT ZAPF

13 **An entangled history of environmental and cultural sustainability: satirical reflections on the German forest and the German oak as resources of cultural energy** 153
EVI ZEMANEK

14 **The cultural sustainability of Victorian waste** 183
URSULA KLUWICK

15 **The moral dilemma of unsustainability: eco-dystopian fiction as cultural critique** 193
ALEXA WEIK VON MOSSNER

16 **A tale of love and fallout: Lauren Redniss' graphic narrative and the ecological imaginary** 208
TORSTEN MEIREIS AND GABRIELE RIPPL

17 **Sustainability, eco-ekphrasis and the ethics of literary description** 220
GABRIELE RIPPL

18 **Ecogames: playing to save the planet** 232
JOOST RAESSENS

PART IV
Where do we go from here? 247

19 **Conclusion** 249
TORSTEN MEIREIS AND GABRIELE RIPPL

Index 256

Figures

7.1	Two perspectives on the relationship between culture and consumption	78
7.2	Examples of two perspectives on the relationship between culture and sustainable consumption	79
10.1	Overview of the sustainability conceptions as paradigm candidates	120
13.1	Crassus (1873). "Elegie". *Fliegende Blätter*, 58(1452), 159	161
13.2	Johnson, A. (1908). "Das Schweigen im Grune-Wald (frei nach Böcklin)". *Beiblatt zum Kladderadatsch*, 61(35), 1 (30 August 1908)	163
13.3	Anon. [signature unidentifiable] (1921). "Nieder mit dem deutschen Wald!" *Kladderadatsch*, 74(42), 2 (16 October 1921)	165
13.4	Anon. (1898). "Die deutsch-österreichische Eiche". *Kikeriki. Humoristisches Volksblatt*, 38(8), 2 (27 January 1898)	167
13.5	Heine, T. (1907). "Ein Blatt aus der deutschen Naturgeschichte". *Simplicissimus*, 11(49), 1 (4 March 1907)	169
13.6	Heine, T. (1915). "Siege in Ost und West". *Simplicissimus*, 20(7), 1 (18 May 1915)	169
13.7	Gulbransson, O. (1919). "Die Ratifikation". *Simplicissimus*, 24(31), 1 (28 October 1919)	170
13.8	Anon. (1896) "Im Hochgebirge". *Kikeriki*, 36(74), 4 (13 September 1896)	172
13.9	Lange-Christopher, K. (1932). "Deutscher Wald im Dritten Reich". *Der Wahre Jacob*, 53(12), 14 (4 June 1932)	174
13.10	Stuttmann, K. (2005). "25 Jahre Die Grünen. Schützt den deutschen Wald!!" *Der Tagesspiegel* (13 January 2005)	176
13.11	Stuttmann, K. (2004). "Wenigstens der deutsche Wald ist gerettet!!" *Der Tagesspiegel* (2 April 2004)	176
18.1	*World Climate Conference*	233
18.2	*Walden, a Game*	237
18.3	Model for the encouragement of sustainable action	240
18.4	*Join the rebellion and turn VW away from the dark side*	241
18.5	A circular Amsterdam	242

Tables

7.1 Three aspects of culture 76
10.1 Overview of the sustainability conceptions as paradigm candidates 111
18.1 Barriers, solutions, properties, and strategies 234

Contributors

Aleida Assmann is Professor Emeritus of English Literature and Literary Theory at the University of Konstanz, Germany. She held guest professorships at Rice, Princeton, Yale, Chicago, Vienna, Mainz and Lucerne and received an Honorary Degree from the University of Oslo (2008). Selection of prizes: Heineken Prize for History of the Royal Netherland Academy of Sciences (2014) and the Balzan Prize (2017, together with Jan Assmann). Main areas of research: historical anthropology, media history and cultural memory (esp. Holocaust and trauma). Recent Publications: *Memory in a Global Age. Discourses, Practices and Trajectories* (ed. with Sebastian Conrad, 2010), *Cultural Memory and Western Civilization. Functions, Media, Archives* (2012), *Shadows of Trauma. Memory and the Politics of Postwar Identity* (2016).

Marius Christen studied Philosophy, Sustainability Sciences and History. His first book analysed the axiological foundations of sustainability (*Die Idee der Nachhaltigkeit. Eine werttheoretische Fundierung*, 2013). Currently, he researches and teaches at the Sustainability Research Group, University of Basel, Switzerland, focusing on sustainability governance and questions of inter- and transdisciplinarity.

Michael Gerber is Switzerland's Special Envoy for Global Sustainable Development. In 2012, the Swiss Federal Council appointed him to prepare Switzerland's position in the UN post-2015 process. In 2015, he was mandated to head the Swiss delegations in the intergovernmental negotiations at the United Nations on the 2030 Agenda for Sustainable Development, as well as on the associated financing and implementation framework, the Addis Ababa Action Agenda.

Wolfgang Huber is Professor Emeritus of Theology and Honorary Professor of Systematic Theology at the Humboldt University Berlin, Germany, at Heidelberg University, Germany and Stellenbosch University, South Africa. He served as Bishop of the Protestant Church of Berlin-Brandenburg-schlesische-Oberlausitz (1994–2009), was chairman of the Protestant Church of Germany (2003–2009) and a member of the National and the German Ethics Council. Publications: *Konflikt und Konsens* (1990), *Ethik. Die Grundfragen unseres*

Lebens von der Geburt bis zum Tod (2013), *Handbuch der Evangelischen Ethik* (2015, co-edited).

Sacha Kagan is a Research Associate at the Institute of Sociology and Cultural Organization, Leuphana University Lüneburg, Germany; Principal Investigator at the research project "The City as Space of Possibility"; Former Chair of the Research Network "Sociology of the Arts" at the European Sociological Association. Dr. Kagan has been researching at the intersection of the arts, culture and sustainability for over a decade. He is the author of *Art and Sustainability: Connecting Patterns for a Culture of Complexity* (2011).

Ursula Kluwick is Assistant Professor of Modern English Literature at the University of Bern. She has published on the nineteenth century, ecocriticism, contemporary and postcolonial literatures and cultures, as well as on Shakespeare. She is the author of *Exploring Magic Realism in Salman Rushdie's Fiction* (2011) and has co-edited *The Beach in Anglophone Literatures and Cultures* (2015).

Anton Leist is Professor Emeritus of Practical Philosophy of Zurich University. He worked on applied and theoretical ethics. Presently he is engaged with the epistemology of pragmatism and topics of social justice. He is co-editor of *J.M. Coetzee and Ethics: Philosophical Perspectives on Literature* (2010) and author of *Action in Context* (2007).

Ulf Liebe is Professor of Sociology and Quantitative Methods at the Department of Sociology, University of Warwick. His fields of interest include environmental sociology, economic sociology and experimental methods. He has published in journals such as *European Sociological Review*, *Sociological Methods & Research*, and *Environment and Behavior*. He is author of *Zahlungsbereitschaft für Kollektive Umweltgüter: Soziologische und Ökonomische Analysen* (2007).

Andreas Losch is a theologian who works with philosophers and space researchers. His current project on "Ethics of a Planetary Sustainability" combines his expertise in the dialogue between science, philosophy and theology. He is a member of the Center of Theological Inquiry (Princeton, NJ) and serves in several societies dedicated to the study of the relation of science and religion. Recent publications: *What is Life? On Earth and Beyond* (2017, ed.) and *"Alles in der Schrift ist echte Gesprochenheit". Martin Buber und die Verdeutschung der Schrift* (2016, co-ed.).

Torsten Meireis serves as Full Professor of Ethics and Hermeneutics in the field of Systematic Theology at the Humboldt University Berlin and as Director of the Berlin Institute for Public Theology. Current research interests include sustainability, political and economic ethics. Recent publications: *Religion and Democracy. Studies in Public Theology* (with Rolf Schieder) (2017); *Handbuch Evangelische Ethik* (with Wolfgang Huber, Hans-Richard Reuter) (2015).

Joost Raessens is Chair and Full Professor of Media Theory at Utrecht University and Scientific Director of the Utrecht Center for Game Research. His research concerns the ludification of culture, focusing in particular on games for change in relation to global climate change and refugee issues. Among his co-edited book publications are the *Handbook of Computer Game Studies* (2005), *Playful Identities* (2015) and *Playful Citizens* (2018).

Gabriele Rippl is Full Professor and Chair of Literatures in English at the University of Bern. Current research interests: cultural sustainability, narrative ethics, intermediality. Recent publications: *Handbook of Intermediality* (2015); *Towards a New Monumentalism? Cultural and Aesthetic Perspectives beyond Postmodernism* (2013); Rippl is co-editor of *Anglia. Journal of English Philology* and the De Gruyter series *Handbooks of English and American Studies: Text and Theory*.

Peter Seele is Professor of Corporate Social Responsibility and Business Ethics at USI Lugano, Switzerland. He publishes on sustainability with regard to corporate reporting, education and recently on digital sustainability and predictive sustainability control. As principal investigator he conducts research projects on behalf of the Swiss National Fund on the credibility of CSR reporting and recently on Greenwashing.

Alexa Weik von Mossner is Associate Professor of American Studies at the University of Klagenfurt, Austria. She has published widely on cosmopolitanism, ecocriticism and affective narratology and is the author of *Cosmopolitan Minds: Literature, Emotion, and the Transnational Imagination* (U of Texas P, 2014) and *Affective Ecologies: Empathy, Emotion, and Environmental Narrative* (Ohio State UP, 2017).

Galit P. Wellner, PhD, is a Senior Lecturer at the NB School of Design Haifa, Israel, and an Adjunct Professor at Tel Aviv University. Her research area is philosophy of digital technologies. She published several peer-reviewed articles and book chapters. Her book *A Postphenomenological Inquiry of Cellphones: Genealogies, Meanings and Becoming* was published in 2015 by Lexington Books.

Hubert Zapf is Professor Emeritus at the University of Augsburg, Germany. Main areas of research: literature and cultural ecology, literary history, literary and cultural theory. Recent publications: *Literature as Cultural Ecology: Sustainable Texts* (2016) and *Handbook of Ecocriticism and Cultural Ecology* (2016). He is co-editor of *Anglia. Journal of English Philology*, of the Book Series of *Anglia*, and of the De Gruyter Handbook Series *English and American Studies. Text and Theory*.

Lucas Zapf combines in his research ethics, economics and studies of religion. He researched the market economy's hidden connections to Luther's work ethics, the deification of the markets and, most currently, the nexus between business ethics and risk management. He works as a PostDoc and Lecturer at the University of Basel.

Evi Zemanek is Associate Professor of German Literature and Media Studies at the University of Freiburg, Germany, fellow at FRIAS (Freiburg Institute of

Advanced Studies) and at the Rachel Carson Center for Environment and Society, Munich. She has published widely in the field of transmedia ecocriticism. She is editor of *Ökologische Genres. Naturästhetik – Umweltethik – Wissenspoetik* (2017) and co-editor of *Ecological Thought in German Literature and Culture* (2017).

Acknowledgements

Cultural Sustainability is a field that, by definition, needs inter- and transdisciplinary approaches. This volume of essays is faithful to this prerequisite and could not have been assembled without joint efforts by people from a range of institutions.

In addition to all contributors to this volume, our heartfelt thanks go to our colleagues and team members, Leona Goop, Camille Gray, Sibylle Hofer, Bettina Schön, Malaika Sutter, Edward Wright, Clemens Wustmans, Hubert Zapf and Anne Zimmermann, without whom the job of bringing this volume to publication would have been so much harder. With unflagging commitment, they discussed our ideas with us, pushed us to develop our topic and helped us with proof-reading and editorial work. A big thank you also goes to our MA and PhD students of the seminar "Cultural Sustainability: Literary Ecology – Normative Structures – Religious Imagery" and the research colloquium "Cultural Sustainability", co-taught at the University of Bern in spring term 2016 and autumn term 2018 respectively. Likewise, the students of the seminar on "Cultural Sustainability and Religion" and the Summer School on "Religion – Sustainability – Politics", taught at the Humboldt-Universität zu Berlin in summer term 2016 and summer term 2017 respectively, have helped us to develop our ideas. We are grateful to the University of Bern and the Humboldt-Universität zu Berlin for fostering interdisciplinary teaching. Our thanks also go to both the Swiss National Science Foundation and the Swiss Academy of the Humanities and the Social Sciences for generously supporting an exploratory workshop held at the University of Bern in November 2016, some results of which have fed into this volume. Last but not least, we would like to thank Charlotte Endersby and Leila Walker at Routledge together with Libby Robin and Paul Warde for their support in editing this volume in the Routledge Environmental Humanities series, as well as our two anonymous reviewers whose helpful suggestions inspired us and helped give the volume its final shape.

Part I

Concepts of culture and cultural sustainability

1 Introduction

Torsten Meireis and Gabriele Rippl

A long-neglected question: the cultural dimension of sustainable development

> Humanity has the ability to make development sustainable to ensure that it meets the needs of the present without compromising the ability of future generations to meet their own needs. The concept of sustainable development does imply limits – not absolute limits but limitations imposed by the present state of technology and social organization on environmental resources and by the ability of the biosphere to absorb the effects of human activities. But technology and social organization can be both managed and improved to make way for a new era of economic growth.
>
> (WCED 1987: 16)

This quote from the famous "Our Common Future" report from the World Commission on Environment and Development stresses the interdependence of the ecological, social and economic dimensions of sustainable development – or in short: sustainability. While the necessity of the relationship between the components of sustainability has been highlighted time and again, the quote also shows that one important aspect tends to be overlooked, to wit, the cultural. This cultural dimension is usually omitted in the so-called triangle of sustainability or the three-pillar model of ecological, social and economic capital propagated by the World Bank (1997).

International debates on the function of culture in sustainable development did, belatedly, gain ground. The UNESCO-organised Intergovernmental Conference on Cultural Policies for Development in Stockholm (UNESCO 1998a) as well as the UNESCO and UNEP 2002 Johannesburg Roundtable on Cultural Diversity and Biodiversity (UNEP 2003) and the Johannesburg Summit's decision for a Decade of Education for Sustainable Development (UNESCO 2002) functioned as stepping stones, but they mostly concentrated on topics concerning respect for cultural diversity and creativity (UNESCO 1998a: 13–14, UNESCO 1998b: 93–104; UNEP 2003). The overall significance of the cultural dimension of sustainability has only recently been acknowledged by the UNESCO's

Hanghzou Declaration, which voices a commitment to "Placing Culture at the Heart of Sustainable Development Policies" (UNESCO 2013). The road leading from the UN division specialising in cultural questions to the UN at large has been rockier still: as Michael Gerber in this volume convincingly demonstrates, corresponding UN-Resolutions confirming this stance have been rather vague and late in appearance (65/166, 2010; 66/208, 2011; 68/223, 2013; 69/230/2014; 70/214, 2015), although admittedly now the 2015 Sustainable Development Goals contain elements expressly pertaining to culture, such as 'quality education'.

In academia, debates on the cultural dimensions of sustainability in the wake of the Brundtland report sprang up in the fields of educational sciences (Sandri 2013; Sorgo 2011; Stoltenberg et al. 1999; cf. also Conclusion). In cultural policy, culture was proclaimed as a "fourth pillar of sustainability" and calls for a concerted effort in public and urban cultural development were made (Davies 2015; Böhm et al. 2014; DifU 2011; Banse et al. 2011; Duxbury and Jeannotte 2010; Nadarajah 2007; Krainer and Trattnigg 2007), because the "engines of cultural production require a singular and co-ordinated setting within government management structures" (Hawkes 2001: 38). However, it also became evident that culture may act as a factor severely impeding sustainability awareness (Hoffman 2015). A recently completed European research network explored different understandings of cultural sustainability (Dessein and Soini 2016). It analysed scientific discourses on policy development (Soini and Birkeland 2014), and showed that culture needs to be understood as an instrumental factor in sustainable development for the very reason that it may act as an additional, separate pillar in fostering sustainability, as a mediating agency between ecological, social and economic aspects of sustainability and as a basic pattern establishing a sustainability-friendly orientation (Dessein et al. 2015: 20–37). Even though cultural sustainability has received growing attention, the ways in which culture works in contexts of sustainability and how it is related to a system of values which, in turn, influences understandings of sustainability, are yet to be scrutinised. It is worthwhile to highlight one important aspect: the sustainability discourse has had a history that started long before the 1987 report from the World Commission on Environment and Development was published. It goes back as far as the eighteenth century: the concept is usually attributed to Saxonian official Hans Carl von Carlowitz, who proposed sustainable forestry within a deeply religious framework (Carlowitz 1713). The concept has since undergone several changes of meaning (Grober 2010). For that reason, a cultural understanding of sustainability has to take this wide range of different conceptualisations into account (Weigel 2010).

The significance of culture for sustainability issues: key concepts and central questions

As a normative idea, the concept of 'sustainability' has profound implications for the day-to-day cultural routines of societies. The way we consume and the way we understand prosperity or 'well-being', what we see as 'the good life' and which values we hold are closely related to our notion of sustainability. If, for instance, the general idea of 'well-being' in a given society is connected to a lifestyle of intensive resource consumption, then the idea of sufficiency, closely related to a global sustainability strategy accepting planetary boundaries (Rockstrøm et al. 2009) will be hard to implement politically in those regions already leaving the largest ecological footprint (Schneidewind and Zahrnt 2014). For that reason, it may be necessary to change those cultural concepts. On the other hand, evolving cultural ideas – like conservation (Engels 2006; Grober 2010) or new concepts of urbanisation (MoMa 2014) – may further global sustainable development. Moreover, the principle of sustainability may affect the cultural organisation of scientific research and the conceptualisation of scientific knowledge (Schneidewind 2014). Cultural studies and the humanities in general not only tackle normative questions but also reflect on visions of (non)sustainable action, which have the potential to influence the public perception of environmental issues that are linked to moral questions, as negotiated in works of fiction (literature, films, graphic narratives, computer games, etc.), religious or other cultural narratives. Thus, cultural sustainability has a double impact. Firstly, ecological, economic and social questions of sustainability can only be understood within the horizon of culture, rooted as they are in the ideas, imagery and concepts employed to describe sustainability. Secondly, and following from this observation, cultural sustainability is a topic in itself, as it necessitates research into traditions, modes of perception, methods of cognition, the conceptualisation of knowledge and the concept of sustainability itself, for instance, when its close connection to the idea of development is challenged as a colonial way of thinking (Mignolo 2011).

To look into the ways culture 'works' with regards to sustainability questions, we heuristically distinguish a wider and a narrower concept of culture. In a wider sense, 'culture' signifies the human activity of symbolising (Cassirer 1953–1957), thus denoting any way of communicating and coming to terms with the world by means of language, images, concepts and so forth, spanning a range from the most elaborate composition to the ordinary interactions of everyday life. In a narrower sense, 'culture' means the act of expressly producing, reproducing or modifying cultural symbols in performance, artistic, scientific or scholarly production (Williams 1989).

While the focus has been on the economic, social and ecological dimensions of sustainability, research on cultural aspects has mostly been limited to praising and protecting diversity and to describing cultural policy. In current political and academic debates, the important contribution of the humanities – theology, literary, media and cultural studies, social sciences as well as philosophy – to a sustainable world has not been adequately acknowledged. The contributions made by the arts and the humanities along with religious and philosophical discourses have received insufficient attention by politicians and social agents. While the externally visible signs of sustainability – scientific evidence – are necessary for long-term solutions to environmental crisis, equally relevant are those interior dimensions (Horlings 2015), i.e. the internal narratives and images that we use to make sense of the world and explore moral space (Nussbaum 1995). Deploying the concept of 'social imaginaries' (Taylor 2004), it is our claim that cultural products such as literary and religious narratives, visual works of art and films have an enormous impact on people's world views, beliefs and values. They function as important means for the representation and mediation of societal problems and cultural anxieties.

The book and its structure

It has become clear that in existing discourses, culture – and with it cultural sustainability – has been conceptualised within narrow frameworks focusing solely on questions concerning cultural diversity, indigenous cultures, the significance of local identity and the role of creativity for community development. In many academic disciplines, however, much wider and more comprehensive definitions of culture are available which include complex concepts of human symbolic systems (see Meireis' Chapter 5 in this volume). Yet to date, these extensive definitions of culture have had almost no impact on discussions of cultural sustainability. This is also true of scientific approaches to values and conceptualisations of sustainability not focused solely on the Brundtland report. Indeed, one upshot of our own research into the term 'cultural sustainability' is the realisation that values are central to the term and should figure prominently in future discussions of it. Actually, it is precisely these values that determine what we may understand as 'sustainable'. Last but not least, historical, diachronic analyses (cf. Assmann's Chapter 3 in this volume; Schliephake 2017) are indispensable when discussing cultural sustainability. While the general understanding of sustainability is oriented towards the future, cultural sustainability needs to take the past into account and thus deal with questions regarding cultural memory, cultural heritage and institutions such as archives, libraries and museums as vessels of conservation which help to hand down knowledge and cultural information to future generations.

All contributions to this book investigate the potential of cultural products to sharpen people's awareness of their own actions and feelings of responsibility for their environment. Literature, visual works of art, film, graphic narratives, philosophical or religious narratives and images, all enable us to try out ideas in our imagination; they stimulate us to imagine possible alternative worlds and to identify with others (Ricoeur 2005). If the political and social benchmarks of sustainability and sustainable development are to be met, ignoring the role of the humanities on the one hand and that of social, cultural and ethical values on the other is highly problematic. After all, societal change toward more sustainable lifestyles is by no means solely caused and implemented by political decisions. Rather, such change is linked directly to people's world views, beliefs and cultural values, which have an immediate impact on how we act – hence the need to study them as cultural dimensions of sustainability. Culture, however, is not a one-dimensional phenomenon easily harnessed towards a certain normative goal, but a multi-dimensional entity in itself. The complexity of the concept demands careful study, if its significance for sustainability discourse is to be explored.

Possibly, the benchmarks mentioned above also need changing. For that reason, this collection of essays maps the field of cultural sustainability, by discussing and bringing together a range of fundamental concepts of culture, and by taking stock of the contributions from across the humanities and social sciences. Of course, the broadening of the discourse on cultural sustainability engenders a wide variety of scholarly and practical approaches, styles and methods.

The book comprises three main parts, the first of which introduces different concepts of cultural sustainability, thus combining different perspectives. *Michael Gerber*, who acts as Switzerland's Special Envoy for Global Sustainable Development, sketches the role of cultural sustainability as a political term in debates at the United Nations, this, from the view of a participant in political processes. Spanning a period from the 1960s up to the 2015 Sustainable Development Goals, Gerber describes the political and conceptual challenges that sustainability has endured. In her seminal contribution, literary scholar and cultural theoretician *Aleida Assmann* argues that the concept of 'sustainability' adds a new important meaning to our understanding of the future in terms of limited resources and a new understanding of human stewardship towards our planet. Other meanings of 'future', however, continue to be applied to political and economic contexts. She draws attention to the pluralisation of our time's regimes and links the term 'cultural sustainability' to the theory of 'cultural memory' which she has developed. Theologian *Wolfgang Huber* then illuminates the ethical background of the debates on cultural sustainability by pointing to the importance of human rights in this context. In fleshing out the concept of 'cultural genocide' and by exploring the cultural dynamics of international law development, he presents an idea of cultural sustainability rooted with respect to diversity. Concentrating on the meaning of 'culture' in the concept of cultural sustainability, ethicist and theologian *Torsten Meireis* claims that an instrumental understanding of culture – common in sustainability debates – is dysfunctional as it treats normative assumptions as self-evident and does not take the agonal and productive aspects of culture into account. Meireis pleads instead for an explicit debate on normative questions and a stronger regard for the production of cultural assets and the traditions maintaining them.

The second part of the book brings together contributions from philosophy and the social sciences that explore the conditions and challenges of cultural sustainability. Philosopher *Anton Leist* questions the popular notion that sustainability issues are firmly rooted in cultural narratives and images pertaining to a politically liberal, progressive or social democratic and green agenda. By sketching the position of the philosopher Roger Scruton, Leist discusses the merits of a distinctly conservative concept of cultural sustainability. Examining consumption behaviour, sociologist *Ulf Liebe* looks at the behavioural mechanisms furthering or hindering sustainability and distinguishes between a culture as a constraint, which works through social norms, from culture as an enabling force which is embodied in rituals and practices. While social norms have only a short-term effect, the operation of rituals and cultural practices tends to be long-term. Philosopher *Galit P. Wellner* explores the intersection of the social and the cultural dimension of sustainability by presenting social happiness as a cultural value, which implies good social relations, a rich cultural environment and fair institutions, thus favouring a long-term eudaimonic rather than a short-term hedonistic perspective. Starting from the NASA's definition of 'planetary sustainability' and the existing practices of space travel to the Moon or Mars, theologian and ethicist *Andreas Losch* argues that cultural sustainability is at the core of any understanding of interplanetary sustainability, dealing for instance with the dangers of forward or backward contamination of microbial life forms. Economists and scholars of cultural studies *Marius Christen*, *Peter Seele* and *Lucas Zapf* investigate cultural sustainability by applying Thomas S. Kuhn's concept of paradigm shift. They identify three paradigms of sustainability that offer solutions to different problems: *resource crisis*, *environmental crisis* and *system crisis*. As a fourth paradigm, they suggest *contingency management*, which responds to the more recent multi-crisis and which carries religious overtones of salvation.

The third part of this essay collection comprises contributions that investigate how the visual arts, literary texts and computer games scrutinise the relationship between social imaginaries and sustainable development. Cultural studies researcher *Sacha Kagan* sketches the development of the sustainability discourse in the performing and visual arts as well as in cultural studies. He argues that aesthetic and art-based practices and discourses of research provide not only instrumental cultural resources for sustainable development, but also imaginative approaches and a transdisciplinary methodology for sustainability studies in general. Americanist and literary theorist *Hubert Zapf* introduces the theoretical concept of cultural ecology, which helps to describe how literature feeds into sustainable development. Zapf argues that literature contributes to a complex, ethically responsive form of cultural sustainability by combining cultural memory and cultural creativity in ever-new scenarios of embodied culture-nature interaction, thus extending the semantic range of 'sustainability' by emphasising the experimental, non-systemic and creative aspects of literary sustainability. German literature scholar *Evi Zemanek* challenges the distinction between environmental and cultural sustainability by focusing on their entanglement in nineteenth-century satirical texts and drawings in German magazines. She discusses how prominent ethnographers and writers established the 'German forest' and the 'German oak' as symbols of cultural identity by popularising these natural-cultural symbols among a greater audience. However, in a climate of growing nationalism, the 'German Oak' lost its innocence when it was associated with racist ideas. The contribution of literary scholar *Ursula Kluwick* approaches cultural sustainability by focusing on debates concerning the utilisation of human excrement as manure in the context of sanitary reform in Victorian Britain. The public health issues relevant in this respect were discussed in a significantly broader cultural domain (for instance in Charles Dickens' novels), thus lending themselves to the exploration of the cultural dissemination of the notion of sustainability. Literary scholar *Alexa Weik von Mossner* takes the recent resurgence of interest in dystopian fiction as a starting point for an investigation into the relationship between speculative storytelling and the cultural discourse on sustainability. Drawing on the work of sociologists Shai Dromi and Eva Illouz, she suggests that eco-dystopian texts such as Octavia Butler's *Parable of the Sower* are cultural critiques in the sense that they present readers with a dilemma and then imbue that dilemma with emotional value. Literary scholar *Gabriele Rippl* and ethicist and theologian *Torsten Meireis* explore the genre of the graphic narrative and its contribution to cultural sustainability. In connection with their example, Lauren Redniss' *Radioactive – Marie & Pierre Curie. A Tale of Love and Fallout*, they investigate how this graphic narrative presents a popular and widely accessible narrative ethics, which provokes an ethical reflection within readers, instead of simply offering particular morals or values. *Gabriele Rippl* then investigates literature's potential to sharpen readers' awareness of their own values and to foster feelings of responsibility for the environment. As an undertaking in narrative ethics, novels such as Margaret Atwood's *The Year of the Flood* are concerned with the underestimated (but crucial) inner, subjective dimension of sustainability. With the help of descriptions of nature and eco-ekphrases, Atwood introduces ecological concepts into the 'social imaginary' (Charles Taylor). Media theorist and game researcher *Joost Raessens* looks into the dynamics of contemporary digital eco-games and how they frame the cultural dimension of sustainability. His analysis aims at offering a conceptual clarification of the strategies these games use to raise awareness about the issue of climate change and to change or reinforce the player's world views, beliefs and values.

Part 4 consists of the editors' Conclusion. *Gabriele Rippl* and *Torsten Meireis* summarise the significance of the concept of cultural sustainability as presented by the volume's contributors, explore some ways to teach the subject and refer to further areas of research.

This volume is bent on fathoming the hitherto neglected 'inner' dimensions of sustainability: the values, beliefs, ideas and images which the various disciplines within the humanities and social sciences investigate and which have consequences for the exterior dimensions, too. The collection of essays addresses research desiderata, trying to ignite more exchange between the disciplines and to contribute to a comprehensive approach to the concept of cultural sustainability. In showcasing a wide range of previously underexplored facets of sustainability, the authors highlight the role of the humanities in achieving and transforming the goals of sustainable development. They focus on important issues such as social and cultural values, well-being, quality of life and happiness, thus encouraging further inter- and transdisciplinary research, the results of which will hopefully reach the relevant social and political agents.

References

Banse, G., Nelson, G. L. and Parodi, O. (eds.) (2011). *Sustainable Development – The Cultural Perspective: Concepts – Aspects – Examples*. Berlin: edition sigma.

Böhm, S., Pervez Bharucha, Z. and Pretty, J. (eds.) (2014). *Ecocultures: Blueprints for Sustainable Communities*. London: Routledge.

Carlowitz, Hans Carl v. (1713). *Sylvicultura oeconomica oder Haußwirthliche Nachricht und Naturmäßige Anweisung zur Wilden Baum-Zucht*, Nebst Gründlicher Darstellung/wie zu förderst durch Göttliches Benedeyen dem allenthalben und insgemein einreissenden Grossen Holz-Mangel/Vermittelst Säe- Pflanz- und Versetzung vielerhand Bäume zu prospiciren. Leipzig: Braun.

Cassirer, E. (1953–1957). *The Philosophy of Symbolic Forms [1923–1929]*. New Haven: Yale University Press.

Davies, W. K. D. (2015). "Background to Sustainable Cities". In W. K. D. Davies (ed.). *Theme Cities: Solutions for Urban Problems*. Dordrecht: Springer Netherlands, 151–205.

Dessein, J., Soini, K., Fairclough, G. and Horlings, L. G. (eds.) (2015). *Culture in, for and as Sustainable Development: Conclusions from the COST Action IS1007 Investigating Cultural Sustainability*. Jyväskylä: The University of Jyväskylä.

Dessein, J. and Soini, K. (2016). "Culture-Sustainability Relation: Towards a Conceptual Framework". *Sustainability*, 8(2), 167.

DifU (Deutsches Institut für Urbanistik) (2011). *Städte für ein nachhaltiges Deutschland: Gemeinsam mit Bund und Ländern für eine zukunftsfähige Entwicklung*. Berlin: DifU.

Duxbury, N. and Jeannotte, M. S. (2010). *Culture, Sustainability, and Communities: Exploring the Myths*. Working Paper No. 353. Coimbra: Oficina do CES, Universidade de Coimbra.

Engels, J. I. (2006). *Naturpolitik in der Bundesrepublik. Ideenwelt und politische Verhaltensstile in Naturschutz und Umweltbewegung 1950–1980*. Paderborn – Munich: Ferdinand Schöningh.

Grober, Ulrich (2010). *Die Entdeckung der Nachhaltigkeit. Kulturgeschichte eines Begriffs*. München: Kunstmann.

Hawkes, J. (2001). *The Fourth Pillar of Sustainability: Culture's Essential Role in Public Planning*. Champaign, IL: Common Ground Publishing.

Hoffman, A. J. (2015). *How Culture Shapes the Climate Change Debate*. Stanford: Stanford University Press.

Horlings, L. G. (2015). "The Inner Dimensions of Sustainability: Personal and Cultural Values". *Current Opinion in Environmental Sustainability*, 14, 163–169.

Krainer, L. and Trattnigg, R. (eds.) (2007). *Kulturelle Nachhaltigkeit: Konzepte, Perspektiven, Positionen*. Munich: Oekom Verlag.

Mignolo, W. (2011). *The Darker Side of Western Modernity. Global Futures, Decolonial Options*. Durham – London: Duke University Press.

MoMa (Museum of Modern Art, New York) (2014). *Uneven Growth: Tactical Urbanisms for Expanding Megacities*, organized by Pedro Gadanho, New York.

Nadarajah, M. (2007). "Culture of Sustainability: Multicultural Reality and Sustainable Localism. A Case Study of Penang (George Town), Malaysia". In M. Nadarajah and A. T. Yamamoto (eds.). *Urban Crisis: Culture and the Sustainability of Cities*. Tokyo: United Nations University Press, 107–133.

Nussbaum, M. C. (1995). *Poetic Justice: The Literary Imagination and Public Life*. Boston, MA: Beacon Press.

Ricoeur, P. (2005). *Das Selbst als ein Anderer*. München: Wilhelm Fink.

Rockstrøm, Johan et al. (2009). "A Safe Operating Space for Humanity". *Nature*, 461, 472–475.

Sandri, O. J. (2013). "Exploring the Role and Value of Creativity in Education for Sustainability". *Environmental Education Research*, 19(6), 765–778.

Schliephake, C. (ed.) (2017). *Ecocriticism, Ecology, and the Cultures of Antiquity*. Lanham, MD: Lexington Books.

Schneidewind, U. (2014). *Transformative Wissenschaft: Klimawandel im deutschen Wissenschafts- und Hochschulsystem*. Weimar: Metropolis-Verlag.

Schneidewind, U. and Zahrnt, A. (2014). *The Politics of Sufficiency: Making It Easier to Live a Good Life*. Munich: Oekom Verlag.

Soini, K. and Birkeland, I. (2014). "Exploring the Scientific Discourse on Cultural Sustainability". *Geoforum*, 51, 213–223.

Sorgo, G. (ed.) (2011). *Die unsichtbare Dimension: Bildung für eine nachhaltige Entwicklung im kulturellen Prozess*. Vienna: Forum Umweltbildung im Umweltdachverband.

Stoltenberg, U., Michelsen, G. and Schreiner, J. (eds.) (1999). *Umweltbildung – den Möglichkeitssinn wecken*. Schneverdingen: Alfred-Toepfer-Akademie für Naturschutz.

Taylor, C. (2004). *Modern Social Imaginaries*. Durham – London: Duke University Press.

UNEP (2003). Cultural Diversity and Biodiversity for Sustainable Development. A jointly convened UNESCO and UNEP high-level Roundtable held on 3 September 2002 in Johannesburg during the World Summit on Sustainable Development, Nairobi.

UNESCO (1998a). *World Culture Report 1998: Culture, Creativity and Markets*. Paris: UNESCO.

UNESCO (1998b). Final Report. Intergovernmental Conference on Cultural Policies for Development, Stockholm, Sweden, 30 March–2 April.

UNESCO (2002). *Education for Sustainability. From Rio to Johannesburg: Lessons Learnt From a Decade of Commitment*. Paris: UNESCO.

UNESCO (2013). *The Hangzhou Declaration. Placing Culture at the Heart of Sustainable Development Policies*. Paris: UNESCO.

Weigel, S. (2010). "ÜberLebensQualität: Kulturwissenschaft und Nachhaltigkeit". In Deutscher Hochschulverband (ed.). *Glanzlichter der Wissenschaft: Ein Almanach*. Saarwellingen: Lucius &Lucius, 153–159.

Williams, R. (1989). "Culture is Ordinary". In *Resources of Hope: Culture, Democracy, Socialism*. London: Verso, 3–14.

World Bank (1997). The International Bank for Reconstruction and Development/The World Bank, Expanding the Measure of Wealth. Indicators of environmentally sustainable development (Environmentally sustainable development studies and monographs series no. 17), Washington, DC.

World Commission on Environment and Development (WCED) (1987). "Report of the WECD. Our Common Future". *WCED*. Transmitted to the General Assembly as an Annex to document A/42/427 – Development and International Co-operation: Environment. www.un-documents.net/wced-ocf.htm [Accessed 15 October 2017].

2 Culture and sustainable development from a UN perspective

Michael Gerber

Introduction

In September 2015, at the United Nations Sustainable Development Summit in New York, all 193 UN member states successfully adopted the 2030 Agenda for Sustainable Development (2030 Agenda) with its 17 Sustainable Development Goals (SDGs). The 2030 Agenda serves as a new global reference framework for sustainable development until 2030. It covers the three dimensions of sustainable development (social, economic and environmental) in a balanced way and is universally applicable to all countries. Not least due to the inclusive elaboration phase, the 2030 Agenda can be considered the latest and politically most widely agreed upon description of sustainable development.

In the respective elaboration and negotiation processes from 2012 to 2015, the concept of culture gained political relevance. Some observers and commentators even defined it as the fourth dimension of sustainable development, in addition to the three dimension concept, politically accepted since the definition given in the Brundtland report ("Our Common Future") in 1987 (World Commission on Culture and Development 1995).

This contribution[1] aims at discussing the role and position the concept of culture assumes in the sustainable development discourse at the multilateral level. After a brief historic overview on the relevance culture was ascribed in relevant UN processes, I will elaborate on a range of specific inputs in the political debate about culture and sustainable development, mainly stemming from important human rights treaties, the most relevant UN conventions and declarations as well as resolutions by the UN General Assembly. Eventually, I will discuss key contributions during the negotiations on the 2030 Agenda, and outline the relevance of culture in the final outcomes document.

Herewith, I contribute a political perspective to the academic debate around culture and sustainable development. I reveal and discuss the existing gap between the academic discourse, which usually tends to be more advanced, and the political reality at an international level, thus spurring an in-depth debate about the notion of *cultural sustainability* in multilateral practices.

The role of culture in (sustainable) development

The understanding of the role of culture in development processes has altered significantly over the last few decades. In conjunction with the steady evolution of development theories, culture has become increasingly important and is nowadays widely acknowledged as a factor that needs to be integrated in development policies and actions (Maraña 2010: 3). This relation between culture and development is highly complex and a detailed historic analysis would require the integration of diverse academic perspectives, an ambition that cannot be met in this short contribution. Instead, a brief overview of the most important milestones, mainly based on the UNESCO Working Paper by Maider Maraña (2010) and the respective primary resources, will be given in the following paragraphs.

The contemporary understanding of the role of culture in development stands in stark contrast to the early theories of development of the 1960s, which strongly focused on economic growth and did not take culture into account. Even worse, culture as a way of life was declared an obstacle to development by many traditional development economists (Streeten 2006). This econometric concept of development was gradually replaced by more complex ones, though not necessarily leading to the acknowledgement of the role of culture. The work of ecologists and environmentalists contributed to the conceptualization of *sustainable development* in the 1970s, which went beyond pure economic growth and also included concerns about the impact of economic activity on the physical environment. At the same time, various experts started pointing to the importance of cultural factors in development processes. Nevertheless, according to Maraña, culture remained side-lined until the 1980s or even 1990s, when the first international bodies and aid agencies started exploring the relations between culture and development, however still with little effect on the work of the international development community (2010: 4).

The difficulty of finding agreement on a clear definition of the term *culture* is likely to have contributed to the exclusion of the concept from development policies for such a long time (Maraña 2010: 4). It is a term that can mean everything and nothing and the problems arising from the lack of a definition are exacerbated in the discourse on *cultural sustainability*, as shown later on. Some understand culture in a narrow sense and thus focus on the promotion of arts or the creative industries in general. This approach differs substantively from trends in the 1990s towards a more human-centred development that stressed the need for culturally sensitive, participatory approaches and capacity building. This humanistic understanding of development, which was inspired by the work of Amartya Sen, led to the creation of the Human Development Index (HDI) by the United Nations Development Programme (UNDP).

It did not however trigger a more systematic inclusion of cultural aspects in development cooperation, despite the advocacy of non-governmental organizations (NGOs).

This exclusion of cultural aspects also applies to the domain of multilateral development cooperation. The Millennium Development Goals (MDGs), which were launched by the UN in 2000 aimed at halving hunger and poverty by 2015, did not include culture in their targets and indicators. This decision was taken despite a growing recognition of culture's role in development during the elaboration phase, which also led to important initiatives and further research. One of the reasons for not including culture in the goal and indicator framework of the MDGs was the difficulty of precisely measuring culture's impact on development. As we shall see later, the same issue was raised in the negotiations of the 2030 Agenda, which was adopted in 2015 and consolidated the two work streams of the Rio-process and the MDGs to constitute a global framework for sustainable development.

In the 2030 Agenda, again no dedicated goal on culture exists. Nevertheless, the concept of culture is strongly integrated and explicitly mentioned several times in the declaration as well as in specific targets of the SDGs. However, in the period between the two agendas, a lot of work on culture and sustainable development was conducted by international development or research bodies and was fed into the negotiations on the 2030 Agenda.

Against this background, the previous political debate will be elaborated on in the following chapter before turning to a more detailed description of the role of culture in the 2030 Agenda.

Political discourse prior to the 2030 Agenda

Human rights documents

An important basis for the debate around culture and development is provided by the most important human rights documents. The "Universal Declaration of Human Rights" by the United Nations General Assembly in 1948 (UN General Assembly 1948) includes two key references to the protection of culture-related individual rights. Article 22 protects cultural rights that are indispensable for the dignity and development of personality. Another aspect of the complex concept of culture is protected in article 27, which states that "[e]veryone has the right freely to participate in the cultural life of the community" and the right to "the protection of the moral and material interests resulting from any scientific, literary or artistic production of which he is the author" (UN General Assembly 1948: Art. 27). These individual rights were further codified in the "International Covenant on Economic, Social and Cultural Rights" from 1966 (UN General Assembly 1966). The most important articles regarding culture guarantee the right of all peoples to self-determination (article 1) and operationalize the provisions made in article 27 of the "Universal Declaration of Human Rights", namely the right of everyone "[t]o take part in cultural life", the right "[t]o enjoy the benefits of scientific progress and its applications" as well as the right "[t]o benefit from the protection of the moral and material interests resulting from any scientific, literary or artistic production of which he is the author" (UN General Assembly 1966: Art. 15(1)).

As described above, these international human rights documents were adopted in an era, when culture did not play an important role in development processes.

It was largely ignored or even considered detrimental to economic development. Nevertheless, the aforementioned documents also laid out the basis for the work of the United Nations Educational, Scientific and Cultural Organization (UNESCO), which is the only United Nations body entrusted with culture in its mandate (Maraña 2010: 6), and which provided much input to the debate.

UNESCO conferences and declarations

UNESCO is mainly concerned with the protection of cultural heritage, but has also been a driving force behind raising attention for the interconnection between culture and development. It was in the 1960s that UNESCO started promoting culture as an important element of development processes, mainly by organizing inter-governmental conferences and preparing reports on the subject matter (Maraña 2010). A first milestone was achieved in 1966 with the adoption of the *Declaration of the Principles of International Cultural Cooperation* (UNESCO 1966), which states in article 1 that "[e]ach culture has a dignity and value which must be respected and preserved" (UNESCO 1966: Art. 1). In 1970, the first Inter-Governmental Conference on the Institutional, Administrative and Financial Aspects of the Cultural Policies took place in Venice (Italy) and enabled a discussion of the cultural dimensions of development. Surprisingly for that time, the final report states that "cultural progress is an essential component of economic and social progress" (UNESCO 1970), even though it must be assumed that this notion was shared within a comparatively small circle of experts and academics, without any real impact on development practice.

In 1982, Mexico hosted the World Conference on Cultural Policies (Mondiacult), where the link between culture and development was explicitly made. The Mexico Declaration stated in paragraph 16 that "[b]alanced development can only be ensured by making cultural factors an integral part of the strategies designed to achieve it" (UNESCO 1982: para. 16). It also established a definition of culture, which is still valid today. It goes beyond the understanding of culture as marketable artistic products and includes

> the whole complex of distinctive spiritual, material, intellectual and emotional features that characterize a society or social group. It includes not only the arts and letters, but also modes of life, the fundamental rights of the human being, value systems, traditions and beliefs.
>
> (UNESCO 1982)

Two further milestones were achieved with the proclamation of the World Decade for Cultural Development from 1988 to 1997 and the elaboration of the report *Our Creative Diversity* by the World Commission on Culture and Development in 1995. Both acknowledged and strengthened the interdependence of cultural aspects and economic and social development. The report *Our Creative Diversity* even states that "[d]evelopment divorced from its human or cultural context is growth without a soul" (UNESCO 1996: 15), thereby pointing to the fact that development efforts had often failed because cultural factors were underestimated (22). This same report started using the term *sustainable development*, based on the definition introduced by the Brundtland Report 1987, which acknowledges culture as an essential determinant thereof (World Commission on Culture and Development 1995).

One of the most important landmarks was reached in 2001 with the adoption of the Universal Declaration on Cultural Diversity (UNESCO 2001). Even though the declaration was of a political nature and not legally binding, it was "pioneering in promoting cultural diversity as world heritage and identified cultural diversity as a development factor" (Maraña 2010: 11). Article 3 states that "[c]ultural diversity widens the range of options open to everyone; it is one of the roots of development, understood not simply in terms of economic growth, but also as a means to achieve a more satisfactory intellectual, emotional, moral and spiritual existence" (UNESCO 2001: Art. 3), and is thus to be seen as a continuation of the 1948 "Universal Declaration of Human Rights". Furthermore, the declaration mentions the concept of *sustainable development* in article 11, affirming that cultural diversity is the key to achieving it (UNESCO 2001: Art. 11).

However, due to the non-binding nature of the declaration, it was necessary to go one step further. It was against this background that the *Convention on the Protection and Promotion of the Diversity of Cultural Expressions* was elaborated and adopted in 2005 (UNESCO 2005). As an international treaty, the convention is of a legally binding nature. It allows states to take measures to protect and promote the diversity of cultural expressions (article 1) and acknowledges the role that international cooperation can have in promoting the relationship between culture and development (Maraña 2010: 12). Furthermore, the Convention makes several explicit references to the relationship between cultural diversity and *sustainable development*. The preamble states that "cultural diversity creates a rich and varied world, which increases the range of choices and nurtures human capacities and values, and therefore is a mainspring for sustainable development for communities, peoples and nations" (UNESCO 2005). It also recognizes "the importance of traditional knowledge as a source of intangible and material wealth, and in particular the knowledge systems of indigenous peoples, and its positive contribution to sustainable development" (UNESCO 2005). Moreover, *sustainable development* is one of the guiding principles of the whole convention (article 2) and is again explicitly mentioned in article 13: "Parties shall endeavour to integrate culture in their development policies at all levels for the creation of conditions conducive to *sustainable development*" (UNESCO 2005: Art. 13).

Given these trends, it has become evident that the number of references to sustainable development in landmark documents constantly increased and that the importance of culture in the discourse on development grew over the course of decades. However, the claim to integrate culture as a defining dimension of sustainable development was not made until 2013, when the Hangzhou Declaration, entitled "Placing Culture at the Heart of Sustainable Development Policies" (UNESCO 2013a), was adopted by over 100 leaders in the field of culture and development on the occasion of the Hangzhou International Congress

(UNESCO 2013b). The declaration states that culture is a fundamental enabler of sustainability and has great potential to be a driver of sustainable development. Therefore, culture should be placed at the heart of future policies for sustainable development and "be included as the fourth fundamental principle of the post-2015 UN development agenda, in equal measure with human rights, equality and sustainability" (UNESCO 2013a). Following this reasoning, the participants of the congress called for the inclusion of a dedicated goal on culture in the post-2015 UN development agenda (later the "2030 Agenda for Sustainable Development").

United Nations General Assembly resolutions

Similar developments, from acknowledging the value of culture in development to explicitly demanding the integration of culture in the post-2015 development framework, can also be tracked in several resolutions adopted by the United Nations General Assembly. The period from 2010 and 2014 was especially remarkable in that regard.

Five resolutions on culture and (sustainable) development were adopted between 2010 and 2014. Four of them (UN General Assembly 2010, 2011, 2013, 2014) belong to a series of reoccurring resolutions, which build on each other, recall earlier commitments and keep the topic on the political agenda. Nevertheless, it can be observed that the statements on the relationship between culture and sustainable development become slightly more pronounced over time. The 2010 and 2011 resolutions ("Culture and Development") recognize that culture is an essential component of human development and that cultural diversity is important for sustainable development. Furthermore they acknowledge culture's important contributions to sustainable development and the Millennium Development Goals (MDGs), which formed the guiding development framework from 2001 to 2015. The ensuing resolutions, which are explicitly entitled "Culture and *Sustainable* Development", explicitly mention that culture is both an enabler and a driver of sustainable development, that it contributes to all three dimensions of sustainability (social, economic and environmental) and that it needs to be given due consideration in the elaboration phase of the post-2015 development agenda. The 2013 resolution even recalled the Hangzhou Declaration and states that its recommendations should be taken into account in the post-2015 process.

The 2012 resolution (UN General Assembly 2012) entitled "The Future We Want", adopted on the occasion of the Rio+20 UN conference on sustainable development, cannot be compared with the other four resolutions on culture and development mentioned earlier. With the adoption of this resolution, the United Nations General Assembly decided to elaborate a post-2015 development agenda, which shall include Sustainable Development Goals as a follow-up framework for the Millennium Development Goals that were about to expire in 2015. "The Future We Want" is much less vocal on the importance of culture for sustainable development than previous and subsequent resolutions. It recognizes the importance of cultural heritage and diversity (UN General Assembly 2012:

para. 30/41/58) and that culture can contribute to sustainable development (para. 41). However, apart from these short references, culture is only mentioned in the context of sustainable tourism (para. 130/131), sustainable cities (para. 134) and biodiversity (para. 197). This is important to note, as the way and degree to which culture is included in the 2030 Agenda on Sustainable Development mainly reflects the language of the 2012 resolution. Interestingly, subsequent resolutions with stronger language on "Culture and Sustainable Development" (UN General Assembly 2013, 2014) had less influence on the 2030 Agenda.

The final outcome document "Transforming the World: The 2030 Agenda for Sustainable Development" is the result of long political negotiations between governments of all UN member states. Yet other groups, such as NGOs, the private sector, academia, minority groups, sub-national and regional actors and even the public had a voice in the process. Thus, mainly non-state actors were advocating for the integration of culture as a central concept in the 2030 Agenda.

Inputs by other groups and networks

Among the most vocal networks were the International Federation of Arts Councils and Culture Agencies (IFACCA), the Agenda 21 for Culture, the International Federation of Coalitions for Cultural Diversity (IFCCD) and Culture Action Europe. Together, these networks of organizations and global cultural actors started advocating for a strong role of culture in the 2030 Agenda in 2014. They demanded to put culture at the heart of a positive transformative change (International Federation of Arts Councils and Culture Agencies et al. 2014). Based on the recommendations made in the Hangzhou Declaration (2013a), they argue that culture is a means to foster development in all three dimensions of sustainability (culture as an enabler) but also a development end in itself. Therefore, they call for a specific goal on culture in the 2030 Agenda, "in order to reinforce the potential of cultural resources for sustainable development and to achieve their long-term sustainable use for current and future generations" (International Federation of Arts Councils and Culture Agencies et al. 2014: 7). The suggested title for a SDG on culture was "Ensure Cultural Sustainability for the Wellbeing of All". Together with this suggestion, the networks proposed ten specified targets that relate culture to all dimensions of sustainable development and focus, among others, on a normative institutional framework, cultural and creative industries, education and training systems as well as cultural heritage (International Federation of Arts Councils and Culture Agencies et al. 2014: 8–9; Swiss Confederation 2014).

During the same period of inter-governmental negotiations, an informal group of countries, the "*Group of Friends on Culture and Development*", was established in New York. It was chaired by Peru and comprised 29 countries from all the different regions. The group worked towards ensuring that the role of culture is fully acknowledged in the 2030 Agenda, however not taking such a strong position as the cultural networks mentioned above. As Switzerland was part of this group of friends and because the author was Switzerland's lead negotiator during the elaboration process of the 2030 Agenda, the Swiss position with regard to culture and sustainable development will be outlined in the following paragraph.

The Swiss position

Although the integration of culture was not a top priority of Switzerland in the negotiations on the 2030 Agenda, the Swiss delegation incessantly argued in favour of a strong integration of culture in the emerging universal goal framework (Swiss Confederation 2014). It thereby referred to the UN General Assembly Resolution of 2013 (UN General Assembly 2013), which stipulated that due consideration be given to culture in the elaboration of the post-2015 development agenda. Switzerland acknowledged the role of culture as an enabler and driver of sustainable development and gave three main reasons for the prominent integration of culture in the 2030 Agenda.

First, culture is understood as a critical component of human rights. Not only is guaranteeing cultural rights (as mentioned earlier in this chapter) central for "forging inclusive and equitable societies, and an effective, people-centred approach to sustainable development" (Swiss Confederation 2014: 2). Culture is also strongly linked to other human rights, such as the right to education, the right to freedom of speech or the right to peaceful assembly (Swiss Confederation 2014: 3).

Second, cultural diversity needs to be protected and promoted, as it contributes to sustainable development. Universal policies and one-size-fits-all approaches to development, which do not take into account the cultural context, are doomed to failure and are to be considered unsustainable. Only people-centred, participative and culturally sensitive development processes can foster endogenous development and empower populations by building their cultural assets. Moreover, cultural diversity also produces social capital and strengthens resistance to unfair treatment by the authorities. It is therefore a source of resilience and stability. The promotion of cultural diversity can even enhance tolerance and prevent conflicts between and within nations (Swiss Confederation 2014: 3). This last point is particularly important, as the 2030 Agenda is the first universal sustainable development agenda that also comprises a dedicated goal on peaceful and inclusive societies (SDG 16), and acknowledges that there is no development without peace and no peace without development.

Third, cultural and creative sectors have to be considered in shaping the sustainable development agenda. They are both resilient and rapidly expanding sectors in developing and developed countries (e.g. cultural tourism) and thus contribute to creating decent jobs and economic growth. They can also catalyse social change in that artistic expression can help fight against exclusion and inequality or support healing the scars of violence (Swiss Confederation 2014: 4).

Despite the importance Switzerland assigned to culture in sustainable development processes, it did not advocate for the inclusion of a dedicated standalone goal in the 2030 Agenda. One reason for this position was the ambition to create an agenda with a limited number of goals covering all three dimensions of sustainability. At the same time, a certain lack of clarity around the concept of culture prevailed, which is why cultural aspects tended to be eclipsed by social, economic and environmental issues. Moreover, many negotiators were sceptical with regard to the *goalability* of culture because of the lack of adequate indicators to measure concrete targets on culture. Lastly, the inclusion of culture as a standalone goal bore the risk of opening doors for cultural relativism with regard to the other goals of the agenda and human rights aspects in particular. For Switzerland, a strong human rights framework was the pre-condition for any discussion on culture.

On these grounds, even Switzerland, a country fully in favour of integrating cultural aspects into the new framework, and recognizing culture as a powerful resource for sustainable development and peace, was therefore not in a position to support a standalone goal on culture. Instead, the objective was to ensure that "the principle of a people-centred, culture-sensitive approach to development cooperation is taken into account within the human-right-based implementation framework of the Post-2015 Agenda" and that "the specific roles of culture, cultural diversity and/or artistic and cultural expressions in contributing to sustainable development are accordingly reflected in the goals or transversal approaches" (Swiss Confederation 2014: 4). The next section will summarize to what extent these objectives have been achieved and how the concept of culture was eventually integrated in the 2030 Agenda for Sustainable Development.

The role of culture in the 2030 Agenda for Sustainable Development

The final outcome document of the 2030 Agenda (UN General Assembly 2015a) was adopted on September 25, 2015 at the Sustainable Development Summit in New York. All 193 UN member states agreed on the text and more than 140 heads of state and government were present in person, along with Pope Francis and celebrities from all over the world. The universally applicable framework with its 17 Sustainable Development Goals (SDGs) and 169 targets covers all three dimensions of sustainability in a balanced way. At the same time, it also puts a specific focus on peace, security and governance as well as on a renewed global partnership to achieve the goals (i.e. the "five Ps": People, Planet, Prosperity, Peace, Partnerships). As described above, inputs influencing the elaboration and negotiations of the 2030 Agenda had been numerous and positions on the integration of culture were diverse.

In the end, a paradigm change was achieved insofar as, for the first time in history, culture was integrated in a global sustainable development agenda. Several references to cultural aspects are made in the political declaration as well as in some of the targets. The declaration explicitly mentions cultural diversity as a goal, along with respect for human rights and human dignity, the rule of law, justice, equality and non-discrimination, as well as the respect for race and ethnicity (UN General Assembly 2015a: para. 8). In paragraph 36, the member states "pledge to foster intercultural understanding, tolerance, mutual respect and an ethic of global citizenship and shared responsibility" and they "acknowledge the natural and cultural diversity of the world and recognize that all cultures and civilizations can contribute to, and are crucial enablers of, sustainable development" (para. 36).

At the target level, four explicit and two indirect references to culture are made. Target 4.7 on education for sustainable development mentions culture's role in contributing to sustainable development by demanding that, by 2030,

> all learners acquire the knowledge and skills needed to promote sustainable development, including, among others, through education for sustainable development and sustainable lifestyles, human rights, gender equality, promotion of a culture of peace and non-violence, global citizenship and appreciation of cultural diversity and of culture's contribution to sustainable development.
>
> (UN General Assembly 2015a: Target 4.7)

Target 8.9 mentions culture as a factor for economic development in the context of sustainable tourism by stating that states have to "devise and implement policies to promote sustainable tourism that creates jobs and promotes local culture and products" (UN General Assembly 2015a: Target 8.9). In the context of sustainable tourism, target 12.b suggests that states "[d]evelop and implement tools to monitor sustainable development impacts for sustainable tourism that creates jobs and promotes local culture and products" (UN General Assembly 2015a: Target 12.b). Furthermore, target 11.4 explicitly calls states to "[s]trengthen efforts to protect and safeguard the world's cultural and natural heritage" (UN General Assembly 2015a: Target 11.4).

Besides these explicit references, two targets indirectly mention specific aspects of culture and their contribution to sustainable development, namely traditional farming knowledge and the creative industries as a source for entrepreneurship and job creation. Within SDG 2 on hunger, food security and sustainable agriculture, target 2.5 requires "fair and equitable sharing of benefits arising from the utilization of genetic resources and associated traditional knowledge" (UN General Assembly 2015a: 2.5). Within SDG 8 on economic growth and decent work for all, target 8.3 asks states to "[p]romote development-oriented policies that support productive activities, decent job creation, entrepreneurship, creativity and innovation" (UN General Assembly 2015a: 8.3).

Despite the progress made by anchoring a culture-sensitive perspective in a universal sustainable development agenda, the ambitious claims of the aforementioned groups were not taken into full consideration. Neither was a standalone goal on culture established, nor could culture be positioned as the fourth dimension of sustainable development (next to the social, economic and environmental dimensions). The level of integration thus largely reflects the position of the 2012 General Assembly resolution "The Future We Want". More recent and comprehensive claims were not fully taken into account. At the same time, the term *cultural sustainability* was not used in the outcome document of the 2030 Agenda.

It does not figure in any important UN declaration or treaty and it is yet to be seen if and how it will make its way into the international political discourse.

Conclusion

The importance of culture in the (sustainable) development discourse has increased remarkably since the 1960s. Culture has come from being considered an obstacle to economic development in early development theories to being recognized as a crucial contributor to, and an enabler of sustainable development in the 2030 Agenda. The main drivers of this evolution at the international political level have been UNESCO and the UN member states in their capacity as actors in the UN General Assembly, not least during the negotiations of the 2030 Agenda. Non-state actors have also been actively involved, as illustrated by the example of several networks that promoted a standalone goal on culture in the 2030 Agenda. However, precisely this objective could not be achieved and the concept of culture is thus still not acknowledged as fourth pillar of sustainable development. On the contrary, the widely accepted definition of sustainability with its three dimensions – social, economic and environmental – was reinforced by the UN member states, and not complemented by a fourth, cultural dimension. In order to establish an even stronger interlinkage between culture and sustainable development in the political discourse, more research as well as active academic engagement is needed.

Among the reasons for not including culture as a dedicated goal in the SDG framework was the lack of clarity regarding its concept, definition and interpretation, as well as doubts with respect to the measurability of progress on cultural goals. Improvement on these two issues would help make the cultural dimension more explicit and prevent it from being eclipsed by the politically agreed social, economic and environmental ones.

Despite the adoption of the 2030 Agenda and its explicit enhancements with regard to the inclusion of culture, the discussion about its role in the context of sustainable development will continue. Only a few months after the Sustainable Development Summit in New York, the UN General Assembly adopted another resolution on "Culture and Sustainable Development" (UN General Assembly 2015b), thereby affirming culture's contribution to all three dimensions of sustainable development and confirming that this contribution shall be taken into consideration in the follow-up and review framework of the 2030 Agenda (UNESCO 2015).

However, the term *cultural sustainability* has up until now not been used in any major policy document at the UN level, revealing a gap between the academic discourse, which tends to be more advanced, and the political reality. This might also be owed to the lack of an accepted definition. It remains unclear what the concept intends to preserve. Does it relate to cultural heritage, cultural liberties and rights, different modes of life, the creative industries or encompass all these aspects?

Yet, the topic seems to have gained relevance in the past few years, not only at international, but also national level. The UN System Task Team on the

Post-2015 UN Development Agenda (2012), which was comprised of experts from over 50 UN entities, found that in 2012, culture was mentioned in 70 percent of all UN Development Assistance Frameworks (UNDAF). Five years before, this number had been at a low 30 percent only. The Task Team also found that by 2012, 18 different UN organizations were working on culture or regularly adopted cultural-sensitive approaches (UN System Task Team on the Post-2015 UN Development Agenda 2012). Precisely because of this momentum, it would be worthwhile to concretize the notion of cultural sustainability in a way that it is both applicable to the international political discourse and relevant to sustainable development practitioners around the world.

Note

1 This contribution is based on a presentation held by Switzerland's Special Envoy for Global Sustainable Development, Ambassador Michael Gerber, in the context of the workshop "Cultural Dimensions of Sustainability" at the University of Bern, 10–12 November 2016. In composing this article, Michael Gerber highly profited from the assistance of Stefan Luegstenmann, Advisor in the 2030 Agenda team, whom he sincerely thanks for his valuable support.

References

International Federation of Arts Councils and Culture Agencies, Agenda 21 for Culture, International Federation of Coalitions for Cultural Diversity, Culture Action Europe (2014). *Culture as a Goal in the Post-2015 Development Agenda.* www.agenda21culture.net/index.php/docman/per-a-circulars/492-cultureasgoal-final/file [Accessed 21 April 2017].

Maraña, M. (2010). "Culture and Development – Evolution and Prospects". *UNESCO Etxea Working Papers*, No.1, 2–28. www.unescoetxea.org/dokumentuak/Culture_Development.pdf [Accessed 21 April 2017].

Streeten, P. (2006). "Culture and Economic Development". In V. A. Ginsburgh, and D. Throsby. *Handbook of the Economics of Art and Culture*. Amsterdam: Elsevier, 399–412.

Swiss Confederation (2014). *Culture in the Post-2015 Agenda*, 1–5. www.eda.admin.ch/content/dam/agenda2030/en/documents/topics/culture-post-2015.pdf [Accessed 21 April 2017].

UN General Assembly (1948). "Universal Declaration of Human Rights", A/RES/217(III). www.un.org/en/ga/search/view_doc.asp?symbol=A/RES/217(III) [Accessed 21 April 2017].

UN General Assembly (1966). "International Covenant on Economic, Social and Cultural Rights". www.ohchr.org/EN/ProfessionalInterest/Pages/CESCR.aspx [Accessed 21 April 2017].

UN General Assembly (2010). "Culture and Development", A/RES/65/166 www.un.org/en/ga/search/view_doc.asp?symbol=A/RES/65/166 [Accessed 21 April 2017].

UN General Assembly (2011). "Culture and Development", A/RES/66/208. www.un.org/en/ga/search/view_doc.asp?symbol=%20A/RES/66/208 [Accessed 21 April 2017].

UN General Assembly (2012). "The Future We Want", A/RES/66/288. www.un.org/en/ga/search/view_doc.asp?symbol=%20A/RES/66/288 [Accessed 21 April 2017].

UN General Assembly (2013). "Culture and Sustainable Development", A/RES/68/223. www.un.org/en/ga/search/view_doc.asp?symbol=A/RES/68/223 [Accessed 21 April 2017].

UN General Assembly (2014). "Culture and Sustainable Development", A/RES/69/230. www.un.org/en/ga/search/view_doc.asp?symbol=A/RES/69/230 [Accessed 21 April 2017].

UN General Assembly (2015a). "Transforming Our World: The 2030 Agenda for Sustainable Development", A/RES/70/1. www.un.org/en/ga/search/view_doc.asp?symbol=A/RES/70/1 [Accessed 21 April 2017].

UN General Assembly (2015b). "Culture and Sustainable Development", A/C.2/70/L.59. www.un.org/ga/search/view_doc.asp?symbol=A/C.2/70/L.59&Lang=E [Accessed 21 April 2017].

UN System Task Team on the Post-2015 UN Development Agenda (2012). "Culture: A Driver and an Enabler of Sustainable Development", 1–10. www.un.org/en/development/desa/policy/untaskteam_undf/thinkpieces/2_culture.pdf [Accessed 21 April 2017].

UNESCO (1966). *Declaration of the Principles of International Cultural Co-operation*. http://portal.unesco.org/en/ev.php-URL_ID=13147&URL_DO=DO_TOPIC&URL_SECTION=201.html [Accessed 21 April 2017].

UNESCO (1970). *Final Report of the Intergovernmental Conference on the Institutional, Administrative and Financial Aspects of the Cultural Policies, Venice*. www.ifa.de/fileadmin/pdf/abk/inter/unesco_venedig.pdf [Accessed 21 April 2017].

UNESCO (1982). *Mexico City Declaration on Cultural Policies*. http://portal.unesco.org/culture/en/files/12762/11295421661mexico_en.pdf/mexico_en.pdf [Accessed 21 April 2017].

UNESCO (1996). *Our Creative Diversity – Report of the World Commission on Culture and Development*. http://unesdoc.unesco.org/images/0010/001016/101651e.pdf [Accessed 21 April 2017].

UNESCO (2001). *Universal Declaration on Cultural Diversity*. http://portal.unesco.org/en/ev.php-URL_ID=13179&URL_DO=DO_TOPIC&URL_SECTION=201.html [Accessed 21 April 2017].

UNESCO (2005). *Convention on the Protection and Promotion of the Diversity of Cultural Expressions*. http://portal.unesco.org/en/ev.php-URL_ID=31038&URL_DO=DO_TOPIC&URL_SECTION=201.html [Accessed 21 April 2017].

UNESCO (2013a). *The Hangzhou Declaration – Placing Culture at the Heart of Sustainable Development Policies*. www.unesco.org/fileadmin/MULTIMEDIA/HQ/CLT/images/FinalHangzhouDeclaration20130517.pdf [Accessed 21 April 2017].

UNESCO (2013b). *The Hangzhou Declaration: Heralding the Next Era of Human Development*. www.unesco.org/new/en/unesco/resources/the-hangzhou-declaration-heralding-the-next-era-of-human-development/ [Accessed 21 April 2017].

UNESCO (2015). *UN General Assembly Adopts a New Resolution on Culture and Sustainable Development*. http://en.unesco.org/news/general-assembly-adopts-new-resolution-culture-and-sustainable-development [Accessed 21 April 2017].

3 The future of cultural heritage and its challenges

Aleida Assmann

Introduction: three meanings of 'future'

Today we are witnessing a radical shift in the structure of our temporality: While the future has lost much of its luminosity, the past has more and more invaded our consciousness. In this context, the meaning of the word 'future' has changed considerably and widened its semantic range. Let me start with three meanings of future that are commonly used today and exist peacefully side by side. If looked at more closely, however, they contradict and exclude each other. As they are all in use, we may conclude that they are all needed.

First meaning of future: progress, utopia and novelty

A first meaning and vision of the 'future' is linked to the narrative of progress. The future is the time when the world is expected to become a better, healthier, more just and freer place. This narrative of progress is bound up with the project of modernization and is based on constant innovation in the fields of science and technology, harbouring the universal promise of transforming the old into the new, with the new always standing for the better, the more satisfactory, the more advanced option. Hegel's philosophy in the nineteenth century was built on this plan of history as a long-term, linear process, in which civilization evolves over centuries from less conscious to more conscious stages of freedom. Even though the end of history is endlessly deferred, the 'telos' of history has already been revealed and contains promises of material fulfilment, aesthetic perfection, or spiritual transcendence.

This positive concept of the future as a time in which our hopes and expectations of a better world will be fulfilled, clearly prevailed after the Second World War. After 1945, the determination to turn one's back to the catastrophic cataclysm of the twentieth century, to mark a new beginning and to embrace a bright future was a general hope in both Eastern and Western Europe. "Risen from rubble and ruins and turning to a bright future" (*Auferstanden aus Ruinen und der Zukunft zugewandt*) were the first lines of the National anthem of the GDR. This first meaning of future is connected with concepts like revelation, discovery and the emergence of the new. To fully grasp the new, we have to forego the old; to embrace the future, we have to let go of the past.

Second meaning of future: uncertainty and risk

In post-war West Germany, I grew up with the first concept of a bright future. By the 1980s and 1990s, however, it was rapidly losing much of its glamour and was replaced with a concept of the future that was less optimistic and far-reaching. This second concept of future is much older than the first, and can be traced back to many ancient cultures. In the second meaning, the word future stands for what is unexpected and unthought of, what comes towards us as a pleasant or unpleasant surprise. Que sera, sera – some of us remember the words of the song in which Doris Day looked into the future in an American film of the 1950s.[1] This concept of the future is linked to fate, fortune, contingency and coincidence. It is often accompanied by a fatalistic attitude towards time: que sera, sera. The old wisdom that the song conveys is that one has no other choice but to wait, see and accept whatever the future brings. There are, however, also other ways of responding to the veiled and hidden dimension of the future. In ancient cultures, techniques of divining were created to be prepared for the unknown and to back up decisions. Already in the early modern age, the negative quality of uncertainty was transformed into the positive quality of risk-taking, transforming the fickle mediaeval goddess Fortuna into the modern goddess of Chance. In the form of challenge, experiment and productive risk, uncertainty was built into the temporal structure of innovation. In modern societies, new methods like opinion polls and financial insurance systems have been devised to reduce the invisibility and unknowability of the future and have achieved some degree of security in anticipating failures and minimizing losses.

Third meaning of future: sustainability

There is, I want to claim, a third meaning of 'future' that is generally accessible and is equally obvious but had totally disappeared from consciousness until it was recovered in the 1970s in the framework of a new ecological awareness. Knowledge about the limits of growth and the finitude of natural resources has radically changed our relation to the world and our orientation in time. With the shift towards ecological thinking the meaning of future changed radically: Nature was no longer considered to be an inexhaustible source of change and renewal or the telos of all our wishes and hopes, but became the object of serious concerns, care and a new temporal ethics of responsibility.

This third concept of future is linked to the term 'sustainability', which dates back to the 1980s. It arose in the ecological discourse and, so far, has only reluctantly been applied to the humanities. Sustainability, however, can be conceived both in terms of natural and cultural resources (cf. Rehling 2014). In both cases, the idea of future is no longer linked to surprise, change or innovation, but applies to objects and values that have already existed for a shorter or longer time and are meant to be preserved. Future in the sense of sustainability refers to the preservation and continuation of what we already have, know, value and cherish. This concept of course also applies to cultural texts, objects, buildings and monuments that have existed for some time and are deemed worthy of being preserved to have a future. The Early Modern poet John Milton, for instance, had hoped that he "might perhaps leave something so written to after-times, as they should not willingly let it die" (1641: 37).[2] And 180 years later, the Romantic poet Percy Bysshe Shelley reminded his readers that "all language, institution and form, require not only to be produced but to be sustained" (1962: 238).

Cultural sustainability

The third meaning of future involves a cultural pact between past, present and future generations, based on an assertion that here is something that we will not give up, lose, forget or ignore but continue to invest with interest, value and affect. Future in this context does not refer to progress or an eternal craving for the new, but is related to cultural memory, heritage and identity. This third meaning of future is no longer the opposite of the past but intimately linked to something in the past, which is to be ensured for the future. While the categorical imperative of the time regime of modernity had been: 'break with the past!', the categorical imperative of the time regime of sustainability is 'care for the past!' applying interest, diligence and responsibility – of course not to all objects of the past but only to what has been selected for further re-use. Such a relation to selected objects of the past can be based on material value, aesthetic value, scientific value or identity value: This object is important because it is part of my extended past, history and tradition, and I understand and respect that the same affective link is active in other communities of culture. This kind of future demands financial resources, attention and curation; it requires specialists who are engaged in preserving an afterlife for certain cultural objects. This form of cultural sustainability has existed for centuries; a more recent variation is the concept of 'cultural heritage' or, 'world heritage' as established by UNESCO, emphasizing a vital link between culture, history, memory and identity.

What can we learn from this short overview of three different concepts of the future? It helps us to realize that the words we use to describe temporal relations are not neutral terms that can be taken for granted, as if invested with stable and unequivocal meaning. Quite the contrary: These terms are related to different time regimes that are connected with different world-views and consequently prescribe different values and forms of action. Past and future, therefore, are not safely situated behind us or before us on a perennial temporal axis. Rather, they are dimensions of time that are created according to cultural patterns and performed within human frames of action. While the natural sciences and technology constantly produce the framework of a new future through their exclusive emphasis on inventions and innovations, the humanities produce a different framework of future by also recovering, maintaining and transforming the cultural memory of a group. Cultures depend on forms of transmission, backed up by memory, they exist in a constant process of transformation through recovering, reworking, revaluing, reanimating and restructuring the collected and collective heritage of the group. But this also means that the future of cultural memory and heritage is always precarious. It relies on renewed acts of attention, interest, remembering, preservation, transmission and discussion. As such, cultural memory can be easily neglected and destroyed.

Until very recently, the topic of cultural heritage and the work of UNESCO were hardly topics of intellectual interest or common concern. Even though there exists today a growing number of university chairs and academic departments with the denomination 'heritage studies', the term is still controversial and unpopular among many historians. For this group, the polemic term 'heritage industry' stands for the exploitation of the past for economic purposes fostered by the state and city tourism, for a world wide competition of national symbols, elitist concerns and identity politics. This sweeping dismissal of a huge sector of global cultural concerns and practices, however, can hardly be considered an adequate approach to these problems. After a long series of brutal assaults by Islamic State (IS) troops on sites of Western and Eastern cultural history, the fashionable criticism of 'heritage industry' seems grossly out of date. Let us take an imaginary visit to an actual site of destruction in order to rethink the concept of heritage in this time of acute heritage-crisis, to recover a glimpse of its history and to try to assess its future.

The discovery of Palmyra by Robert Wood

The sites of antiquity that we cherish today have not always been part of European cultural memory. After ostensibly valuable materials had been robbed and plundered from ancient temples in antiquity, such sites were abandoned to the slow destructive work of time. For many centuries, they existed in a state of complete negligence and dilapidation. But during all these years, the lack of attention and sheer bulk of these monuments were also a reason for their survival.

It was in this ruinous state that Robert Wood and two of his friends discovered the ancient city of Palmyra in the Syrian dessert with its temples, theatres, arches and amazing streets lined with pompous columns. An enchanted reader of Homer, Wood had gone looking for the geographic sites of former civilizations. Writing in the middle of the eighteenth century, he belonged to the Age of Enlightenment, but was also a forerunner of Romanticism. He combined both sentiments, the classical spirit of Hellenism and a Romantic passion for the sublime quality of picturesque ruins. He was the first to do what many were to do after him: He travelled to the historical sites of the ancient Mediterranean world, described in detail what he saw there and produced and circulated images of the site. Wood was convinced that the legacy of Homer and the Greeks was not only transmitted through words but that there were also material places and palpable traces in geographical space that could be visited, recovered and estimated. His fascination with ruins was animated by the idea that what was far away and lost in time could be rediscovered and accessed in space.

We owe our vision of Palmyra to Robert Wood, the pioneer who rediscovered the historic site and shaped its imaginary recovery. His description and the images of the early artists who travelled with him created the cultural pattern through which succeeding ages and generations accessed these ruins after him (cf. Wood 1753).[3] The love for ancient Greek civilization, its history and culture became a living tradition and an influential part of Western cultural memory. Wood's visual discovery of Greek art and his enthusiasm are at the roots of Western historical consciousness, identity construction and the canon of its art.

The rise of the concept of cultural heritage

The concept of 'cultural heritage' emerged in the European nations in a process throughout the long nineteenth century. In the context of this new historical consciousness, the interest in heritage became a growing occupation, involving official administrative efforts from above as well as civic movements from below. The concept ranged from simple retro-fictions and mythical fantasies to modern endeavours valuing 'authenticity' and guarding the monuments of the past. It was rooted in the spirit of historicism, a new branch of historical scholarship established at nineteenth-century universities including the disciplines that we sum up today under the label 'humanities'. Within the context of historicism, an empirically grounded interest in and investigation of a nation's culture emerged, becoming the object of scientific scrutiny that was focused on topics such as the nation's history, its art, landscape, local traditions and folklore. This academic interest in one's own culture and its past transcended also the elitist aesthetic canon and included more and more local and vernacular elements. It was propelled by scholarship, measures for public education and museum presentation. It was by no means confined to these institutions, however, but soon spilled over and reached a growing public through popular novels, local commissions, exhibitions, courses and excursions.

The central motor behind this new interest in national cultural heritage was the French Revolution that had violently cut off the French past by destroying the institutions of the monarchy, clergy and aristocracy. Together with these institutions it destroyed its traditions, abolishing the past in the name of radical modernization. This abrupt change had a huge impact on forms of life and local milieus. The past was cut off in acts of violent 'culturoclasm', but it was not destroyed and abolished altogether. It was preserved as 'history' and delegated to the care of new professions such as historians, archaeologists and curators of museums. In co-evolution with a forceful break, the past was reconsidered as an object of professional scholarship and public veneration. The revolutionary origin of the concept of cultural heritage explains its complexity, involving a number of tensions and paradoxes such as that between

destruction and preservation
top down and bottom up processes
modernist and traditionalist interests and
the national and the international.

While the nineteenth-century concept of heritage helped to invent, demarcate and highlight national differences, it also turned into a common European and even global project built on a consensus and practical rules that transcended the aims and purposes of the national collective. From its very beginning, it was a process that developed on a local, a national and an international level, and as such was not only a European but also a global movement, creating transnational networks that promoted common standards on the basis of shared values and concerns. In other words, there is such a thing as a common culture of heritage that is only being discovered and historicized today.

A first manifestation of a universalizing spirit of heritage consciousness was the term 'vandalism' as coined by the Abbé Henri Grégoire during the French Revolution. The term summed up a new historical sensibility, criticizing violent acts against material monuments, irrespective of national points of view. There were two new ideas in particular that became influential after the French Revolution and developed in co-evolution: "that nations had a heritage" and "that preservation was a sign of civilized governance" (Swenson 2013: 46). National heritage, in the long run, came to be seen as the particular share of a common world heritage, owned, valued and protected by 'mankind'. Thus, from the start, patriotism and internationalism were intimately linked in the creation and promotion of this new concept, in which the respect for time and the past became a new universal religion, promoting, as it was hoped, "peace and goodwill among the nations" (Swenson 2013: 194).[4]

Only "barbarians and slaves", Grégoire had written, "hate the sciences and destroy the monuments of art. Free men love them and conserve them" (Swenson 2013: 34). The act of vandalism was understood as a lapse from the newly shared and internationally ratified principle of the sanctity of cultural heritage. Violent seizures of monuments and works of art had been condemned as 'crimes against humanity' as early as 1800 (Swenson 2013: 39). To consciously act against these standards meant a serious breach in the common bond of civilized nations (or 'Kulturstaaten').

The concept of heritage was not only built on values and sentiments, but also on an evolving process of international legislation. In parallel with rules protecting civilians and wounded soldiers in times of war (The Geneva Conventions), rules were drafted for safeguarding cultural property in armed conflict (The Hague Conventions). We have experienced many times that these laws proved sadly inadequate to the task of protecting cultural heritage, but we have also witnessed that every new atrocity was answered with further statutes involving new legal measures, values and standards.[5]

War on culture

It is quite obvious that the two world wars have dramatically ignored and acted against the humanitarian principles established by the various conventions. We have to add, however, that these conventions were also confirmed and consolidated through the devastating losses. Until recently, many states in the world were ready to act against these principles, but none of them would go so far as to expressly veto them and to proudly present transgressive acts to the global public.

This is, however, exactly what the IS is doing. It is leading a symbolic warfare not only against Western civilization but also against the normative and regulative idea of an inclusive concept of humanity based on a shared consensus of values. The recent IS attacks on world cultural heritage are painfully spelling out through the channels of global communication that such an inclusive concept of humanity based on universal values and supported by all states and nations no longer exists.

The role of culture in war is complex and changing, but culture has always been implicated in one form or another. Imperial wars were aimed at conquering other countries and incorporating the culture of the other as a trophy into one's own treasure, exploiting and exposing it symbolically as a representation of the superior power. Religious and nationalist wars aimed at destroying the culture of the other as a blasphemous contradiction to one's own beliefs (Loewenthal 1987). In colonial wars, the so-called 'civilizing mission' played an important part not only in camouflaging the scramble for new territories and material resources, but also for forcefully exporting religion and culture and imposing it on indigenous populations. More recently, civil and religious wars have continued this bloody legacy of war against culture in the Balkans with the destruction of ancient cities and the burning of the library of Sarajevo.

"The destruction of cultural heritage of world-historic value by the IS is breaking with everything that we hitherto knew" (Schmickler 2015: n.p.). This comment by Sönke Neitzel, historian at the London School of Economics, is perhaps a bit rash. Unfortunately, there is a long record of the intentional destruction of world-historic objects and buildings of culture. Leo Loewenthal has written an important essay on the burning of books and libraries, in which he listed various motives behind the destruction of cultural heritage.[6] According to Loewenthal, the central motive driving this kind of violence against cultural objects is *the extinction of history*. After a political regime change, there is often the desire to eliminate the past and to replace it by a new foundational narrative. As history is always long, complex and composed of different strands and layers, it is also polyvocal and complex, undermining the construction of a single legitimizing myth. To wipe the slate of history clean by creating a *tabula rasa* was a common strategy for a new political power to prevent contradictions and claims to another reality in the act of starting over and inventing itself as new. Under these circumstances, the destruction of cultural relics goes hand in hand with *acts of purification*. Whatever jars with the absolute truth imposed by the regime in power has to disappear. These politics of *damnatio memoriae* have been fostered by all regimes built on absolute truth and totalitarian power. Purification was also the strong motive behind acts of iconoclasm in the early time of the Reformation.

The Western veneration for antiquities and heritage was a secular and modern project starting with the Enlightenment around 1750 and was consolidated on an institutional basis around 1800 after the French Revolution. The cult of antiquities, the value of heritage and the high esteem for extinct cultures became part of modern Western civilization. We may even refer to the veneration of this complex of aesthetics, art and historical consciousness as a new secular religion. Libraries, theatres and museums became the modern temples of this religion, archaeologists, art historians and literary scholars became its priests, and the touristic sites of historic relics and ruins became the sacred places of modern pilgrimages.

It is exactly this secular heritage religion of the West against which the 'blasphemous acts' of IS terror are aimed. This is indeed a new dimension of age-old vandalism. We are witnessing today violent acts of 'purification' and the destruction of history with brutal force, deleting in one stroke Western aesthetics and centuries of historical thinking and heritage consciousness. But this is not only the time of nameless destroyers but also of the courageous preservers who deserve to be named, such as Abdel Kader Haidara, curator of one of the most important libraries in Timbuctoo, a position handed down in his family for generations. It was he who clandestinely organized the evacuation of 350,000 precious manuscripts to the south of Mali to safeguard them against Jihadist violence. The collection contains books on art, philosophy and science, documenting the mediaeval Arabic enlightenment. The Syrian archaeologist Khaled al-Asaad (1934–2015) was director and devoted curator of antiquities in Palmyra for 50 years, before he was publicly executed at the age of 82 by IS militants on 18 August 2015. His successor in office, Maamoun Abdel-Karim and his staff had managed to transfer before the arrival of the IS in Palmyra thousands of antique works of art to the national museum in Damascus, including many statues and precious funerary objects.

The world is in a continuous state of shock watching the slow and inevitable work of destruction produced by the threatening and alien violence of the IS. Since UNESCO is finally bringing together the West and the rest of the world in agreeing to value and protect cultural heritage for future generations, we are faced with a radical non-synchronicity of the synchronous ('die Gleichzeitigkeit des Ungleichzeitigen'). Humanity, as now becomes painfully manifest, is living in a globalized world that is unified through the channels of media of communication, but internally torn apart in its values. We are no longer talking about 'cultural differences' but about living in different worlds and time zones, the values of which have drifted apart and become utterly untranslatable and non-negotiable. Today, we do not just live in a world characterized by a 'clash of civilizations' but by a temporal chasm between the 'archaic' and the 'modern'. The new war on culture is being conducted without a minimum shared consensus.

Conclusion: the future of cultural heritage

The great shock that we are experiencing today is well captured in the words of another famous song, Joni Mitchell's "Big Yellow Taxi".[7] We have all made the experiece that loss heightens the value that we attach to things. The words of Mitchell's song can be equally applied to ecological resources and cultural heritage, emphasizing once more the double meaning of the term 'sustainability'. This new world-historical situation also demands a new perspective on cultural heritage in the past, present and future.

To come back to Palmyra: This is an ongoing story. During the Syrian civil war, the ancient site changed hands several times. Eventually, the Syrian Army retook the city, but considerable parts of the site had been destroyed between May 2015 and March 2017. These new acts of vandalism had been visually staged by the IS as a symbolic triumph. Just as the destruction of Palmyra had been very public, so is the debate about its reconstruction. It even reached my own town Konstanz, in which in July 2017 an exhibition opened under the title 'Rebuild Palmyra?' (Hauser and Reiterer 2017). It was prepared by students of two universities and three professors from different disciplines such as communication design, architecture, history and computer science.[8] The exhibition was planned and designed by the students themselves. It was highly mediated, using the most advanced equipment of interactive digital technology including holograms to engage the visitors in a reflection about the past, present and future of cultural heritage. It was displayed on four floors: The ground floor presented a plethora of media reports on Palmyra, indicating its importance as a public global issue. The first floor showed the history of Palmyra from its beginning in the first century B.C. as a node of trade for the exchange of luxuries, mediating between East and West in the outskirts of the Roman Empire, and its rediscovery in the eighteenth century. The second floor focused on the destruction itself by offering a digital architectural model and a telescope for 3D visualizations that allowed a virtual walk through the undestroyed archaeological site. The last and third floor rounded up the presentation by initiating a discussion about the future of this site. Here the question 'Rebuild Palmyra?' was raised again and the visitors were invited to vote yes or no for a reconstruction of the site.

The fate of Palmyra is far from unique. The site has become a symbol for the precarious state of cultural heritage in general as it is oscillating between pride and terror.[9] Sites of cultural heritage are exceptional in that they are venerated and attract many visitors, but they can also become the target of violence that turns them into sites of trauma and destruction. In the age of terror, a paradigm shift may occur in the approach to Cultural Heritage from a policy of conservation to a policy of reconstruction. As any site can become the target of *destruction*, curators and managers of the UNESCO now face the new responsibility to carefully document a site to provide evidence for a possible future assault and reconstruction. We are dealing here also with a new twist in our concepts of 'future', as 'cultural sustainability' may include the temporary absence of what we wish to keep continuously present. This has also an impact on our concept of 'authenticity'. "Today, authenticity is back on the agenda and turns out to be a complex topic of different definitions and perspectives brought to bear on a variety of contemporary reconstructions of the past" (ICOMOS 2017: n.p.).

So what will the future of cultural heritage be like? Que sera, sera? If we take the precarious status of these sites seriously, there are two scenarios. The first is digitization. When Palmyra will be totally destroyed, one may argue, it will not have been erased from Western cultural memory. Such sites from which we have no material relics can survive in texts, in stories, in discussions and in the circulation of images. There is a paradox involved here: While the city of Palmyra is dwindling in material substance, its archive of images is growing and growing. There is always a revival on the Internet.

The second paradox concerns reconstruction. If not only natural disasters but also violent human destruction are now more and more inscribed into the concept of cultural heritage itself, new guidelines are needed concerning a careful documentation for post-trauma appraisal and interventions. Whatever the option will be in a specific case, with their exhibition 'Rebuild Palmyra?', the students of Konstanz have spelt out these pressing problems and given us much food for thought for both scenarios.

Notes

1 The song was written by Jay Livingston and Ray Evans and made popular through Alfred Hitchcock's film *The Man Who Knew Too Much* (1956), starring Doris Day and James Stewart (Roberts 2006). Due to copyright restrictions, a stanza from the song 'Que sera, sera' had to be eliminated from the text.

2 James Joyce picked up this quote in his short story "The Dead" in which the protagonist Gabriel nostalgically meditates on the fragility of cultural memory: If the dead "are gone beyond recall let us hope, at least, that in gatherings such as this we shall still speak of them with pride and affection, still cherish in our hearts the memory of those dead and gone great ones whose fame the world will not willingly let die" (Joyce 1976: 201).

3 See also Veyne (2017).

4 This slogan had been coined for the Great World Exhibition of 1851 and was taken up and repeated for the preservation of national cultural heritage (cf. Swenson 2013: 195–197).

5 There was a rapid fall of esteem for the status of a civil nation after the beginning of the Great War. Marinetti and Mussolini both praise the new cultural value of barbarism. In his book *Dynamic of Destruction. Culture and Mass Killing in the First World War*, Alan Kramer cites German sources that prefer barbarism over civilization: "That may be barbarism, but then it is a piece of healthy barbarism for which we will never want to be ashamed" (2007: 27–26).

6 Loewenthal, L. (1987). "Caliban's Erbe. Bücherverbrennungen und kulturelle Verdrängungsmechanismen". In Aleida und Jan Assmann (eds.). *Kanon und Zensur*. Munich: Fink, 227–236.

7 The relevant lines come from the song "Big Yellow Taxi", written, composed and recorded by Joni Mitchell in 1970 and first released on her album *Ladies of the Canyon*, had to be eliminated from the text due to copyright reasons.

8 The exhibition in the Konstanz 'Bildungsturm' from 30th June to 17th September 2017 under the supervision of Professors Eberhard Schlag, Harald Reiterer and Stefan Hauser. See also the film that shows a walk through the Konstanz exhibition: http://rebuild-palmyra.de.

9 The exhibition was transferred to Braunschweig, where it was shown until 29 April 2018. It is hoped that this innovative experiment will make it all the way to Berlin where it could attract and engage many more visitors.

References

Hauser, S. and Reiterer, H. (2017). *Rebuild Palmyra?* (30.06.2017–17.09.2017). http://rebuild-palmyra.de [Accessed 08 February 2018].

ICOMOS (2017). *Introduction to the ICOMOS University Forum Workshop on Authenticity and Reconstructions (13–15 March 2017)*. Paris: (s.n.).

Joyce, J. (1976). *Dubliners*. Harmondsworth: Penguin.

Kramer, A. (2007). *Dynamic of Destruction: Culture and Mass Killing in the First World War*. Oxford: Oxford University Press.

Loewenthal, L. (1987). "Caliban's Erbe. Bücherverbrennungen und kulturelle Verdrängungsmechanismen". In Aleida und Jan Assmann (eds.). *Kanon und Zensur*. Munich: Fink, 227–236.

Milton, J. (1641). *The Reason of Church Government Urg'd against Prelaty*. London: Printed by E. G. for Iohn Rothwell.

Rehling, A. (2014). "'Kulturen unter Artenschutz'? – Vom Schutz der Kulturschätze als Gemeinsames Erbe der Menschheit zur Erhaltung kultureller Vielfalt". *Jahrbuch für Europäische Geschichte*, 15, 109–137.

Roberts, D. (2006). *British Hit Singles & Albums* (19th ed.). London: Guinness World Records.

Schmickler, B. (2015). "So gefährdet ist das Weltkulturerbe". *Tagesschau.de* (24.08.2015). www.tagesschau.de/ausland/faq-zerstoerte-kulturgueter-101.html [Accessed 8 February 2018].

Shelley, P. B. (1962). "A Defence of Poetry (1821)". In D. J. Enright and E. De Chickera (eds.). *English Critical Texts: 16th Century to 20th Century*. London: Oxford University Press.

Swenson, A. (2013). *The Rise of Heritage: Preserving the Past in France, Germany and England, 1789–1914*. Cambridge: Cambridge University Press.

Veyne, P. (2017). *Palmyra: An Irreplaceable Treasure*. Chicago: Chicago University Press.

Wood, R. (1753). *The Ruins of Palmyra, otherwise Tedmore, in the Desart*. London: (s.n.).

4 Cultural sustainability in times of cultural genocide

Wolfgang Huber

Introduction: culturecide

"Culturcide" is the name of a Punk Band based in Houston, Texas. But "Culturecide" is at the same time an abbreviation for "Cultural Genocide". This new term is not yet very well established. The only article I could find under the heading "Culturecide" was published in the *Encyclopedia of Race and Ethnic Studies*, edited by Ellis Cashmore (Stein 2004: 99). Just as the condensed form and the longer "cultural genocide" are yet to take hold in the Anglosphere, the German translation, "kultureller Völkermord", is also rarely used.

There are good reasons to approach the term with hesitation. I myself had previously felt that "Cultural Genocide" was an exaggeration. Its use carries with it the danger of diminishing the exceptionality of physical genocide and its associated atrocities, not only characteristic for the twentieth but also for the twenty-first century. Yet physical and cultural genocides are inevitably intertwined. Historically, the burning of books, the destruction of synagogues and many comparable attacks against cultural values were all too often immediately followed by murderous attacks against those for whom those cultural artefacts formed a part of their personal or collective identity.

The interconnectedness of cultural and physical destruction suggests that they are a part of the same phenomenon, and therefore ought to be treated collectively through the use of the one term, for it is clear that the term "genocide" means the murderous aggression against groups whose unity has ethnic, religious, political or cultural roots. And the extermination of the cultural self-expression of individuals or groups violates them deeply in their identity as well as in their integrity. Cultural genocide is an intentional attempt to negate human dignity. But at the same time, it negates culture as such, because *every* expression of culture forms a part of the cultural heritage of humankind, and not just those cultural witnesses explicitly acknowledged as parts of the world cultural heritage by UNESCO.

Indeed, the proof of this is before our eyes. Today we see the intertwining of forms of physical and cultural genocide in acts perpetrated by the terrorist organization "Islamic State" (IS), which include not only brutal violence against groups that do not belong to the Umma in the fundamentalist Sunnite understanding, which is taken by IS as the only legitimate form of Islam, but also the demonstrative destruction of buildings, statues and other witnesses of former non-Islamic cultures and religions in, for example, Mossul, Nimrud, Hatra and Palmyra. The exclusion of the other refers not only to persons and their physical integrity but also to the symbolic cosmos of other convictions, world views or forms of cultural expression. The negation of any kind of legitimacy, worth or dignity of persons as well as of cultural entities finds its expression in brutal violence against subjects as well as against objects. The destruction of culture and the destruction of human lives are inextricably intertwined.

What we experience is not a clash of civilizations but a struggle over culture. This struggle is the form, or at least one of the forms, in which the problem of cultural sustainability becomes apparent. And there is no reason for the assumption that cultural sustainability, compared with the seemingly classical elements of "Brundtland-sustainability", namely economic stability, ecological balance and social security, is only of secondary relevance. Humans are cultural beings. This is something we realize when their cultural world is destroyed, or they become alienated from the symbolic means of their possible self-interpretations. The genocidal atrocities in Syria, Iraq and other places are dramatic demonstrations of this. But of course, there are many other places in which people suffered, and continue to suffer, under this twofold form of genocide.

'Genocide' as a concept: a brief history

The evidence of suffering moves us and its contemplation may stir us into action, but it does not lead to conceptual clarity. Despite a protracted debate over whether or not to include the destruction of culture in the moral and legal concept of genocide, little progress has been made. People lament what they see as a cultural decline, and they use the term "cultural genocide" as a means of dramatization. Or they want to defend a specific kind of cultural separatism or immobility, and they call the challenge for cultural adaptation or transformation simply "cultural genocide". The terms "cultural genocide" and "culturecide" can easily be used to characterize and accentuate all kinds of cultural integration, not the least of which would be "assimilation". The term is inclined for a more or less ideological use, working to immunize prejudice by attacking opposing positions with a politically loaded term.

At this point a brief history of the term "genocide" and its relationship to "cultural genocide" is warranted. A useful starting point would be a court case heard in Berlin 1921. The Armenian student Soghomon Tehlirjan was accused of murdering Mehmet Talaat Pascha, the former grand vizier and Minister of the Interior of the Ottoman Empire on Hardenberg-Street in Berlin- Charlottenburg on March 15, 1921. Talaat Pascha was well known as one of the organizers of the Armenian massacres of 1915. In his absence he was sentenced to death by an Ottoman court in Istanbul in 1919. But in Germany he could live without any restrictions – and even be killed on an open street.

The procedures before the court had many implications. I name only three of them. They refer to three different persons, namely Johannes Lepsius, Robert M.

W. Kemper and Raphael Lemkin. Johannes Lepsius, the Christian advocate of the Armenians, was one of the experts before the court. He succeeded in convincing the jurors that the cruelties against the Armenians, including 89 members of the Tehlirjan-family, influenced the psychological situation of the defendant so deeply that he was not responsible for his actions. The jury therefore decided for an acquittal, because Tehlirjan had committed his act in the heat of the moment and was mentally not responsible for what he did. The fact was rather, that Tehlirjan executed an order that he had received already months before in a Boston coffeehouse. But under the influence of Lepsius, the court's judgement addressed the Armenian massacre rather than the assassination of Talaat Pascha.

That was at least the impression of another person present in the courtroom, namely Robert M. W. Kempner, a 21-year old of Jewish descent studying law and political science. He later interpreted the Berlin legal proceeding as the first case in the history of law in which a crime against humanity committed by a government is combatted by a foreign state. He drew the conclusion that this ought not to be understood as an intervention into the internal affairs of another country. That was of course an exaggerated interpretation of the action of the judiciary in Berlin. But for Kempner his participation in the Tehlirjan court case was evidently a preparation for his later role as deputy chief prosecutor in the Nuremberg trials between 1945 and 1948, the first trials on crimes against humanity executed by foreign countries. Born in 1900 as the son of a Polish-Jewish family, Raphael Lemkin was in 1921 a young student of linguistics in Lemberg, where he became aware of the Tehlirjan court case. His reaction differed from that of Kempner. He asked himself, why it was not possible to prosecute somebody who committed crimes against humanity outside of his own country. Lemkin changed the subject of his studies to law and committed himself to the question how these crimes could be made justiciable internationally.

'Ludobójstwo', or: terms matter

Lemkin's first effort in this direction was a proposal that he presented to an international conference of the League of Nations in Madrid in 1933. Lemkin himself summarized his proposal as follows: The intention of his two articles was

> that actions aiming at the destruction and oppression of populations . . . should be penalized. The author formulated two new international law crimes to be introduced into the penal legislation of the thirty-seven participating countries, namely, the crime of *barbarity*, conceived as oppressive and destructive actions directed against individuals as members of a national, religious, or racial group, and the crime of *vandalism*, conceived as malicious destruction of works of art and culture because they represent the specific creations of the genius of such groups. Moreover, according to this draft these new crimes were to be internationalized to the extent that the offender should be punished when apprehended, either in his own country,

if that was the situs of the crime, or in any other signatory country, if apprehended there.

(Lemkin 2005: 91)

Lemkin's proposal did not succeed. The author had much reason to regret that in the following years. He was convinced that the lack of an appropriate term was one of the reasons for his failure. At the same time, he experienced the Nazi atrocities himself, having not persuaded his family members to emigrate from Poland early enough. He himself went first to Sweden, then to the USA. In both places, he continued to work for the introduction of crimes against humanity into the international penal law (cf. Mayers 2015). In 1943, he gave advice to the Polish government in exile to use the Polish term "ludobójstwo", newly coined by himself, for the German crimes against the Polish people, a compound word from *lud* (people) and *zabójstwo* (murder). In the following year, he finally found the equivalent "genocide" from the Greek *genos* (people) and the Latin *caedere* (kill) and he used it consistently in the following years (cf. Lemkin 2005: 79). He ceaselessly campaigned for his idea to develop a legal instrument for the international penalization of what he now called "genocide". Consistent with his earlier proposal, his definition of *genos* was not restricted by ethnicity or nationality. Moreover, it included religious and cultural as well as political groups. In Lemkin's understanding the facts of genocide were not only realized by policies of outright extermination of Jews, Gypsies or other groups, but also by policies of demographic reconstruction or – as it was called later during the Balkan wars – by "ethnic cleansing". And he included the cultural aspects of genocide, namely the destruction of the "culture, language, national feelings, religion" of the respective groups (Lemkin 2005: 79). His tireless activity was especially motivated by the conviction that the victims of Nazi barbarism were killed twice, once by the Nazi-Regime itself and a second time by the Anti-Hitler coalition, because the member governments knew about Hitler's crimes but refused to publish their knowledge or publicly decry them. Thus, Lemkin became the architect of the first covenant related to human rights and crimes against humanity. It was adopted by the General Assembly of the UN, namely the "Convention on the Prevention and Punishment of the Crime of Genocide", on December 9, 1948, the day before the promulgation of the "Universal Declaration on Human Rights".

The 'Convention on the Prevention and Punishment of the Crime of Genocide' (1948)

The draft for the Genocide-Convention explicitly included cultural genocide. It defined genocide as such, as "a criminal act directed against any one of the aforesaid groups of human beings, with the purpose of destroying it in whole or in part or of preventing its preservation or development" (*Draft Convention on the Crime of Genocide*, 5). Under this general definition it introduced the distinction between physical, biological and cultural genocide. Physical genocide was defined as "causing the death of members of a group or injuring their health or physical integrity" (ibd.), biological genocide was understood as restricting births by different means.

Cultural genocide was explained as

> [d]estroying the specific characteristics of the group by:
>
> (a) forced transfer of children to another human group; or
> (b) forced and systematic exile of individuals representing the culture of a group; or
> (c) prohibition of the use of the national language even in private intercourse; or
> (d) systematic destruction of books printed in the national language or of religious works or prohibition of new publications; or
> (e) systematic destruction of historical or religious monuments or their diversion to alien uses, destruction or dispersion of documents and objects of historical, artistic, or religious value and of objects used in religious worship. (*Draft Convention on the Crime of Genocide* [28 March 1947: 6f.])

This distinction between physical, biological and cultural genocide disappeared on the way to the final version of the convention. Now Article 2 defines genocide as follows:

Genocide means any of the following acts committed with the intent to destroy, in whole or in part, a national, ethnical, racial or religious group, as such:

a) Killing members of the group;
b) Causing serious bodily or mental harm to members of the group;
c) Deliberately inflicting on the group conditions of life calculated to bring about its physical destruction in whole or in part;
d) Imposing measures intended to prevent births within the group;
e) Forcibly transferring children of the group to another group. (*Convention on the Prevention and Punishment of the Crime of Genocide* [9 December 1948: 309])

The difference to Lemkin's draft is remarkable. Political groups are now excluded from the definition. Outside of the national, ethnical or racial coherence only religious groups are mentioned. Cultural coherence disappears under the categories of national, ethnical or racial unity. The former category of "biological genocide" is included in the item "Imposing measures intended to prevent births within the group". The former category of "cultural genocide" appears only in the item "[f]orcibly transferring children of the group to another group". All other aspects of violent attacks against the cultural identity and self-expression of a group disappeared during the deliberations and debates on the Convention. That was perhaps helpful for the clarity of the convention, but the double restriction – the elimination of the political as well as the cultural aspect of group identities – remains regrettable and, regarding the forms in which these kinds of crimes developed over time, even deplorable.

The failure to include cultural genocide into international human rights law

In this case there is good reason to look at the fact that in the further development of human rights instruments, cultural rights are only seen as rights of the individual. The Universal Declaration of Human Rights of 1948 is a good example for that. Its Article 27 says: "Everyone has the right freely to participate in the cultural life of the community, to enjoy the arts and to share in scientific advancement and its benefits" (Its Article 27 says). The *International Covenant on Economic, Social and Cultural Rights* of 1966 has cultural rights in its title, but is more specific only with respect to education. In all other dimensions it repeats in slightly different words simply the Declaration of 1948 in saying:

1 The States Parties to the present Covenant recognize the right of everyone:

 (a) To take part in cultural life;
 (b) To enjoy the benefits of scientific progress and its applications;
 (c) To benefit from the protection of the moral and material interests resulting from any scientific, literary or artistic production of which he is the author. (Art. 15,1)

The Article adds some duties of the member states in stating:

2 The steps to be taken by the States Parties to the present Covenant to achieve the full realization of this right shall include those necessary for the conservation, the development and the diffusion of science and culture.
3 The States Parties to the present Covenant undertake to respect the freedom indispensable for scientific research and creative activity.
4 The States Parties to the present Covenant recognize the benefits to be derived from the encouragement and development of international contacts and co-operation in the scientific and cultural fields. (Art. 15,2–4)

There is no mention of the prevention or punishment of cultural genocide and therefore also of a common task of the international community of law in this respect.

The rights of indigenous peoples – another chance missed

An occasion to revitalize the concept of cultural genocide was the preparation of the *United Nations Declaration on the Rights of Indigenous Peoples* of September 13, 2007. During the 24 years of preparation for that Declaration the term was introduced. And indeed the preservation and protection of the cultures of indigenous peoples formed an especially important aspect of the work on this declaration. Already the Vienna Declaration that summarized the outcome of the United Nation's Vienna Conference on Human Rights of 1993, had emphasized respect for the value and diversity of the cultures and identities of indigenous peoples, and had emphasized the elaboration and protection of the rights of indigenous people and other minorities as a high priority. However, the prevention or even the punishment of cultural genocide is not addressed in the final text of the Declaration on the Rights of Indigenous Peoples as adopted in 2007. Importantly, the issue is not dealt with in a Covenant, but only in a Declaration. It has only been adopted by the General Assembly and not ratified by the member states, and lacks, therefore, a legally binding force, comparable with the Covenants of 1966. That may be a reason for a concentration on rights on the one hand and obligations of the respective states on the other. It is a declaration of those rights and obligations without any instruments of legal enforcement. No measures of prevention or punishment regarding violations of those rights are mentioned. That may be one of the reasons why the topic of cultural genocide or culturecide does not appear.

The issue of culture itself, however, plays an important role in the 46 articles of this rather lengthy declaration. The relevance of culture for the identity of indigenous peoples is mentioned repeatedly (Art. 8, 11–16, 24, 25, 31). Although they are not restricted to the traditional cultural forms in which indigenous peoples preserve and represent their identity, those forms play an important role, as Art. 31.1 shows in an exemplary manner:

> Indigenous peoples have the right to maintain, control, protect and develop their cultural heritage, traditional knowledge and traditional cultural expressions, as well as the manifestations of their sciences, technologies and cultures, including human and genetic resources, seeds, medicines, knowledge of the properties of fauna and flora, oral traditions, literatures, designs, sports and traditional games and visual and performing arts. They also have the right to maintain, control, protect and develop their intellectual property over such cultural heritage, traditional knowledge, and traditional cultural expressions. (United Nations Declaration on the Rights of Indigenous Peoples [13 September 2007]: 301)

It is worth mentioning that the chairperson of the Global Indigenous Caucus, Les Malezer, after the adoption of the Declaration by the General Assembly, emphasized that the intention of the declaration was not to separate indigenous cultures off from other cultures but to understand them as a part of the cultural diversity – evidently understood in parallel to biological diversity or biodiversity – of humankind. And he added: The "message is not about secession . . . but about co-operation and partnership to ensure that all individuals, regardless of race or beliefs, are truly equal and that all peoples are respected and allowed to develop" (Malezer 2007: 2).

Conclusion: human rights universalism – the basis of cultural sustainability

Culture is about diversity. Cultural sustainability has to do with the interaction between cultures in their differentness. The debate on the rights of minorities, especially of indigenous people is a good test case for that insight. But the diversity of cultures can be used either in a spirit of co-operation and togetherness or in a spirit of separation and opposition. In this respect, it is important to address for a moment new forms of cultural separatism applied to the issue of human rights themselves.

Recent decades have seen new kinds of cultural exceptionalism with regard to human rights. Binding the idea of human rights exclusively to "Western" values became a good argument for constructing exceptions from the obligatory character of human rights on cultural grounds.

The debate on "Asian values" was of specific importance in this respect. With growing intensity after 1990, Asian participants in the debate opposed the idea that human rights could really appeal for universal validity. They saw in them a one-sided emphasis on individual rights and freedoms to the detriment of family values and the rights of communities. They wanted to emphasize the role of local leaders replacing the autonomy of the individual. Or, to quote the president of Singapore, Wee Kim Wee, who became a spokesperson for "Asian values" in the early nineties:

> If we are not to lose our bearings, we should preserve the cultural heritage of each of our communities, and uphold certain common values which capture the essence of being a Singaporean. These core values include placing society above self, upholding family as the basic building block of society, resolving major issues through consensus instead of contention, and stressing racial and religious tolerance and harmony. We need to enshrine these fundamental ideas in a National Ideology. Such a formal statement will bond us together as Singaporeans, with our own distinct identity and destiny.
>
> (Wee Kim Wee 1989: 12)

That at least is a clear understanding of 'cultural sustainability': the formal establishment of a "National Ideology" or of 'Asian cultural values'. In a comparable manner representatives of Islam argue for the father's right to decide on questions regarding his wife or his children, especially his daughters. Arranged marriages for instance are interpreted as not interfering with women's rights. You have to address comparable problems when it comes to the inclusion of vernacular rights into the rule of law. The way in which different cultures deal with the problem of human rights is a good litmus test for the understanding of 'cultural sustainability'. The idea can easily be used as a legitimation for cultural exceptionalism putting into question the universality as well as the indivisibility of human rights.

In critically discussing this kind of exceptionalism on cultural grounds it is necessary to look back to the challenges that determined the atmosphere in which the first international human rights documents were formulated after World War II. Genocides on the one hand and the situation of refugees on the other were the most challenging experiences. Hannah Arendt summarized those experiences in asking for the right to be treated as human beings, or more precisely "the right to have rights" (Arendt 1949: 754–770; Arendt 1993: 452–470). The Convention on the Prohibition and Punishment of Genocide of 1948 was followed in 1951 by the Convention Relating to the Status of Refugees (Brownlie and Goodwin-Gill 2010: 312–326). It is true that the idea of universal human rights emerged from the abysses of the moral history of the twentieth century (Glover 2012). It is this history that lends Jonathan Shay's statement its severity: "The understanding of trauma can form a solid basis for a science of human rights" (Shay 1994: 209).

From a theological perspective there are strong reasons to support this approach to the universality of human rights. A Christian understanding of the human person takes into account the vulnerability of humans and looks on the status of a society from the perspective of its most vulnerable members (Koopman 2007). It considers the inviolability of human dignity with the eyes of those whose dignity is endangered by hunger and illness, poverty and loneliness, flight and migration, violence and war. The preferential option for the vulnerable and the suffering is mandatory for a Christian perspective on human dignity and human rights. This option forms the theological starting point for a critical evaluation of human rights exceptionalism on cultural grounds.

Seyla Benhabib, in her essays on human rights (2011), keeps up Arendt's formula of the "right to have rights" and brings her into dialogue with Lemkin. Whereas Arendt understands the "right to have rights" primarily as a "right to membership in a political community," Benhabib conceives this basic right as "the claim of each human person to be recognized as a moral being worthy of equal concern and equally entitled to be protected as a legal personality by his or her own polity, as well as the world community" (Benhabib 2011: 62). She sees the basis of moral universalism rooted in the demonstration of "equal respect for the other as a being capable of communicative freedom" (Benhabib 2011: 64). And she endorses a statement by Rainer Forst, who writes: "Human rights secure the equal standing of persons in the political and social world, based on a fundamental moral demand of respect" (Forst 2010: 718). If cultural sustainability is to remain distinct from justifications for cultural separatism and exceptionalism, then its proponents have to ask how the fundamental moral demand of respect, mutual recognition and equal treatment of moral beings may be used as a basis for the interaction of cultures – i.e. diversity – on common ground.

But it is also the other way around. Moral universalism is only taken seriously if it opens the way to respect for diversity, including cultural diversity. Arendt learned from her experience as a refugee from a totalitarian regime what Lemkin learned from his confrontation with genocide: Whoever wipes plurality off our life-world destroys this world itself. This is because our perception of the world per se depends on a plurality and diversity of perspectives. In extinguishing one of those perspectives we damage this world. Just as we might mourn the victims of violence or the casualties of war, we should also lament the loss of a perspective on the world, a part of the world as such.

Therefore, we ought to defend a plurality of perspectives on the world, represented by people both in their individual diversity as in the variety of their cultures and religions. Equality, therefore, means the equality of the different. The inherent dignity of every human person provokes our recognition of difference (Huber 2011). Culture is nothing else than the ensemble of symbolic forms through which we represent those diversities on the ground of mutual recognition.

References

Arendt, H. (1949). "Es gibt nur ein einziges Menschenrecht". *Die Wandlung*, 4, 754–770.

Arendt, H. (1993). *Elemente und Ursprünge totaler Herrschaft*. 3rd Ed. München: Piper.

Benhabib, S. (2011). *Dignity in Adversity: Human Rights in Turbulent Times*. Cambridge – Malden: Polity Press.

Brownlie, I. and Goodwin-Gill, G. (eds.) (2010). *Brownlie's Documents on Human Rights*. 6th Ed. Oxford: Oxford University Press.

Convention on the Prevention and Punishment of the Crime of Genocide (9 December 1948). GA Res. 260 (III). In I. Brownlie, and G. Goodwin-Gill (eds.). (2011). *Brownlie's Documents on Human Rights*. 6th Ed. Oxford: Oxford University Press, 308–311.

Draft Convention on the Crime of Genocide (28 March 1947). UN Doc. E/447. www.un.org/ga/search/view_doc.asp?symbol=E/447 [Accessed 17 January 2017].

Forst, R. (2010). "The Justification of Human Rights and the Basic Right to Justification: A Reflexive Approach". *Ethics*, 120, 711–740.

Glover, J. (2012). *Humanity: A Moral History of the Twentieth Century*. 2nd Ed. New Haven, London: Yale University Press.

Huber, W. (2011). "The Dignity of the Different: Towards a Christian Ethics for Pluralistic Societies". In L. Hansen, N. Koopman, and R. Vosloo (eds.). *Living Theology: Essays Presented to Dirk J. Smit*. Wellington: Bible Media, 427–440.

International Covenant on Economic, Social and Cultural Rights (16 December 1966). GA Res. 2200A (XXI), 21 UN GAOR Supp. (No.16) at 52, UN Doc. A/6316 (1966), 993 UNTS 3, entered into force January 3, 1976. In I. Brownlie, and G. Goodwin-Gill (eds.) (2010). *Brownlie's Documents on Human Rights*. 6th Ed. Oxford: Oxford University Press, 370–379.

Koopman, N. (2007). "Some Theological and Anthropological Perspectives on Human Dignity and Human Rights". *Scriptura*, 95(2), 177–185.

Lemkin, R. (2005). *Axis Rule in Occupied Europe: Laws of Occupation, Analysis of Government, Proposals for Redress (1944)*. Clark: The Lawbook Exchange.

Malezer, L. (2007). "Statement by the Chairman, Global Indigenous Caucus to the United Nations General Assembly" (13 September 2007). https://www.humanrights.ch/upload/pdf/170130__07-09-13IPCaucusStatementAdoptionDeclaration.pdf [Accessed 29 June 2018].

Mayers, D. (2015). "Humanity in 1948: The Genocide Convention and the Universal Declaration of Human Rights". *Diplomacy & Statecraft*, 26, 446–472.

Shay, J. (1994). *Achilles in Vietnam: Combat Trauma and the Undoing of Character*. New York: Scribner.

Stein, S. (2004). "Culturecide". In E. Cashmore (ed.). *Encyclopedia of Race and Ethnic Studies*. London, New York: Routledge, 99–100.

United Nations Declaration on the Rights of Indigenous Peoples (13 September 2007). A/Res/61/295. In I. Brownlie, and G. Goodwin-Gill (eds.) (2010). *Brownlie's Documents on Human Rights*. 6th Ed. Oxford: Oxford University Press, 293–303.

Wee Kim Wee. (1989). "Government's Aim Is to Create a Better Life for all Singaporeans. President Wee's Address at the Opening of the Seventh Parliament". *The Straits Times*, 10 January, 12.

5 Sustainable development and the concept of culture – an ethical view

Torsten Meireis

Introduction: cultural sustainability as an ethical problem

Sustainability, or more precisely: sustainable development, is an inherently normative concept as it implies the presupposition and evaluation of desirable ends. The key sentence of the 1987 Brundtland report defines sustainable development as a process that "meets the needs of the present without compromising the ability of future generations to meet their own needs" (WCED 1987: 27). While the fulfilment of needs might be understood as an articulation of particular interests – even though the language of interest is probably already transgressed by a proclamation of the obligation to look to the needs of future generations, the report makes use of even stronger moral language: "Poverty is not only an evil in itself, but sustainable development requires meeting the basic needs of all and extending to all the opportunity to fulfil their aspirations for a better life" (WCED 1987: 16). The second part of that proposition might even call Amartya Sen's version of the capability approach (1999) to mind.

However, from an ethical viewpoint, moral claims need to be argued for as ethics is a critical reflection upon morals and ethos. The sustainability discourse, however, seems to be riddled with implicit moral claims that are not fully examined, thus contributing to the vagueness of the concept (cf. Diefenbacher 2001: 58–72). In the following paper I'll try to show that this also applies to the discourse on cultural sustainability that has been developed since the turn of the Millenium. However, my aim is not the obliteration of moral language from the sustainability discourse, but to reflect upon its application. As a remedy to the ills I detect, I will propose an understanding of culture as an agonal arena, an instrument and a horizon of the sustainability discourse that allows for ethical distinction and implies measures to further the discourse on cultural sustainability.

Tainted love: moral bias in sustainable culture discourse

The first thesis I want to defend states that concepts revolving around the idea of cultural sustainability often imply strong moral convictions. However, due to their implicit nature, they usually tend to be taken for granted, thus tainting the concept of culture in a problematic way. To exemplify this, I'd like to consider an early example and a more recent one.

For the early discourse on cultural sustainability, the contribution by Hildegard Kurt and Michael Wehrspaun (2001) may be paradigmatic. It appeared in the Gaia journal's 2001 volume and argued succinctly for more consideration to be given to the cultural aspects of sustainability discourse, especially in public policy, thus arguing in the same vein as cultural policy advocates elsewhere (cf. Hawkes 2001). Kurt, a scholar in cultural studies, and Wehrspaun, then an official in the German government department for environmental affairs, first observe a lessening of public interest in the environment, as indicated by a common ignorance of the concept of sustainability in Germany. They then connect this development to the absence of cultural topics in the official sustainability discourse and the absence of sustainability topics in cultural policy. In their analysis, this gap is rooted in cultural history leading back to the Age of the Enlightenment when the concepts of culture and nature became separated. Although the reasons differed, the concepts converged in outcome: While for the Kantian tradition, the cultural emancipation from nature was the precondition of freedom, Rousseau's nature-friendly philosophy could be read as a program for the liberation of an alienated nature through human culture. For that interpretation, Kurt and Wehrspaun turn to the works of Marx and the Frankfurt School, while relying on ethicists Günter Altner and Klaus Meyer-Abich for their diagnosis that the cultural 'forgetfulness of nature' and the 'homelessness of modern man' must be overcome and the question of culture therefore move to the centre of the sustainability triangle. Consequently the authors claim the necessity of a new image of the world, a new understanding of man's role in the universe, the facilitation of which they expect from the arts. Even though the authors dismiss any instrumental use of the arts, their plea boils down to an instrumental understanding of culture itself:

> If the thesis presented here contains any plea, it's directed at the intensified dialogue with those segments of the art world that can be seen as . . . avant-garde . . . of a culture of sustainability. For isn't the search for sustainable ways or styles or a new prosperity an essentially aesthetic challenge?
>
> (Kurt and Wehrspaun 2001: 21)

This chapter is not the place to discuss the claims in philosophical history or the positions of the ethicists mentioned. The point in question seems to be that culture here is invoked as a normatively loaded concept instrumental to furthering sustainability. But this instrumentalism comes at a price: Kurt and Wehrspaun reduce the reach of the cultural dimension of sustainability by focusing on a specific narrative, which, albeit supposedly universal, is actually based upon Western paradigms of philosophy. The fact that all cultural narratives consist of situated knowledges (Haraway 1988) and are therefore particular does not imply that striving for the universal significance of a certain narrative by feeding it into discourse is prohibited. But the identification of the dimension of culture with one certain narrative, to my mind, does not allow for the plurality of culture. This identification proceeds a little too fast, ignoring other narratives and their impact, let alone the critique of universality mounted by postcolonial cultural theory (Mignolo 2011: 118–120).

A second example is more recent – it has been devised in the context of a European research project expressly directed at cultural sustainability (cf. Dessein et al. 2015) and uses a very different, discourse-analytical and policy-oriented approach. Starting out from a wide concept of culture, understood as "the meaning content of human communities, the symbolic patterns, norms and rules of human communities that divide humans from nature" (Soini and Birkeland 2014: 214), the authors screen scientific, peer reviewed English articles from disciplines like geography, tourism management, agricultural studies, political science and so on, which are making use of the term 'cultural sustainability'. They try to discern the storylines within these research papers by posing the following three questions: What representations of culture are being made, what problems are being addressed, and which interests are involved? Subsequently, they describe seven storylines which are then contextualized politically. Storylines focusing on the conservation of cultural heritage against globalization (1) and the preservation of such heritage in contexts of globalization (2) are assigned to a conservative pattern. The use of culture as a way of sustaining economic viability (3) is characterized as neoliberal, while articles stressing cultural diversity (4) or local cultural development (5) are attributed to communitarian tendencies and ones which focus on the interdependence of cultural and ecological practices (6) or demand an ecological turn in values and behaviour (7) are filed under an environmentalist header (Soini and Birkeland 2014: 219–220). Whereas the conservative uses of cultural sustainability are understood as furthering the preservation of certain particular cultures, the neoliberal and communitarian ones are described as using culture as instrument to achieve different aspects of sustainability. Environmentalist storylines in particular are characterized as treating culture as the foundation of sustainability (Soini and Birkeland 2014: 220–221). As the focus on policy development with the search criterion 'cultural sustainability' suggests, all the approaches include normative principles, but as the study concentrates on the description of the storylines in a political context, those normative dimensions are not explicitly discussed. Due to the policy perspective, culture tends to appear as a means to an end, that is, ecological, economical or social sustainability. This is not only true for the storylines expressly labelled as seeing culture as instrumental, but also for those who are understood to see culture as a foundation of sustainability. The conservative storylines, on the other hand, are understood to use culture as a way of preserving particular identity. Even if the merits of that approach to the survey of how the concept of cultural sustainability is actually used in policy reflection are granted, still the concept of culture appears as a means to an end: political identity, social, economical or ecological sustainability or all of those at once.

Now we might ask: What is so bad about using culture as a means to achieve worthy goals? My answer would be twofold. Firstly, understanding culture solely or even predominantly as a means to an end might be seriously flawed considered the nature of what we usually understand by the concept. Secondly, normative evaluation of political goals necessitates the discussion of the moral principles involved, implying an open consideration of what might count as a 'worthy goal' – which of course is a necessary step in the implementation of such goals. Failure to stand up to moral debate boils down to moral resignation or moral domination, both of which are problematic for policy development in democratic civil societies. Consequently, my thesis then would be: Understanding culture as a means to an end and thus implying normative positions that are somehow taken for granted or not discussed at all taints sustainability discourse on culture with a claim-overload that cannot possibly be made good on.

The concept of culture in contexts of sustainability

But how should we conceive of culture in contexts of sustainability? First of all, a look at the classical theories of culture may be helpful. As suggested prominently by Ernst Cassirer, culture may be seen as a symbolic system proprietary to humans (and up to a point to certain primates) involving the use of language to represent, signify and thus symbolically reconstruct not only the physical world, but also poetic fantasies, myths and religious ideas, emotions (Cassirer 1944: 44) and abstract relations (Cassirer 1944: 59). This symbolic structure has to be seen as dynamic: "A genuine human symbol is characterized not by its uniformity but by its versatility. It is not rigid or inflexible but mobile" (Cassirer 1944: 57). In that line of thinking, culture therefore cannot be reduced to a certain social sphere or to the elaborate products and performances of art but is deeply embedded in ordinary life. The ordinary is culture and culture is ordinary, it encompasses "a whole way of life – the common meanings" as well as "the arts and learning – the special processes of discovery and creative effort" (Williams 1989: 5). The social nature of that dynamic symbolizing has been explained by Clifford Geertz in his essay on thick description. We can draw here on his famous example of the winking eye. Given a room with more than two people, the winking of an eye triggers a cascade of interpretation. The person who winks not only moves a muscle, but in doing so draws on a culturally given set of symbols, a meaning attached to that gesture. Usually, the winking is addressed to another person and may indicate some secret understanding. People watching, however, through their interpretation of its meaning, may add to the stock of symbols culturally belonging to that gesture. A second person may wink, and, in performing the same gesture, parody the action of the first as clumsy. The meaning, then, is not conspiracy, but ridicule (Geertz 1973: 10–12). A third may then understand the wink of the second not as parody, but as an unintentional twitch triggered by the cognizance of a conspiracy and so on. Another person may then practice her satirical abilities at home before a looking glass, so the gesture becomes part of a rehearsal of an expressive repertoire. A fifth person may add to the interpretational network by twitching the eye in response to a physical irritation, its gesture, however, being understood by a sixth as parody of the clumsy satirist and so on. A wink of the eye may thus symbolize a signal of affection, a token of ridicule, the idea of conspiracy, an element of rehearsal and so on; it draws on a cultural pool of meanings which are continuously interpreted and becomes independent from the intention of the author instantaneously. What applies to gestures most certainly applies to more complex or abstract symbols: to concepts like 'love', or 'faith', or 'sustainable development' or 'cultural sustainability', for that matter. If we then understand culture as a network of symbols, the content and relation of which are continually interpreted and reinterpreted by all people involved intentionally and unintentionally, the idea of a fixed meaning becomes something of a chimera. Of course, the durability of certain interpretations is a matter of gradual difference depending not least on resources. If I have the means to repeat a certain interpretation of a certain symbol a thousand times a day, reaching a million people and relying on everyday use, on psychotechnical tools like emotional bonding and mnemotechnic instruments like rhyming, I may make that interpretation stick for a while. Exercises in commercial branding rely not least on those effects, making 'cellotape' or 'kleenex' household words. However, even such powerfully amplified interpretations are subject to time and reinterpretation: The German detergent brand name 'Persil' was used to ironically label the slip granting the dropping of charges in the denazification trials after World War II in Germany, insinuating that the person in question had been washed white – on the outside.

If culture is thus to be understood as an ongoing and dynamic process of symbol-interpretation and reinterpretation in which everybody in the context of given cultural manifestations – personal or medial – takes part in the measure of his or her symbolic, social and economic capital (Bourdieu 1983), and if the durability of interpretations gradually differs according to various parameters, then at least three different sociological dimensions of culture can be discerned (Nassehi 2011: 145–161). Firstly, a given – if dynamic – pool of symbols may be used as an instrument of ascription, assigning certain traits to groups of people identified by certain index markers. A woman's headscarf worn in summer in a certain way may in a Western European context act as a signal which leads agents to ascribe certain traits to the person wearing it: She's a Muslim, she's of Turkish origin, she's very restricted by a patriarchal set of customs, subject to her husband's rule, doesn't go to public pools and so on. Stereotyped markers like this may well lead to false conclusions. The woman may in fact be a Christian of self-proclaimed German ethnicity and cultural belonging, having recently emigrated from Kazakhstan. Or such markers may of course be faulty in themselves: Muslim women of Turkish origin may understand the headscarf as a sign of personal identity in a Western context which goes well with the lifestyle of an independent professional. What is more, they are subject to change: A woman with a headscarf in summer would have been perfectly usual in 1950s Germany or maybe even understood to just have stepped out of an open sports car. 'Culture' is then used as an ascriptional marker of difference and understood to comprise group traits adhering to ethnicity, stratum, class, lifestyle and so on. It is relatively recent, as it implies a plurality of legitimate pools of symbols and cultures existing alongside each other.

Secondly, the concept of culture is used to denote a pool of symbols in the context of a given group or society that are taken for granted, even though they're subject to change, including changes in their concepts, rituals and images. Rather than bowing, Western Europeans ritually shake hands when they meet formally. They take for granted that the question 'what do you do' at informal gatherings refers to activities related to gainful employment in a market environment, just as they use the term 'work' in a way that implies positive recognition rather than a manifestation of a lower class membership. Knowledge of those pools of symbols relevant for interpretation is usually latent. Assumptions about the nature of those symbols are only disrupted only when the group of reference, the 'we' is changed: either by travelling to places where a majority of people adhere to significantly different latent codes or by changes in a given society itself.

Because cultural codes work as ascriptional signals and latent knowledge in relation to given groups, they can also work as signals of distinction. Certain aesthetic preferences may be the index marker of a social milieu characterized by disproportional social power or wealth, or, vice versa, of a lower stratum. If the powerful declare their codes to be the normative ideal ('high culture'), that pool of symbols may also be used as a basis to discard values and preferences of subcultural groups as crude or even immoral, thus at the same time discharging that group's claims to equal social power or prosperity – thus popular music derived from African pentatonic musical traditions was understood as immoral or even destructive and sensual spectacles were often discarded as the cheap thrills of the lowly. On the other hand, subcultural symbolic codes may be utilized to draw emancipatory or identity-related distinctions between a cultural minority and a majority culture or the codes of those perceived to be in power. In such cases, 'discarded' markers, values or concepts become signals of a counter-culture striving for socially countervailing power which may even try to achieve cultural hegemony (Kebir 1991:74–89, 187–217). Slogans of this kind are familiar: "Black is beautiful", "no future" – or just see Roland Barthes (2010) on wrestling. In this sense, culture can be understood as an agonal arena, where cultural codes are means of distinction used for the purpose of identity formation, gaining recognition, manifesting social power and accumulating material resources.

If culture is properly understood as a dynamic symbolic social system and constantly changing interpretational network in which every person capable of the use of language, symbolic thinking and symbolic actions is actively and passively involved, and if it works as an ascriptional mechanism, as latent knowledge and as an agonal arena of distinction, then the idea of using it as a means to an end, to me, has a very reductionist flavour. Of course, it is possible to strive for cultural hegemony in the name of a certain normative concept of sustainability, but we need to understand that any such concept will be contested on various grounds:

1 Any concept will be understood very differently depending on the respective *cultures of reference*. To use a somewhat imperfect example, while in the 1970s and 1980s ecological sustainability was seen by many as a crucial issue in the global North, the global South suspected it to be a ruse to keep it from developing economically and in terms of political power – an issue still visible for instance in South African debates (cf. de Gruchy 2007).

2 The chance of creating a universal, monocultural notion of sustainability ingrained in everyday *latent knowledge* across cultures is rather remote, because cultures are inherently different and changing fast.
3 Any hegemony in the *agonal arena of distinction* will – in the context of social, political and economical inequality – most certainly produce counter-cultures. Take, for example, the 'coal rollers' in the US; people who deliberately modify their truck engines so as to emit unburnt diesel in thick clouds of smoke to protest against what they understand to be a liberal and well-to-do, ecologically minded establishment (Tabuchi 2016), may be only an extreme version of the anti-elitist stance manifested in Trump supporters who associate the idea of ecological sustainability with comparably rich, well-connected and intellectual liberals threatening their way of life.
4 As culture is an ever-changing network, any such hegemony will be temporal at best.

That doesn't mean that any attempt at sustainable development or cultural sustainability is futile. The human rights discourse is an example of the challenges, but also of the opportunities inherent in such an effort. But conceiving of culture as but a means to an end (1), associating it to a set of fixed moral principles that are taken to be self-evident (2) or dissociating cultural sustainability from the sustainability of cultural traditions (3), all fail to account for the complexity of culture.

Culture as horizon, agonal arena and instrument of sustainability discourse – a proposal

To avoid the traps of instrumentalist or undercomplex concepts of culture in contexts of sustainability, I would like to propose that culture be understood in three ways – as horizon, agonal arena and instrument in the sustainability discourse. Cultural sustainability then means turning to the *horizon* of the interplay of preserving, importing, interpreting, reinterpreting, inventing and testing cultural patterns that take the form of moral or ethnic traditions, of literary fiction, religious narrative or poetic imagery. It furthermore necessitates entering the *agonal arena*, facing the cultural, social and moral struggle, accepting the conflict that goes with it and understanding it not only as a nuisance but also as a creative resource for cultural sustainability. Finally, it refers to the *instrumental* use of particular cultural elements to further contested moral principles, procedures, institutions and practices of sustainability.

As a theologian and ethicist it is that idea of *culture as a horizon for sustainability discourse* that I will turn to first. As Sigrid Weigel (2010), a German scholar in literature and cultural studies has contended, concepts like sustainability not only have a normative basis which needs to be explored and cannot be supplanted by seemingly objective quantification, but are also rooted in historical contexts, bodies of traditions and experiments in literary fiction exploring the implications of cultural patterns. Thus Weigel shows that the moral idea of each generation's right to its own decisions, which is closely connected to today's sustainability discourse, is a quite recent one related to the Western modern shift towards civil societies with capitalist economies. Bodies of tradition and cultural patterns, however, are not to be understood as monolithic blocks or fixed entities, but are constantly on the move as they are interpreted and reinterpreted in the light of changing situations, new ideas imported from other cultures or according to the views of groups new to the tradition. For that matter, culture as a horizon isn't limited to traditions of the past and can't be reduced to questions of heritage, but also implies literary imagination or technical invention. Just one example from the Christian realm: The concept of a 'responsibility towards creation' which was quite effective in spurring on environmentalism, has an immense scope depending on the persons asked. European theologians may dismiss it as imprecise and misleading because it insinuates that humans can preserve creation (Honecker 1999: 353); Swiss citizens may see it as an appropriate expression for the notion that there is more to nature as a resource pool as the notion has been integrated into the Preamble of the Swiss constitution (BV 1999); African Christians may understand it as God's statement against a capitalist sell-out of their fields (LenkaBula 2000: 18–19) and some American born-again Evangelicals may take it to imply a strong stance against evolution theory. Principles and concepts, narratives and images, however, are not proprietary to a certain culture or specific groups of people, but are continuously on the move and form the pool from which the social imaginaries (Taylor 2004) are fed. To care for that pool, a preservation of bodies of tradition or the nurturing of particular cultures is necessary for the whole of cultural sustainability, but it's not enough by far. The creative processes dealing with the preservation, interpretation and reinterpretation of cultural patterns, including their concrete and fictional application in different settings, need to be furthered. That in turn means that moral principles relating to sustainability must be reconstructed and tested in different cultural and religious languages. As there is no abstract human religion, morality or culture, we cannot suggest an objective and universal 'sustainability paradigm' that then only has to be translated in different particular languages. The upside is that we may expect creative and fruitful inputs from different traditions and languages.

Incorporating such an inclusive definition of culture into sustainability discourse would require, first of all, that the cultural patterns we – as respective members of a particular tradition and agents with a particular history – understand as guiding and normative, be considered in the notion of sustainability. That entails asking ourselves why we find sustainability – and the notion of intra- and intergenerational justice attached to it – plausible and attractive on the basis of our respective normative traditions and patterns and how, in turn, those patterns may be influenced by the idea of sustainability. Institutionally, this entails the preservation of bodies of tradition or the nurturing of particular cultures as well as sustaining creative processes dealing with the preservation, interpretation and reinterpretation of cultural patterns and their concrete and fictional application in different settings.

As culture must also be understood as the *agonal arena for sustainability discourse*, the creative processes, debates, dialogues and discourses in which different moral principles, concepts, images, motives and narratives are compared, contested and held against each other also need attention. This not only applies to the furthering of public debate and competition, but also to critical reflection on the role of political or economic power, and on cultural ascriptions and distinctions. The coal rollers might be dismissed as fools, but if the proliferation of environmentally friendly behaviour is understood as a moral priority, it might be a good idea to analyse the driving powers behind their distinctive strategy, in order to test the stereotypes being employed in describing this group, as well as to find ways to engage in some sort of civil dialogue – which implies that everyone should remain open to learning experiences. The search for terms upon which to conduct a dialogue, of course, means engaging not only international bodies, but also the state and all levels of civil society.

In that vein, cultural patterns of sustainability must be understood as debatable and brought to the fore not only in international organizations but in civil society and politics proper on all levels, including faith based organizations, lobby organizations, unions, clubs, public libraries, the media and of course also in everyday life and practices. Debates, dialogues and practical experiments are important even before becoming official policies.

In the context of culture as the semiotic horizon and as the agonal arena of sustainability discourses, cultural activities and formations may then also form an *element of sustainability strategies*. This instrumental aspect has been pointed out by the concept devised in the context of the aforementioned European research (Dessein et al. 2015). Relying heavily on the research of Soini and Birkeland (2014), the COST group has developed a framework rooted in an instrumental view of culture for sustainability. In that framework, culture is conceived of in three ways, as culture in, for and as sustainability. Culture **in** sustainability signifies culture as a separate pillar of sustainable development besides the social, the economical and the ecological one, thus forming an instrumental aspect as cultural means are used to further sustainability ideas (Dessein et al. 2015: 20–37). Culture **for** sustainability sees culture in the focus of the sustainability triangle, thus supposedly acting as an instrument to mediate the ecological, economical and social dimension, for instance by granting a common symbolic language to protagonists of the different dimensions. Culture **as** sustainability signifies a concrete sustainability-friendly cultural pattern including for instance models and images of social cohesion, economic instruments that further effective and efficient sustainable production, and a widespread ecological awareness expressed for instance in a sufficiency orientation. Culture in that vein signifies a constitutive instrument for sustainable development.

In that frame, cultural patterns may then be seen as an important aspect, a fourth pillar, and may function as mediating principles between the classical three dimensions and even work as foundations of sustainability as a whole. But we need to keep in mind that those instrumental functions are volatile and subject to constant questioning, debate, interpretation and reconstruction.

Case study: a Protestant religious normative argument for sufficiency

To give a very short and sketchy example of which questions need to be addressed in the perspective of cultural sustainability, let me turn to the idea of sufficiency. The idea that there must be boundaries to consumption and production has been propagated by authors like Bob Goudzwaard and Harry de Lange (1990), John Cobb and Herman Daly (1994), Thomas Princen (2003), Manfred Linz et al. (2002), Tim Jackson (2011) and Robert and Edward Skidelski (2012). A recent definition that tries to circumvent normative language of 'the good' positions sufficiency as involving the "modification of consumption patterns that help to respect the Earth's ecological limits, while aspects of consumer benefit change" (Fischer and Griesshammer 2013: 10). Of course, the crux lies in the comparison of benefits: To get from A to B, a car and a bike are feasible functional equivalents. On the bike you will get sweaty and wet, will probably need more time and your transport possibilities are limited. The car, however, will keep you cosy and warm, and transport you and your belongings swiftly, albeit not doing a lot for your health or for the environment. To decide which to use, normative orientation is needed.

In the *horizon of culture*, as a Christian theologian I would of course turn to Christian tradition – others may turn to contemporary popular movies like James Cameron's *Avatar* (2009), literary works like Margaret Atwood's *The Year of the Flood* (2009) or an inquiry into legal philosophy like Capra's and Mattei's *Ecology of Law* (2015). In the Protestant tradition that I locate myself in, elements corresponding to the idea of sufficiency can be discerned (cf. Meireis 2016). First of all, the particularly Protestant concept of calling – in its ethical aspects – is directed towards the ultimate aim of serving one's neighbour which carries the promise of fulfilment. In the hierarchy of goods, consumption or recognition are means to achieve that sort of service rather than ends in themselves. Distinctive luxury or excessive consumption is therefore rather spurned, consumption has a measure which allows for the respect of planetary boundaries. Secondly, the Protestant concept of God's merciful justification without works implies the notion that everybody is a loved child of God, no matter what he or she achieves in terms of success or reputation. The necessity of self-aggrandizing through distinctive consumption or the amassing of power may be diminished by this idea of equal worth, which in turn implies another set of boundaries to growth, consumption and accumulation. Thirdly, Christian anthropology considers humankind to be the stewards of the world, not its masters: God's assignment as depicted in the creation myths implies a basic solidarity of the created that forbids limitless exploitation or destruction.

However, the concept of sufficiency has to be perceived in the *agonal arena of culture*, where questions of distinction, ascription and debate are always present – I will not go into all possible issues, just a few controversial aspects to make the point: First there could be a problem with religious traditions in general. The otherwise flawless research of Soini and Birkeland (2014) on cultural sustainability never even mentioned the contribution of religions toward sustainability – drawing on religious tradition might be seen as diminishing the scientific value of a sufficiency orientation. Also, different religious or philosophical traditions might disagree as to the worth of sufficiency. Secondly, sufficiency might very well be a debated point in a certain faith based community – let's say: Protestantism – itself. Religious antagonists to the sufficiency argument might suggest that religious thinking proper has nothing to contribute to secular problems of well-being but is only concerned with salvation of the soul and should therefore refrain from that kind of moral language (Reuman 1992: 103–113). Thirdly, the value of a sufficiency orientation on a general scale may well be debated on the grounds that it lacks universalism and effectiveness. Such an argument might suggest that the economic growth paradigm has been largely effective without putting consumers into restraining moral or legal reigns, and that where such restraints have been tried through economic planning, they failed miserably.

Asserting such claims concerning Protestant tradition and theology and not avoiding debates in the agonal arena of culture, may bring us to an understanding of sufficiency as a *cultural element of sustainability* discourse: Sufficiency could be understood as part of a 'Protestant way of life' – which in turn needs to be preserved. This religious dimension would then be part of the cultural as aspect of sustainability, maybe forming an overlapping consensus with other religious or secular reasons for sufficiency. Protestant ethics being a part of the cultural discourse as a whole could offer models of mediating the ecological, economic and social aspects of sufficiency, for instance arguing for an economic shift towards publicly funded care work that authors like Daly, Princen or Jackson recommend. And of course, a Protestant ethos on the whole might contribute to a moral basis for sufficiency, as it did in the 1970s and 1980s, when it was one of the sources of the environmental movement in the German-speaking world.

Conclusion

To some, this description of possible Protestant contributions towards a culture of sustainability may seem appealing, while others might find it appalling. Here, it serves as an example of models based on a certain religion or world view. It is, of course, only one religiously oriented model among many, all of which might approach sustainability in equal-but-different ways, according to their own ethics. Their contribution to cultural sustainability might be substantial, if they present their moral claims for open debate. A prerequisite of that debate, however, is a concept of cultural sustainability that takes the dynamics of culture into account and understands culture not only as an instrument, but also as a horizon in an agonal arena, thus opening the cultural field to sustainable discourse.

References

Atwood, M. (2009). *The Year of the Flood*. London: Bloomsbury.

Barthes, R. (2010). "Die Welt des Catchens". In *Mythen des Alltags*. Berlin: Suhrkamp, 15–28.

Bourdieu, P. (1983). "Ökonomisches Kapital, kulturelles Kapital, soziales Kapital". In R. Kreckel (ed.). *Soziale Ungleichheiten* (Soziale Welt Sonderband 2). Göttingen: Otto Schwarz & Co, 183–198.

Bundesverfassung der Schweizerischen Eidgenossenschaft (BV) (1999). vom 18. April 1999. www.admin.ch/opc/de/classified-compilation/19995395/201801010000/101.pdf (accessed 01/03/2018).

Cameron, J. (2009). *Avatar*. 20th Century Fox.

Capra, F. and Mattei, U. (2015). *The Ecology of Law: Toward a Legal System in Tune with Nature and Community*. Oakland, CA: Berret-Koehler Pubs.

Cassirer, E. (1944). *An Essay on Man: An Introduction to a Philosophy of Human Culture*. Yale, New Haven: Yale University Press.

Cobb, J. and Daly, H. (1994). *For the Common Good: Redirecting the Economy toward Community, the Environment and a Sustainable Future*. Boston, MA: Beacon Press.

De Gruchy, J. (2007). "An Olive Agenda. First Thoughts on a Metaphorical Theology of Development". *The Ecumenical Review*, 59, 333–345.

Dessein, J., Soini, K., Fairclough, G., and Horlings, L. (2015). *Culture in, for and as Sustainable Development: Conclusions from the COST Action IS1007. Investigating Cultural Sustainability*. Jyväskylä: University of Jyväskylä.

Diefenbacher, H. (2001). *Gerechtigkeit und Nachhaltigkeit: Zum Verhältnis von Ethik und Ökonomie*. Darmstadt: Wissenschaftliche Buchgesellschaft.

Fischer, C. and Grießhammer (2013). "Mehr als nur weniger. Suffizienz: Begriff, Begründung und Potenziale". *Öko-Institut Working Paper* 2/2013.

Geertz, C. (1973). "Thick Description: Toward an Interpretive Theory of Culture". In *The Interpretation of Cultures: Selected Essays*. New-York: Basic Books, 3–30.

Goudzwaard, B. and de Lange, H. M. (1990). *Weder Armut noch Überfluss: Plädoyer für eine neue Ökonomie*. München: Kaiser.

Haraway, D. (1988). "Situated Knowledges: The Science Question in Feminism and the Privilege of Partial Perspective". *Feminist Studies*, 3, 575–599.

Hawkes, J. (2001). *The Fourth Pillar of Sustainability: Culture's Essential Role in Public Planning*. Altona, Australia: Common Ground Publishing Pty Ltd PO.

Honecker, M. (1999). "Art. Schöpfung IX. Ethisch". G. Müller, H. Balz and G. Krause (eds.). *Theologische Realenzyklopädie*. Vol. 30. Berlin: De Gruyter, 348–355.

Jackson, T. (2011). *Wohlstand ohne Wachstum: Leben und Wirtschaften in einer endlichen Welt*. München: Oekom Verlag.

Kebir, S. (1991). *Gramsci's Zivilgesellschaft: Alltag, Ökonomie, Kultur, Politik*. Hamburg: VSA.

Kurt, H. and Wehrspaun, K. (2001). "Kultur: Der verdrängte Schwerpunkt des Nachhaltigkeits-Leitbildes: Überlegungen zur Notwendigkeit und den Chancen einer stärker kulturpolitischen Fundierung der Umweltpolitik". *Gaia*, 10/2001, 16–25.

LenkaBula, P. (2000). *Choose Life, Act in Hope: African Churches Living out the Accra Confession. A Study Resource on the Accra Confession: Covenanting For Justice in the Economy and Earth*. Geneva: World Alliance of Reformed Churches.

Linz, M., Bartelmus, P., Hennicke, P., Jungkeit, R., Sachs, W., Scherhorn, G., Wilke, G., and von Winterfeld, U. (2002). *Von nichts zu viel: Suffizienz gehört zur Zukunftsfähigkeit*. (WP 125) Wuppertal: Wuppertal Institut.

Meireis, T. (2016). "Schöpfung und Transformation: Nachhaltigkeit in protestantischer Perspektive". In T. Jähnichen, T. Meireis, J. Rehm, H.-R. Reuter, S. Reihs and G. Wegner (eds.). *Nachhaltigkeit: Jahrbuch Sozialer Protestantismus, Band* 9. Gütersloh: Gütersloher, 15–50.

Mignolo, W. D. (2011). *The Darker Side of Western Modernity: Global Futures, Decolonial Options*. Durham, London: Duke University Press.

Nassehi, A. (2011). *Soziologie: Zehn einführende Vorlesungen*, 2nd Ed. Wiesbaden: Verlag für Sozialwissenschaften.

Princen, Thomas (2003). "Principles for Sustainability: From Cooperation and Efficiency to Sufficiency". *Global Environmental Politics*, 3(1), 33–50.

Reumann, J. H. (1992). *Stewardship and the Economy of God*. Eugene: Wipf and Stock.

Sen, Amartya (1999). *Development as Freedom*. Oxford, New York: Oxford University Press.

Skidelski, R. and Skidelski, E. (2012). *How Much Is Enough? Money and the Good Life*. New York: Other Press.

Soini, K. and Birkeland, I. (2014). "Exploring the Scientific Discourse on Cultural Sustainability". *Geoforum*, 51, 213–223.

Tabuchi, H. (2016). "'Rolling Coal' in Diesel Trucks, to Rebel and to Provoke". *The New York Times Online*. www.nytimes.com/2016/09/05/business/energy-environment/rolling-coal-in-diesel-trucks-to-rebel-and-provoke.html?_r=0 [Accessed 17/05/15].

Taylor, C. (2004). *Modern Social Imaginaries*. Durham, London: Duke University Press.

Weigel, S. (2010). "ÜberLebensQualität: Kulturwissenschaft und Nachhaltigkeit". In Deutscher Hochschulverband (ed.). *Glanzlichter der Wissenschaft: Ein Almanach*. Saarwellingen: Lucius & Lucius, 153–159.

Williams, Raymond (1989). "Culture is Ordinary". In *Resources of Hope: Culture, Democracy, Socialism*. London: Verso, 3–14.

World Commission on Environment and Development (WCED) (1987). *Report of the WECD. Our Common Future*. Transmitted to the General Assembly as an Annex to document A/42/427 – Development and International Co-operation: Environment. www.un-documents.net/wced-ocf.htm [Accessed 15 October 2017].

Part II

Philosophy, sociology, economics and cultural sustainability

6 Sustainability and modernisation in conflict

On Roger Scruton's conservative green philosophy

Anton Leist

Introduction

Politically ecological aims are traditionally wedded to the left-liberal or social-democratic part of the political spectrum. The philosopher Roger Scruton is out for correcting the expectation toward such an alliance. According to him, conservatism is a more proper basis for ecological aims. If we want to discuss sustainability within the expanded frame of 'cultural sustainability', it is worthwhile addressing Scruton's thought.

The conflict inherent in 'sustainability'

Beginning with the Brundtland Report (1987), 'sustainability' shifted from being a technical term used in ecological debates to a household word. Today many things, whether products or practices, are routinely labelled 'sustainable'. Often this means little more than conceiving of a certain self-imposed restriction, a cap on the expenditure of resources. If there is any consensus on its meaning given the widespread use of the term, it is that it signals ecological consciousness and self-restraint, even if only in a multivariate and jumbled way. The difficulty in using the concept precisely has been ecologists' complaint since its linguistic debut.

Nevertheless, with the Brundtland Report's initiation, a principled core for the definition of the word was included. The Report gave priority to the 'needs' of 'future generations' and also highlighted 'development', widely understood to be a sort of incoherent concession to North-South sensitivities. To many eyes, 'sustainable development' was and is a contradiction, exposing the basic conflict between an optimally high standard of living of European proportions and a reduction in overburdening the environment – both on a global scale.

At the time of the Brundtland Report, 'sustainability' was considered to be a new concept, even if, as it turned out, it was a rediscovery of a forestry principle developed by Hans Carl von Carlowitz, a long forgotten Saxonian *Oberberghauptmann* of the early eighteenth century.[1] Post-Brundtland, ecologists, confronted with the expanded concept of sustainable development, tried to situate it in a wider normative context, and in order to do this, they had to fall back on two political moral traditions, liberalism and socialism. Accordingly, they tried to rephrase sustainability either by referring to concepts of justice or by linking it to a critique of capitalism, or both.[2] Mirroring the muddle of an unsustainable economy, the concept only became more opaque through these transformations.

Given the multigenerational and international dimensions of ecological conflicts, resorting to distributive justice, for one, seemed the method of choice, even if a problematic one. Not being able to 'cooperate' with future generations – one of the basic conditions for justice – in the face of accelerated climate change, the only possibility is cooperation with the *next* generation. Also, whether economic and social justice, normally conceptualised as being implemented in a single national society, can be extended to the global sphere, is open to debate. A controversy about real conditions underlies the latter dispute: whether motivational forces are needed and if so, whether the necessary motivation is in place to implement a concept like justice. Moreover, in the rhetoric of sustainability, the usually harmless philosophical penchant for abstractions turns dangerously ideological, highlighting an impossible ideal.

To be clear, sustainability seems a Western apologetic idea as it suggests the conservation of *our* level of welfare – why should we mount opposition to sustaining *our* way of living? Often 'sustainability' is used in the literature, accordingly, to mean the continuance of effective business, good markets and increasing wealth. One wonders whether *ecological* sustainability would not need to be something different. In order to disentangle its internally conflicting forces, I suggest distinguishing between *productive* and *conservational* sustainability. 'Conservation', being etymologically linked with conservatism, refers to a third position that politically has been relegated to the background for some time. Due especially to the impact of globalisation, the political conservatism of the early years after World War II has been thoroughly forgotten, and the label 'conservatism' has little appeal today. But if we want to gain a clearer view of the systematic alternatives, we should not overlook it.

Both liberalism and socialism centre on the values of freedom, equality and justice. These ideals attempt to account for how scarcity is to be regulated, but they are silent on how scarcity is to be mitigated in the first place. The tradition of conservatism can be helpful for just this reason: because liberal and socialist traditions are presently dominant in Western societies, it is no small wonder that explicitly conservative defences are rare. In the following, I will engage with the philosopher Roger Scruton who has paddled for many years 'against the stream' as a spokesman for conservatism, including for a conservative Green position.

Environmentalism and conservatism

In order to prepare the ground for an engagement with the values central to Scruton's conservatism, I will, following him, outline three areas of convergence between environmentalism and conservatism: a conservative attitude towards the environment, a scepticism about technological-capitalist solutions, and a preference for community-based solutions over state-organised ones.[3] Scruton is an eloquent spokesman of all three of these conservative tendencies.

First, the socially aware conservative maintains a *cultural* concept of nature. She appreciates perhaps more than others the historical development of our environments, such as the built environment in our cities, as well as the cultural manifestations of our landscapes. While raw, or pure, nature may exist – for example, the deep sea or the North Pole – for humans there is no such thing in our everyday activities. The natural environment is always already subjected to human use, whether that be agriculture, landscaping or construction. Conservatives value tradition and are committed to maintaining the existing level of human 'interference' in the environment. Here conservative ideology converges with imperatives of conservational sustainability, without invoking an esoteric concept of 'deep ecology' (Naess 1973; Merle 1994; de Jonge 2004).

Second, the conservative is sceptical of techno-capitalist solutions. This scepticism will become more appreciable once we understand the deeper reasons for conservatism, because at face value one would think 'why not conserve capitalism as well?' Techno-capitalist solutions are ones that are based on individualism and instrumental rationality. With techno-capitalist solutions (e.g. solar panels, energy-saving light bulbs, electric cars) short-term problems are solved but underlying social habits remain largely intact. Conservatives, by contrast, focus precisely on affecting social change. They propose tapping the social potential of communities and making intrinsic values visible instead of employing instrumental rationality. Community for conservatives is not a narrow social association or union, but an entity associated with a territory, a tradition and a specific social glue that holds together its members. Scruton uses the German term *Heimat* for this kind of environmentally advantageous commitment (2012: 25–27; ch. 7).

Third, the conservative attitude towards the state is ambiguous. Conservatives typically favour a strong state and a certain amount of regulation. Adverse to market-driven individualism and also direct democracy, conservatives may in some situations seek help from a strong state, even one with oligarchic tendencies. Conservatives believe in intellectual and moral hierarchies on the basis of ineradicable natural and cultural differences among individual persons, and thus are cautious of social equality. In other contexts, however, conservatives can be strongly critical of bureaucracy and regulation. Scruton, for one, points to the ecological disasters socialist regimes left behind, and he sees this as an endemic problem within democratically controlled bureaucracies. He is also extremely critical of bureaucratically structured NGOs like Greenpeace (2012: 30–38). Instead, he favours small-scale communities where opposition and solutions are organised at a personally engaged local level (10, 20–27, 36, et passim).

This sketch of three ways in which the conservative agenda supports environmental aims brings into focus the extent to which conservatism contrasts with liberalism and socialism. Liberals have much more flexibility in their view of the environment than conservatives in that they subsume ecological aims to the individual's needs, either narrowly as preferences (neoliberalism) or for policy reasons (civic liberalism). Conservatives would instead reclaim the worth of traditions, customs and virtues and assume that the community and its organic needs *restrict* what can be subject to individual choice. For example, the priority of the community's need is universally accepted in the community when it comes to cemeteries, and for conservatives, this would be the ideal orientation in general.

Green environmentalists often turn to bureaucratic solutions, something the conservative shuns. Green environmentalists work to balance justice-driven aims with environmental ones which frequently results in conflicting priorities. For example, the *Pendlerpauschale* in Germany, a tax subsidy for commuters, is at odds with environmentalists' call for clean air. Green environmentalists are torn between these two aims. Conservatives, in contrast, are not that much concerned about justice issues, and they would opt for what is best in terms of addressing climate change.[4]

Following Scruton, I will now examine to what extent the moral resources in our societies can help to distance ourselves from a one-sided, 'productivist' use of the sustainability claims, prevalent in Western capitalist societies.[5] Scruton is helpful in making us aware that in searching for an alternative, we are unavoidably confronted with the more basic moral and political levels of modernisation. In the following, I will engage less with concrete environmental problems than with the social attitudes from which they arise, and which is the only relevant medium of possible change.

Metaphysics or a sense of realism?

Different from everyday conservatives who either take tradition to be important as such or favour it as an essential component of political stability, a philosophical conservative like Scruton exploits philosophical traditions to provide a foundation for his ideas. He is uncompromising in his insistence that conservatism must have a metaphysical basis, which I place alongside another layer of conservative thought, to be called the *realist sense*. A metaphysical and a realist sense often go hand in hand in Scruton's analysis, not unlike other conservatives' analyses. To me it seems preferable to hold on to the realist sense only, and I will explain this by help of critical arguments against metaphysical claims.[6] Metaphysical beliefs conflict with liberal and socialist thinking, and the extent Scruton is fixated on them points to a deeper conflict of values.

Should one not expect *every* political and moral doctrine to be 'realistic'? The point here is one of pervasive importance for our reflective thinking, both for academics and laity. The opposite of a realist sense is a utopian sense. With the help of the realist sense, we evaluate others in light of social behaviour, including a realistic sense of whether social behaviour can be developed in one direction or another. Alternatively, in practising the utopian sense, a thinker orients herself to values and ideals, with only a minimal regard for actual behaviour. Of course, both realist and the utopian elements have to be taken into account if our evaluations will, in some way, be adequate to their object.[7] But different practical policies result depending on which end of the realist-utopian continuum takes precedence. Conservatives orient themselves most often to the realist side, whereas socialists, or socio-liberals, are always in danger of situating themselves at the utopian end of the spectrum.[8]

Social relations are constituted by a combination of empirical psychology and normative claims, and here is the heart of the realist/utopian distinction. Utopians rely on judgement alone, whereas realists orient their judgements to the kind of motivation that is endemic in people. To find a balance between both is a duality problem, of which Scruton is well aware. He presents it in the form of a duality of evolutionary and rational explanations (2012: 213; 2014: ch. 1). He suggests that practical rationality does not come in a pure form but is always combined with a sense of accountability for what we do to others (2012: 212–214). This feeling is itself embedded in, or arises out of, typical moral emotions like indignation, resentment, envy, admiration, praise etc. Responsibility would be the larger moral attitude into which 'feeling oneself accountable' fits.

Scruton follows here the 'imperative of responsibility' of Hans Jonas, which he reads in a more down-to-earth, psychologically realistic way than Jonas did. Pure thought – floating free of concrete motivation – is not what creates responsibility, including such typically academic thoughts as those of 'future generations' or of 'humanity'. Instead, responsibility "arises from our attachments to others: it is our ancestors, our successors in title who awaken our concern" (2012: 207). Responsibility leads to sacrifice, and sacrifice is sustained only if there is attachment to others. This is most effective – the Humean clue – in regard to family and the narrow circle of friends, but hopefully also includes a potential for wider social circles. Here is where the term *oikophilia* (along with *Heimat*) becomes relevant (2012: 26). It means identifying with a local community in a way that is interconnected enough to develop responsibility for the following generations of the community.

There is one tradition of political (and moral) philosophy, the social contract, that also offers insight into the problem of duality and that is also, to some extent, realistic. Social norms of obligation could be derived from the fiction of a social contract of all members of a society with each other by way of a hypothetical reflection on the basis of empirically grounded motives. Such a contract is based on the idea that there are advantages to all members agreeing on moral and political norms, such as opposition to anarchy, civil war or amoral behaviour. Would this not be a take on the duality problem as well, suggesting another combination of accountability and responsibility? Scruton offers a Hegelian critique of this combination, objecting that the contract, if the basis of something like morality, must presuppose just what it purports to explain. For Scruton, family and state "depend on *unchosen* obligations, and . . . will collapse if people lose the instinctive gratitude and piety that enable them to identify kin and country as personal assets, to be cherished and protected in return for what they give" (2012: 223; italics added).

If the illusion exists in the modern era that moral attitudes can be embraced by decisions in light of our informed desires or interests, this would be wrong because such decisions are not available to us. In order to make decisions, an amount of presupposed cultural conditions have to be in place, and the contractualist is naive about the extent of these communal givens.

Scruton on piety and the spiritual

Similar to other critics of modernity, Scruton applies the use/end value distinction in a metaphysical way to modernity. Whereas the distinction is unavoidable in general, the end result need not be read, as metaphysically inclined interpreters do, as having value entirely beyond human control. Similar to critics of 'technotopia' and 'instrumental reason' (late Heidegger, old Frankfurt School), Scruton is eager to retain a sense of value that is beyond the reach of humans and their influence. 'Beauty' and 'sacredness' are the two values he considers most important for that. "Since the Enlightenment, aesthetic taste and natural piety have stood vigil over our surroundings, and held back the hand that was raised to destroy them" (2012: 253). Beauty and the sacred are barriers against use, instrumentality and exploitation, and because of this they are part of *oikophilia* and active components of stopping the disruption of natural habitats. In the modern world, beauty and sacredness are under threat of being adapted by, and drawn into, the very instrumental culture they are meant to hold at a distance. Accordingly, it is the task of the conservative philosopher to lay bare the original power of beauty and sacredness: "It seems to me that many of the worst environmental depredations of recent times have come about because beauty has been displaced from the agenda, and utility elevated at its place" (2012: 256).

Scruton rejects the widespread notion of aesthetics as merely subjective and instead attaches the experience of beauty to achieving a common social identity (2012: ch. 8, esp. 256–258, 264). In one telling example he contrasts a purely utility-oriented way of individually gulping down junk food with the aesthetically laden event of a family dinner. He describes the latter as a synthesis of social connectedness, aesthetically intrinsic value and giving meaning to oneself by relating to things eternal: things that "remain forever the same" (2012: 262).

Even if only vaguely articulated, Scruton's social idea concerning beauty seems plausible to me. It is a promising suggestion that "judgments of beauty are bids for socially recognised *presence*" (2012: 264). In agreement with other remarks, this could mean that aesthetic standards are forms of expressed social identities, a common way to interpret and symbolise our everyday activities. All aesthetic expressions have, even if at most times unintentional, to some extent a social function. Let me call this, in line with Scruton's intentions, the *community conception* of aesthetics and beauty. The community conception sees art as exemplifying ideal ways of living together, in the form of shared living.

Now, if art had such a function in an extended way, it would certainly help in the worst cases of environmental despoliation. To what extent and to what end however, depends on what exactly is meant by 'community'. A community could be a present community, realistically conceived, or an idealised community, depending again vaguely on metaphysical ideals. In his comments about the family dinner, Scruton makes clear that he has a metaphysical conception of community in mind.

Look, for example, at his advice to keep eternal traditions alive for the family by enjoying 'willow-patterned plates' as reminders of ancient China (Scruton 2012: 262). The community has to value the sharing of meals in the first place, and through this, the relation of beauty to sacredness.

> You notice (in the case of the family meal) another continuity too, between aesthetic values and piety, which is the recognition that the world is in other hands than ours. Hence the gods are present in mealtimes.
>
> (Scruton 2012: 263)

Oikophilia thus in the most substantial sense fosters *three motivations* for environmental action, one of communal relationship, one of a sense of beauty, and one of piety developed in the face of sacredness. If we orient ourselves to psychological realism, we can appreciate the first two motives but will be hesitant about the third. The point here is not one of total rejection; even the thinker adverse to the metaphysical has to admit that Scruton puts his finger on something universally human (2012: 284). The occasions for piety that Scruton has in mind are ones like birth, illness, death and events related to these, like marriages, funerals or baptisms. Feelings of sacredness and piety arise out of, or in the face of, such existentially outstanding and often unsettling experiences. The message behind these events is that crucial parts of our life are beyond our control. If we open ourselves to such experiences, something like piety arises, which might, in its most raw form, be a universally human feeling.

Several open-ended questions follow, however, in the face of such an experience of existential shock or boundary feeling. Among these are, first, whether adequate terminology is available, and second, what exactly is the content of this existential response. As for the first one, we could ask whether the term 'piety' is the right one in the event of such an existential boundary experience when there is no religious belief involved. A less religious word would be 'respect'. Another, quasi-religious word would be 'awe' or, in German, '*Ehrfurcht*'. I think we should not tie ourselves to any one word too narrowly, as that may result in a transformation of meaning. Why could 'piety' not be used in a religiously neutral way? On the other hand, not to distinguish between a secular and a religious use would blur the dimensions of definition, something that clearly happens in Scruton's writings. With this distinction in mind, I will use 'respect' for religiously neutral existential phenomena, even if a purely secular use of 'piety' might be convenient as well.[9]

To his credit, Scruton too wants to disengage 'piety' from its Christian background and adopt the meaning of the ancient Romans.[10] Unfortunately, this does not offer us a secular definition of piety. The difference between a secular existentialist attitude of piety and a religious one (including the Ancient one) depends on the presupposition of the existence of a right or good order of things beyond humanity. From a secular perspective, one feels respect in the face of existential events of the type mentioned earlier, and out of this, one develops an attitude of care, humbleness and circumspection concerning one's life and of life in general – ends that are open to further reflection and decisions. There is not only one form of practical response, for example awe or self-denial. The religious perspective, on the other hand, leads to just such dramatic responses, because along with the feeling of piety is a presumption of a pre-established right order.[11] This presumption claims that at a minimum the transcendent order is right in and of itself or, in addition, that it has been created by one of the Gods of the monotheistic religions. Consequently, imperatives are imposed on us on behalf of this right order, sometimes of an absolute quality. We are not allowed, for example, to forego our life by suicide.

The upshot of this is that as non-believers we must *not* accept Scruton's claim that sacredness is a human universal (2012: 284), even if we should accept a respectful attitude arising from existential experiences of the boundaries of human life. Such a non-religious existential respect is for non-believers an adequate response. Scruton's writings are biased rhetorically at this point, as he regularly tries to convince his readers of the 'sacred reality' by dramatically painting the shortcomings of a purely 'rational', 'instrumental' or 'utilitarian' perspective (see also his 2012a). This either/or is too crudely put by Scruton, as existential respect is another alternative between such stark extremes.

If one favours a 'religious' attitude of piety (presupposing a 'right order'), one should have deep concern for nature beyond making it instrumentally useful for the fulfilment human needs. If, alternatively, for the non-believer, piety arises out of existential events, it is not usually or easily extended to all living things. We may feel respect (based on our existential experiences) for unknown humans, but not similarly for animals, plants and landscapes. Scruton holds the human person to be 'sacred', a term the non-religious thinker cannot adopt. He restricts sacredness to humans, arbitrarily because he does not base his argument on Christian revelation, and relegates animals to a lower tier by denying them rights (Scruton 1996: ch. 8). Also, it is less sacredness he builds on here than established philosophical reason, which is reserved for humans and distances the moral status of animals. The premium 'sacredness' is attributed to whatever is of human interest.

Traditional and modern morality

I will now try to explain the extent of my sympathy with Scruton's conservatism. I am obviously hesitant to embrace his ideas in so far as they arise out of the belief in sacredness and piety. These I think to be private attitudes, and even if I would share them personally, I would not see fit to advocate them as public attitudes. Scruton seems to think otherwise because he takes these values to be a part of the anthropological frame of man, whereas I doubt that such a claim can be justified. A closer look into the bases of morality may help to make my reasons for hesitation clearer.

First of all, Scruton's belief in sacredness is not an idiosyncratic whim on his part, so much as part of a coherent conservatism. An underlying affinity between conservatism and religion can be seen from the empirical work by the social psychologist Jonathan Haidt (2008, 2012). Haidt keeps distinct several groupings of, as he calls them, 'value modules'. The one grouping I see as *traditional* morality includes values like loyalty, authority and sanctity, and the competing *modern* one centres on values like care, fairness and freedom. If we make use of this distinction, the ideal-typical *conservative* morality could be defined by giving priority to the three traditional values when there are conflicts in values. Somehow, it seems, the value of sanctity coheres with the social attitudes of authority and loyalty and relegates to the background the modern values of care, fairness and freedom, without necessarily being entirely at odds with them. If we take the two value sets to be internally coherent, sanctity, or the belief in sacredness, is not an arbitrary addendum for the conservative attitude.

If we ask ourselves which ideal-typical variant of morality to follow, in the sense of explicating to ourselves what it means to live in Western culture, we will, in all likelihood, situate ourselves on the side of modern morality, which at its core, centres on fairness and freedom, or a belief in individual autonomy which in conflict trumps authority and loyalty. The believer in autonomy will not easily accept the intrinsic value of tradition and will be critical of attributing sacredness to things or practices. Families, old buildings and even outstanding natural objects will not have a premium above all interest-based alternatives.

There is a distinction to be made, however, among the positions within modern morality. Again ideal-typically, the apt distinction is the one between the *realist* and the *utopian* liberal.[12] And here is where the realist liberal and Scruton's conservative converge in part. The realist liberal agrees in many ways with Scruton's sceptical criticism of self-serving bureaucracy, non-legitimised political cosmopolitanism, and suppression by utopian politics. Scruton is right in seeing ethical reasoning to be based on motives, albeit often in a hidden form. His tendency of letting polemics run into diatribe notwithstanding,[13] he is right in his critique of the disruption of well-functioning communal networks, crafts and customs by economic and technological forces (Scruton 2012: 313).

To generalise, the expansion of local economic activities into a sphere of ever wider geographic and political extension, particularly transnational, notoriously implies signs of suppression by means of politics and law. In many cases, this changes the situation for the worse, as the mismatch of ambitious political schemes and available levels of motivation disrupts what was effective earlier. If this criticism of politics and law in their expansion beyond engaged morality is right, conservatives have another argument beyond what at first looks like only a nostalgic adherence to community. They have a piece of practical proof of the social glue of communal living, and realist liberalism shares this.

Conclusion: towards a conserving (cultural) sustainability

Conservatives have a point when they highlight the importance of traditional practice. Traditional practice can be a way of doing business, a way of work, of festivities or religion, or of whatever one could think. Conservatives claim these practices have been proven effective by history and represent a well-established way of doing things. And this is indeed true: long-term traditions incorporate a social experience that should not be overlooked or be left to erode by one-sided, economically driven developments. Conservatives should, therefore, point out what they think the treasures of tradition are and hold on to them.

However, what is also needed is a critical adaptation of traditional practice. A traditional practice in most cases is at the core a social one, which gives moral values an extraordinary role within the practice. This also merges the practice with traditional values. For liberal realism, this suggests a two-pronged method to deal with Scruton's more material attitudes. First, in sharing a dependence on empirical relevance with Scruton, liberal realism acknowledges the communitarian side of environmentalism, as long as it does not openly contradict liberal values. Compromises in values are to be sought after. Second, these compromises are those that Haidt explains with his contrast between traditional and modern morality. If loyalty, authority and sanctity are related to long-standing social practices and institutions, like the family and fathers or the church and the priest, liberal realism will try to appreciate the attitudes that undergird these but try to critically adapt them by subsuming them under care, fairness and freedom.

Traditional marriage is a case in point, if authority and loyalty in it are practiced one-sidedly, and fairness and freedom are invoked against this. It is an idle question whether empirical modernisation – the disruption of traditional practices by (especially) economic development – is the origin of modern values, or the other way around. In any case, modern values arise when traditional practices weaken or break down. Marriage is not breaking down, of course, but turned into a contract-based relationship with mutually equal duties and rights (a development that Hegel and Scruton regret).

A critical attitude towards a traditional practice could be developed in the following way: identify within a traditional practice what is good and the traditional values undergirding it in order to hold onto what is good but substitute modern values for the traditional ones, without, if possible, destroying the entire practice! Also let all presently available experiences and techniques add to improve the practice, but keep what is good in it alive. We could call this the 'modernising change' of traditional practices. This way of proceeding is not what a conservative like Scruton would acclaim. But if we do not want to dogmatically bind ourselves to tradition, reworking it in light of our modern values is surely the only way to proceed.

Notes

1 Von Carlowitz lived in the small Saxonian city of Freiberg and published his book *Sylvicultura Oeconomica* in 1713 in Leipzig, one year before his death. He served as an employee of the Saxonian King August the Strong, not himself an exemplar of frugal living!

2 I followed the first lead myself in Leist (2010). But see also Tremmel (2009: chs. 5–6). Comment on ecology and socialism falls either in the category of self-directed criticism (Grundmann 1991; Benton 1996) or in the category of rescue operations by way

of social justice (Pepper 2002). Of course, socialism is notoriously known for laying waste to the environment, not for sustaining it.

3 Scruton (2012: 9); see also Pilbeam (2003).

4 Or better: for the preservation of his local environment from climate change induced destruction, making the climate topic only instrumental. See below.

5 I call claims to sustainability 'productivist' if they are wedded to technical methods making human behaviour environmentally less damaging, without changing the aims of behaviour. The alternative of 'conservational' sustainability comes into view towards the end of this chapter. This is not, of course, meant to be a quid pro quo.

6 In philosophical parlance the predicate 'realist' has a metaphysical connotation as well. I, however, use the term in the everyday practical meaning of addressable empirical facts, adequately theorised, and not in the sense of an irreducible ontology of values or duties.

7 J. Rawls (1999: 7, passim) coined the term 'realistic utopia' for this.

8 Most academic moral philosophers are inclined to lean towards utopianism, without being outspoken socialists. Libertarians, for example, are also exemplars of utopian thinking.

9 In the dispute following up on the first article of the German constitution, it has become customary to make use of the term 'dignity' ('*Würde*') as a partial substitute for 'sacredness'. If dignity is based in respect alone, this would come to the same end. I suggest something like this in Leist 2005.

10 "I wish to divest it of its specifically Christian connotations in order to return it to something like the use it had in late antiquity, when it expressed an idea of permanent validity in moral thinking" (Scruton 1996: 64).

11 The use of 'religious' is a difficult one here. Christianity would not at all endow the whole of nature with sacred value, as Buddhism does. In consequence, its ethic of nature is more in line with the secular one.

12 The 'liberal,' to make things easier, includes the 'socialist'.

13 For heavy polemics, see Scruton's comparison of cosmopolitan intellectuals with the Jacobins during the French Revolution (2012: 102). For his recurrent plea for motivation-based reasoning, see Scruton (2012: 17–19, 22, 101, 187, 198, 200–201, 206–8, 214, 229, 284, 290).

References

Benton, T. (ed.) (1996). *The Greening of Marxism*. New York: Guilford Press.

Brundtland, G. H., et al. (eds.) (1987). *Our Common Future*. Oxford: Oxford University Press.

De Jonge, E. (2004). *Spinoza and Deep Ecology: Challenging Traditional Approaches to Environmentalism* [2004]. London: Ashgate; second ed. 2016 Milton Park: Routledge.

Grundmann, R. (1991). "The Ecological Challenge to Marxism". *New Left Review*, 187, 103–120.

Haidt, J. (2008). "Morality". *Perspectives on Psychological Science*, 3(1), 65–72.

Haidt, J. (2012). *The Righteous Mind: Why Good People Are Divided by Politics and Religion*. London: Allen Lane.

Leist, A. (2005). "Menschenwürde als Ausdruck". In *Ethik der Beziehungen*. Berlin: Akademie.

Leist, A. (2010). "Nachhaltigkeit, aufgelöst in Gerechtigkeit". *FIPH Journal*, 16.

Merle, J. (1994). "Sustainable Development and Deep Ecology: An Analysis of Competing Traditions". *Environmental Management*, 18(4), 477–488.

Naess, A. (1973). "The Shallow and the Deep, Long-Range Ecology Movement". *Inquiry*, 16, 95–100.

Pepper, D. (2002). *Ecosocialism: From Deep Ecology to Social Justice*. London: Routledge.
Pilbeam, B. (2003). "Natural Allies? Mapping the Relationship Between Conservativism and Environmentalism". *Political Studies*, 51, 490–508.
Rawls, J. (1999). *The Law of Peoples*. Cambridge: Harvard University Press.
Scruton, R. (1996). *Animal Rights and Wrongs*. London: Continuum.
Scruton, R. (2012a). *Green Philosophy*. London: Atlantic Books.
Scruton, R. (2012b). *The Face of God*. London: Bloomsbury.
Scruton, R. (2014). *The Soul of the World*. Princeton: Princeton University Press.
Tremmel, C. (2009). *A Theory of Intergenerational Justice*. London: Earthscan.

7 Culture, consumption and sustainability in a sociological perspective

Ulf Liebe

Introduction

Overconsumption is one of the central challenges of modern societies aiming for sustainable development. So-called developed countries in particular use many more resources than is tolerable for meeting current needs and not affecting the possibilities of future generations to meet their needs (a common definition of sustainability). Most countries already have an ecological footprint that is considerably above biocapacity (Dietz et al. 2007; also www.footprintnetwork.org/). Ethical and political consumption, including the purchase of organic and Fair Trade products and using repair cafes, are forms of social action and features of social movements seeking sustainable societies (Rössel and Schenk 2017; Andorfer and Liebe 2012; Stolle et al. 2005). However, the impact of these sustainable consumption practices remains rather low (FAO 2009; Krier 2008). Many argue that, to secure a sustainable society, individuals must change their lifestyle and consumption patterns (Leng et al. 2017), which is where culture – including values, norms and beliefs – comes into play. Ideally, these cultural components would have to change, but processes of value, attitude and norm change require long-term commitment and can be difficult to enact.

Against this background, I use a sociological perspective to analyse the interplay of culture, consumption and sustainability. Because these are very broad terms and are conceptualized very differently even in sociological research, this contribution cannot do justice to the full range of perspectives. Rather, after presenting some practical conceptualizations of culture, I will highlight two perspectives on the interchanges between culture and consumption that appear contradictory at first glance and argue that, upon closer inspection, both perspectives can be effectively drawn upon to enhance our understanding of how societies can shift towards greater sustainability.

Linking culture and consumption

Similar to other concepts in sociological research, culture is defined in many different ways and is not always easy to distinguish from other phenomena (i.e. "everything is culture"). In the following paragraphs, I refer to some of the conceptualizations discussed in relation to economic processes, including (sustainable) consumption. As shown in Table 7.1, culture can be defined by three aspects (see DiMaggio 1994). Culture has a cognitive component, as expressed in beliefs such as justice in the trading system between developing and developed countries. It also has an expressive component, for example, as indicated through symbols such as the Fair Trade mark. Culture also maintains a valuative component that refers, for example, to values such as universalism (Schwartz 1992), which can be linked to sustainable consumption, including the purchase of fair trade products (Doran 2009, 2010).

To forms of culture like scripts, schemata, classifications, preferences, attitudes and opinions, DiMaggio (1994: 27) adds several other useful distinctions with respect to the study of culture as it relates to the economy. Culture can be conceptualized as a "source of values and goals" that motivate individual behaviour (in line with works by Talcott Parsons and Max Weber, see DiMaggio 1994; Vaisey 2009), as well as a source of "strategies or means" that shape capacities from which strategies of action – ways of organizing action – are constructed (Swidler 1986). According to this framework, culture is represented as a heterogeneous "toolkit" or "repertoire" (DiMaggio 1997: 267, referring to works by Ann Swidler and Charles Tilly) compared to representations as a coherent, latent variable. Research in cognitive psychology supports the "toolkit" perspective (DiMaggio 1997), but also explicitly or implicitly assumes that culture works as a conscious and deliberate process. This conclusion, in turn, tends to neglect automatic processes. Therefore, some researchers argue in favour of a dual-process theory of cultural cognition, which differentiates between fast/automatic and slow/effortful behaviour (Kahneman 2003) and seems more plausible:

> actors are driven primarily by deeply internalized schematic processes ('the elephant'/practical consciousness/ habitus), yet they are also capable of deliberation and justification ('the rider'/discursive consciousness) when required by the demands of social interaction.
>
> (Vaisey 2009: 1687)

Table 7.1 Three aspects of culture

Aspect	*Example of a cultural form*	*Example of manifestation regarding Fair Trade (FT) consumption*
Cognitive	Beliefs	Justice concerns on fair prices for small-hold farmers are related to FT purchases.
Expressive	Symbols	FT marks on clothing, shopping bags, etc. can express support of the Fair Trade movement.
Valuative	Values	Universalism = "protection for the welfare of all people and for nature" (Schwartz 1992: 12) – as a foundation of supporting FT

Source: Ulf Liebe

In considering the correlation of culture to consumption – the "acquisition and use of goods and services" – one may argue that consumption without culture is not possible because "(s)hared understandings and their representations – the components of culture – undergird all economic life" (Zelizer 2005: 337, 348). Many questions concerning the relationship between culture and consumption are also relevant for sustainability (DiMaggio 1994 in the following); among them is the classic question of whether poor people have different preferences, their own culture, compared to "non-poor" people, or whether differences between poor and non-poor people are driven by structural factors (somewhat in line with Pierre Bourdieu's *habitus* concept, which is partly explained by structural constraints DiMaggio 1994: 42). The culture-of-poverty debate offers valuable insights for questions about social sustainability, for example, in terms of reducing social inequality and poverty. Other topics at the nexus between culture and consumption include: How do goods become the objects among which we choose (e.g. ethical products)? Where do preferences come from (e.g. for Fair Trade products)? How does cultural capital (Bourdieu 1984) emerge and lead to in-group effects and distinct consumption patterns (e.g. ethical consumption such as the purchase of organic food, Agovino et al. 2017)? How does status competition work and lead to conspicuous consumption (Veblen 1899, purchase of green products meant to be seen, Griskevicius et al. 2010)? How does consumption constitute identity and vice versa (Simmel 1907; Andorfer and Liebe 2013)? Sociological analysis at the macro-level of research addresses, among others, issues of consumer societies (Schor and Holt 2000).

Keeping the above-mentioned concepts in mind, in the following I will argue that a simple distinction pointed out by DiMaggio (1994) and shown in Figure 7.1 is helpful to understanding (parts of) the relationship among culture, consumption and sustainability:

> Those who treat economic behaviour as *analytically distinct* from culture stress the ways in which norms and conventions constrain the individual's untrammelled pursuit of self-interest. Those who view culture and economic behaviour as *mutually generative* tend to emphasize the former: culture provides the categories and understandings that enable us to engage in economic action.
>
> (DiMaggio 1994: 28, emphasis added)

As the following paragraphs elucidate, research on sustainable consumption features both perspectives and I will argue for their complementarity.

Culture in action towards sustainable consumption patterns

I begin by assuming the perspective that the relationship between culture and (sustainable) consumption is analytically distinct, then change course by adopting the perspective that sees this relationship as mutually generative. Figure 7.2 demonstrates the examples with respect to the two perspectives.

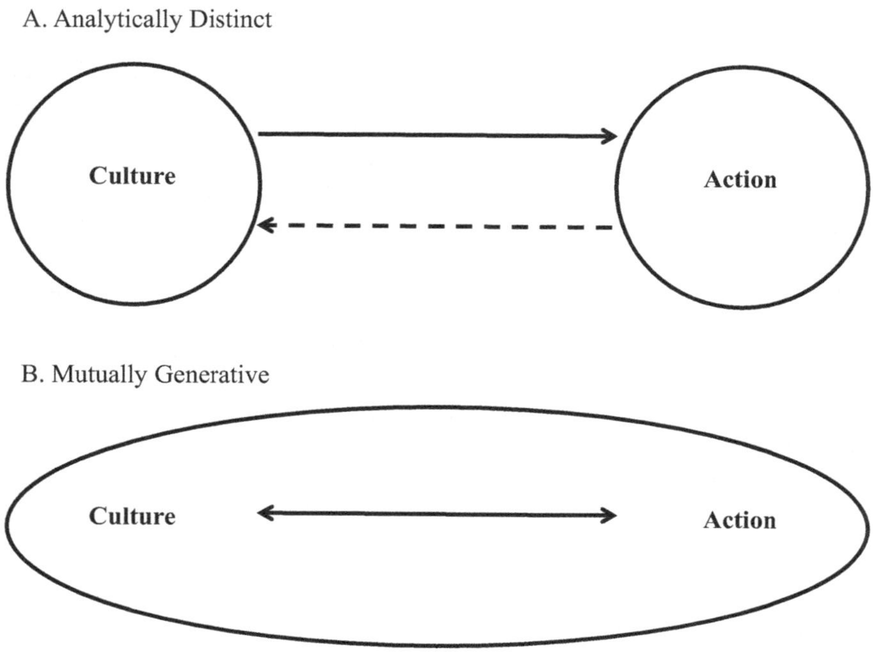

Figure 7.1 Two perspectives on the relationship between culture and consumption
Source: Ulf Liebe

An "Analytically Distinct" perspective

Explicitly or implicitly, based on the conceptualization that culture is analytically distinct from consumption behaviour, behavioural research shows that aspects of culture can be "used" to promote sustainable consumption such as purchasing green products and fair trade products, and participating in other forms of pro-environmental behaviour. These studies suggest that "nudges" – cues in a decision environment – can promote sustainable behaviour like green energy use and consumption patterns (Lindenberg 2012; Thaler and Sunstein 2008). While System 1 nudges, including default options (e.g. defining organic products as standard options), are directed towards automatic processing, System 2 nudges, such as the provision of information about "what most people are doing" (Thaler and Sunstein 2008: 69), a descriptive social norm (Cialdini 2005), target deliberative processing (Kahneman 2003, 2011). Nudges, especially the System 1 type, function well because they are built upon on the general behavioural tendencies of humans such as the desire for relative social status (Griskevicius et al. 2012: 120), perhaps best elucidated in Thorstein Veblen's discussion of "conspicuous consumption" in his *Theory of the Leisure Class* (1899).

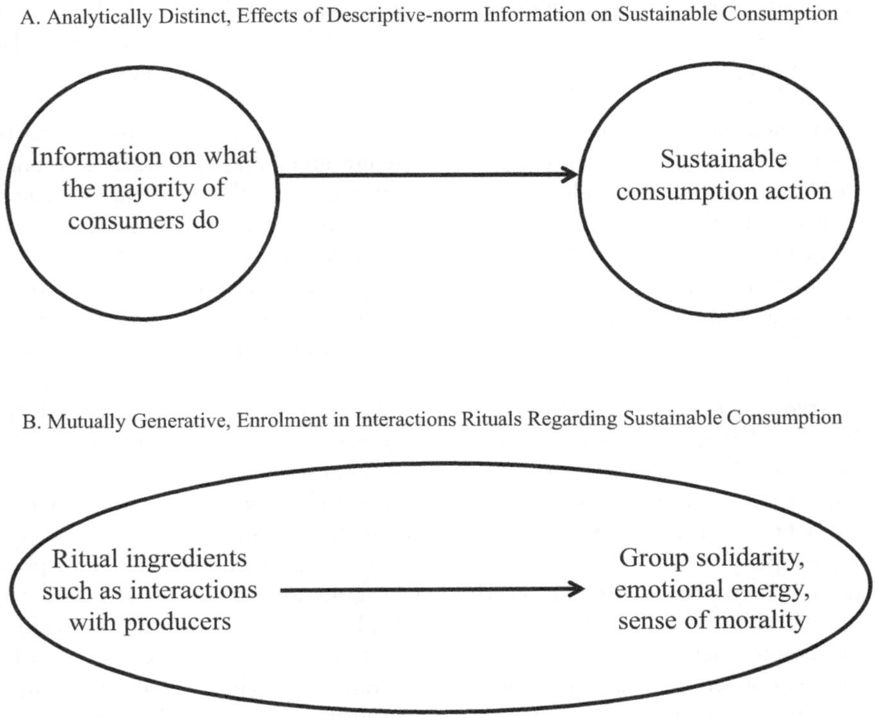

Figure 7.2 Examples of two perspectives on the relationship between culture and sustainable consumption

Source: Ulf Liebe

Griskevicius et al. (2012: 120) summarize how marketers can manipulate these human behavioural tendencies to promote sustainability:

> Strategies for reducing overconsumption that do not account for the importance of relative status are often fighting an uphill battle. . . . Rather than being the root of the problem, savvy marketers are simply exploiting people's innate desires. It will also be difficult to persuade people to be content with their current status or behave in ways that lowers their status. For example, imploring Westerners to consume less because they are wealthier than most people in the world is likely to do little to slow consumption. . . . Rather than trying to eradicate the drive for status, each theory [Costly Signalling, Competitive Altruism, Status, Competitive Environmentalism, added by UL] suggests how this tendency can be harnessed to motivate people to behave in self-sacrificing and pro-environmental ways.

But how do marketers prompt and promote these consumer patterns? Priming experiments have shown how individuals, often unconsciously, can be affected by behavioural cues that activate certain attitudes, norms and so forth (variation in the context of behavioural choices, Sniderman and Grob 1996: 390). For example, in a study conducted by Griskevicius et al. (2010), university students read a short story about a successful university graduate who obtained a high-status first job and was, relative to others, promoted quickly. Students in the control group read a neutral story of similar length. In subsequent experimental tasks, students could (hypothetically) choose between different green and non-green consumer products (car, household cleaner, dishwasher). Each time the products were equal in price but the non-green product was more luxurious and had better performance than the green product, which in turn was pro-environmental. The study showed that, in the treatment group, in which an activation of status motives occurred, 55% chose the green car; in the control group, 37% chose the green car. The corresponding figures for the household cleaner and dishwasher were 42% versus 26%, and 49% versus 35%, respectively. Therefore, activating status motives – exploiting the cultural tendency of a "desire for status" – increased the likelihood of pro-environmental product choices. In further experiments, Griskevicius et al. (2010) showed that the activation of status motives with respect to green product choices was only effective when students were shopping in public (in a brick-and-mortar store versus online at home), and that green products were more desirable when they cost more than non-green products. This phenomenon can be explained by costly signalling theory (Griskevicius et al. 2010, also Fehrler and Przepiorka 2013; Trivers 1971), which suggests that individual pro-environmental behaviour demonstrates not only the willingness to contribute to a collective good (e.g. improved environmental quality) but also the ability to bear the costs for the corresponding contribution. Thus, it might be the desire for status and not the desire for environmental protection that drives these forms of pro-environmental behaviour.

The effectiveness of communicating a descriptive social norm in promoting pro-environmental behaviour was indicated in a field experiment on the reuse of towels in hotels conducted by Goldstein et al. (2008). They randomly assigned 190 hotel rooms in medium-sized hotels in the United States to one of two experimental groups. The rooms for the first group displayed the standard message regarding towel reuse on washroom towel racks: "HELP SAVE THE ENVIRONMENT. You can show your respect for nature and help save the environment by reusing your towels during your stay." The message to the second group included the descriptive norm – that is, what the majority of other guests do:

> JOIN YOUR FELLOW GUESTS IN HELPING TO SAVE THE ENVIRONMENT. Almost 75% of guests who are asked to participate in our new resource savings program do help by using their towels more than once. You can join your fellow guests in this program to help save the environment by reusing your towels during your stay.

Goldstein et al. (2008) observed the behaviour of the hotel guests over a time span of 80 days where there have been 1,058 instances of potential towel reuse.

The data were collected by the hotel room attendants, and hotel guests were not aware that they were part of an experiment. The results show that the towel reuse rate increased by 9 percentage points when the descriptive norm message (44% reuse) was used compared to a standard message, which saw a 35% reuse. In subsequent experiments, Goldstein et al. (2008) varied the reference group of the norm. They found, for example, that referring to the previous hotel guests of the same room (49% reuse) increased the towel reuse rate by 5 percentage points compared to referring to hotel guests in general (44% reuse) and that referring to any reference group (average of 45% reuse) increased the towel reuse rate compared to using the standard message (37%). These results were partly replicated in a similar study in Germany (Bohner and Schlüter 2014).

Nudging citizens into specific behaviour can be a powerful tool for promoting sustainable behaviour. Recognizing the effectiveness of nudging, governments, for example in the United States and the United Kingdom, have begun using these tools widely. Although nudging has been criticized as being potentially manipulative (Wilkinson 2013), many studies have shown that the majority of citizens in Europe and the United States have accepted nudging – especially with regard to pro-environmental behaviour and healthier food choices – when they were in line with values in a population and not involved in illicit purposes (e.g. Sunstein 2015; Reisch and Sunstein 2016).

A "*Mutually Generative*" *perspective*

Works on practice theories offer a perspective that is in line with the assumption that culture and consumption are mutually generative. Spaargaren sees the potential for these theories to promote sustainable consumption patterns:

> A cultural perspective on socio-environmental change should first of all be applicable at the level of situated consumption practices. Second, it should allow for a thorough analysis of the role of objects and symbols in the reproduction of practices. Third, to complement the existing emphases on the negative aspects of (un)sustainable consumption, the cultural approach should make conceptual room for analysing the positive experiences and elements of sustainable consumption.
>
> (2011: 818)

Following Spaargaren (2011), promoting sustainable consumption patterns can be studied by looking at interaction ritual chains following Collins' (2004) theoretical work. Collins' theory considers situations as starting points. These situations – common events – have ritual ingredients such as bodily co-presence, barriers to outsiders, mutual focus of attention and shared moods, which are interlinked and lead to "collective effervescence," which in turn "produces" ritual outcomes such as solidarity, symbols, standards of morality and emotional energy ("strength, enthusiasm" etc.) in the individual (Spaargaren 2011: 819; Collins 2004: 48). While these processes start in face-to-face interactions, they extend to interaction ritual *chains*, in which emotional energy flows and is stored in objects and symbols. Through interaction ritual chains these processes of intersubjectivity can bridge the micro- and macro-level.

With respect to promoting sustainable consumption patterns, Spaargaren explicates:

> With Collins' analysis in mind, we can reframe the issue of green commitment, excitement and awareness in a non-individualist, positive manner. The more frequent, intense and dense the enrolment of individuals in sustainability-related interaction rituals, the higher the chance that their commitments and levels of awareness will increase. This increased commitment can be studied for the kinds of interaction rituals at play and the particular green symbols and objects used to develop and carry along 'green emotional energies'. . . . When green objects and technologies start to perform well across a range of different interaction rituals they can be said to be of strategic (policy) significance for the greening of interaction ritual chains within specific consumption domains.
>
> (2011: 820)

Brown (2011) studied interaction ritual chains related to Fair Trade consumption. For two years, he conducted interviews and ethnographic research in a retail store and a cooperative of Fair Trade (FT) coffee shops in Philadelphia. Among other efforts, he worked as a volunteer and retail clerk in order to study "how meanings of FT emerge through face-to-face interactions with consumers" (Brown 2011: 127). His research showed that when shopping (ritual), consumers learn about Fair Trade by interacting with retailers. He classified the process as involving low-to-moderate emotional energy, which motivates shopping related to aesthetics, quality and ethics. The corresponding symbols include hand-made objects and high-quality coffee. The sense of morality is hardly affected and mainly includes a warm glow by buying Fair Trade products. Other rituals, processes and symbols studied include (sacred) Fair Trade certification logos commended by artisans and farmers. These processes involve low-to-moderate emotional energy, raise awareness and encourage consumers to consider ethics.

Practice theory in general and Interaction Ritual theory in particular can enhance our understanding of consumer situations and consumption processes. It also has the potential to show how sustainable consumption patterns can be promoted and spread in society.

Conclusion

It is clear that culture matters for both consumption and sustainability as well as for the relationship between the two phenomena. However, sociology's great variety of cultural conceptualizations may lead to the deleterious assumption that when "everything" is affected by culture, culture becomes an empty and meaningless concept. Going beyond a "culture matters" perspective in sociological research requires defining culture in research studies so as to uncover variations in culture in or across populations and to investigate the relationship between different forms of culture and action (DiMaggio 1994: 28). By pointing to some forms of culture and distinctions made in the literature, I hope I have exemplified the difficulties of doing cultural research in general and with regard to consumption and sustainability in particular.

Limitations on length prevent discussion of many other valuable insights into conceptualizations and theoretical ideas in sociological research on culture (see Patterson 2014; DiMaggio 1997) and consumption (Zelizer 2005; DiMaggio 1994). A deeper look at (sustainable) consumption would include the broader relationships among production, distribution and consumption (Zelizer 2005: 332). A promising theoretical and empirical approach on research on culture – which might also be fruitful for the study of the nexus between culture, consumption and sustainability – would focus on social networks (Pachucki and Breiger 2010; Emirbayer and Goodwin 1994). As demonstrated by Edelmann and Vaisey (2014), network analysis can, for example, uncover mechanisms and social processes that link cultural tastes to a focus on consumption and non-consumption to social relationships and reveal, in addition to a perspective of "cultural skills" (Swidler 1986), culture as a "system of distinctions" in line with works by Pierre Bourdieu (1984).

While research on culture (partly) seems to move towards more integrative theorizing such as dual process theory (Vaisey 2009) and specific ideas about network analysis (Pachucki and Breiger 2010), I highlighted two analytically distinct perspectives, which can also be found in the literature on sustainable consumption: "culture and action are analytically distinct" and "culture and action are mutually generative" (DiMaggio 1994: 28). I tried to demonstrate that both perspectives can help us understand and explain consumer behaviour and behavioural change towards sustainability. Research on the activation of social norms and desires for social status show how cultural forms prevalent in society can be used to promote sustainable consumption, without changing these cultural forms per se (e.g. values, attitudes, social norms). This strategy seems to be a powerful tool that is already used by decision makers and governments (e.g. in so-called nudging units). Some of these "cultural cues" in the decision environment are so strong that they have remarkable effects and can change behaviour immediately – an attractive tool for gaining improvements in sustainability in the short-term.

While such cultural cues might work short-term, sustainable societies require a substantial change in values, attitudes, lifestyles and so on, which is typically achieved in long-term processes. Practice theories (Spaargaren 2011) offer a framework to understand how such substantial changes might happen by bridging the micro-level and macro-level of society. One approach is interaction ritual chain theory (Collins 2004), which can explain the microfoundations of cultural change related to sustainable consumption. It can also be used to promote sustainable consumption by reframing consumption as a positive, intersubjectively shared practice (Spaargaren 2011).

Therefore, both the "analytically distinct" perspective and the "mutually generative" perspective can be of use in (actively) promoting sustainable consumption. In my view, these perspectives complement each other in that the first perspective offers a way to reach increases in sustainable consumption with immediate impact, which is needed given current levels of overconsumption in many societies; the second perspective can help to reach persistent long-term changes towards cultural forms in line with the goals of sustainable development.

References

Agovino, M., Crociata, A., Quaglione, D., Sacco, P. and Sarra, A. (2017). "Good Taste Tastes Good. Cultural Capital as a Determinant of Organic Food Purchase by Italian Consumers: Evidence and Policy Implications". *Ecological Economics*, 141, 66–75.

Andorfer, V. A. and Liebe, U. (2012). "Research on Fair Trade Consumption – A Review". *Journal of Business Ethics*, 106, 415–435.

Andorfer, V. A. and Liebe, U. (2013). "Consumer Behavior in Moral Markets. On the Relevance of Identity, Justice Beliefs, Social Norms, Status, and Trust in Ethical Consumption". *European Sociological Review*, 29, 1251–1265.

Bohner, G. and Schlüter, L. E. (2014). "A Room with a Viewpoint Revisited: Descriptive Norms and Hotel Guests' Towel Reuse Behavior". *Plos One*, 9(8), e104086.

Bourdieu, P. (1984). *Distinction: A Social Critique of the Judgement of Taste*. Cambridge: Harvard University Press.

Brown, K. R. (2011). "Interaction Ritual Chains and the Mobilization of Conscientious Consumers". *Qualitative Sociology*, 34, 121–114.

Cialdini, R. B. (2005). "Basic Social Influence Is Underestimated". *Psychological Inquiry*, 16 (4), 158–161.

Collins, R. (2004). *Interaction Ritual Chains*. Princeton: Princeton University Press.

Dietz, T., Rosa, E. A. and York, R. (2007). "Driving the Human Ecological Footprint". *Frontiers in Ecology and the Environment*, 5(1), 13–18.

DiMaggio, P. J. (1994). "Culture and Economy". In N. Smelser and R. Swedberg (eds.). *Handbook of Economic Sociology*. New York: Princeton University Press and Russell Sage Foundation, 27–57.

DiMaggio, P. (1997). "Culture and Cognition". *Annual Review of Sociology*, 23, 263–287.

Doran, C. J. (2009). "The Role of Personal Values in Fair Trade Consumption". *Journal of Business Ethics*, 84, 549–563.

Doran, C. J. (2010). "Fair Trade Consumption: In Support of the Outgroup". *Journal of Business Ethics*, 95, 527–541.

Edelmann, A. and Vaisey, S. (2014). "Cultural Resources and Cultural Distinction in Networks". *Poetics*, 46, 22–37.

Emirbayer, M. and Goodwin, J. (1994). "Network Analysis, Culture and the Problem of Agency". *American Journal of Sociology*, 99(6), 1411–1454.

FAO (2009). "The Market for Organic and Fair-Trade Coffee". *FAO Rome*. www.fao.org/fileadmin/templates/organicexports/docs/Market_Organic_FT_Coffee.pdf [Accessed 27 February 2018].

Fehrler, S. and Przepiorka, W. (2013). "Charitable Giving as a Signal of Trustworthiness: Disentangling the Signaling Benefits of Altruistic Acts". *Evolution and Human Behavior*, 34, 139–145.

Goldstein, N. J., Cialdini, R. B. and Griskevicius, V. (2008). "A Room with a Viewpoint: Using Social Norms to Motivate Environmental Conservation in Hotels". *Journal of Consumer Research*, 35(3), 472–482.

Griskevicius, V., Cantu, S. and Van Vugt, M. (2012). "The Evolutionary Bases for Sustainable Behavior: Implications for Marketing, Policy, and Social Entrepreneurship". *Journal of Public Policy & Marketing*, 31(1), 115–128.

Griskevicius, V., Tybur, J. M. and Van den Bergh, B. (2010). "Going Green to Be Seen: Status, Reputation, and Conspicuous Conservation". *Journal of Personality and Social Psychology*, 98, 392–404.

Kahneman, D. (2003). "A Perspective in Judgment and Choice: Mapping Bounded Rationality". *American Psychologist*, 58(9), 697–720.

Kahneman, D. (2011). *Thinking, Fast and Slow*. New York: Farrar, Straus and Giroux.

Krier, J.-M. (2008). "Fair Trade 2007: New Facts and Figures from an Ongoing Success Story. A Report on Fair Trade in 33 Consumer Countries". DAWS – Dutch Association of Worldshops.

Leng, M., Schild, K., Hofmann, H. and Hammer, T. (2017). "Enough Is Good Enough: Sufficiency to Curb Resource Overconsumption". *CDE Policy Brief*, 11. CDE Bern.

Lindenberg, S. (2012). "How Cues in the Environment Affect Normative Behaviour". In L. Steg, A. E. van den Berg and J. I. M. de Groot (eds.). *Environmental Psychology: An Introduction*. New York: Wiley, 119–128.

Pachucki, M. A. and Breiger, R. L. (2010). "Cultural Holes: Beyond Relationality in Social Networks and Culture". *Annual Review of Sociology*, 36, 205–224.

Patterson, O. (2014). "Making Sense of Culture". *Annual Review of Sociology*, 40, 1–30.

Reisch, L. A. and Sunstein, C. R. (2016). "Do Europeans Like Nudges?". *Judgment and Decision Making*, 11, 310–325.

Rössel, J. and Schenk, P. (2017). "How Political is Political Consumption? The Case of Activism for the Global South and Fair Trade". Submitted to *Social Problems*.

Schor, J. and Holt, D. (eds.). (2000). *The Consumer Society Reader*. New York: New Press.

Schwartz, S. H. (1992). "Universals in the Content and Structure of Values: Theoretical Advances and Empirical Tests in 20 Countries". *Advances in Experimental Social Psychology*, 25, 1–65.

Simmel, G. (1907). *The Philosophy of Money*. Boston, MA: Routledge & Kegan Paul.

Sniderman, P. M. and Grob, D. B. (1996). "Innovations in Experimental Design in Attitude Surveys". *Annual Review of Sociology*, 22, 377–399.

Spaargaren, G. (2011). "Theories of Practices: Agency, Technology, and Culture. Exploring the Relevance of Practice Theories for the Governance of Sustainable Consumption Practices in the new World-order". *Global Environmental Change*, 21, 813–822.

Stolle, D., Hooghe, M. and Micheletti, M. (2005). "Politics in the Supermarket: Political Consumerism as a Form of Political Participation". *International Political Science Review*, 26, 245–269.

Sunstein, C. R. (2015). "Which Nudges Do People Like? A National Survey". Working Paper.

Swidler, A. (1986). "Culture in Action: Symbols and Strategies". *American Sociological Review*, 51, 273–286.

Thaler, R. H. T. and Sunstein, C. R. (2008). *Nudge: Improving Decisions about Health, Wealth, and Happiness*. New Haven: Yale University Press.

Trivers, R. L. (1971). "The Evolution of Reciprocal Altruism". *The Quarterly Review of Biology*, 46, 35–57.

Vaisey, S. (2009). "Motivation and Justification: A Dual-Process Model of Culture in Action". *American Journal of Sociology*, 114(6), 1675–1715.
Veblen, T. (1899). *The Theory of the Leisure Class*. New York: Macmillan.
Wilkinson, T. M. (2013). "Nudging and Manipulation". *Political Studies*, 61, 341–355.
Zelizer, V. (2005). "Culture and Consumption". In N. Smelser and R. Swedberg (eds.). *Handbook of Economic Sociology*. 2nd Ed. New York: Princeton University Press and Russell Sage Foundation, 331–354.

8 Social happiness as a cultural value

An analysis of shared values for ecosystem assessment

Galit P. Wellner

Introduction

In 2005, the UN via the World Health Organization published the *Millennium Ecosystem Assessment: Ecosystems and Human Well-Being*. This report initiated a series of National Ecosystem Assessments (NEA),[1] which were intended as tools to assist policy makers ensure their decisions support sustainable development. In Israel the NEA's objective was defined as "the advancement of knowledge-based management of open spaces and natural resources, via continuous production of scientific knowledge on the state of ecosystems and biodiversity in Israel" (HaMaarag 2018). Two objectives were set: to increase awareness as to the multifaceted value of nature and our dependence on functioning ecosystems; and to assist managers, decision- and policy-makers to incorporate the value of ecosystem services, and the biodiversity on which their provision depends, into planning processes, land management and policy.

The first Israeli report is scheduled to be published in 2018. The report will be structured around three main research sections: The first and primary section will include six chapters, each of which will review the local ecosystems: marine, urban, agricultural, Mediterranean, inland water and desert. The second section will analyse the ecosystem services that derive from these ecosystems. The third section will discuss the contributions of the ecosystems to human well-being from economic, health, cultural and social perspectives. I am a contributing author of the social chapter, with Dr Daniel Mishori and Aviram Sariel.

The social chapter is the "Benjamin" of the Israeli NEA report – it is the last, but nevertheless requires attention. The positioning as the last chapter can be explained by the order in which the pillars of sustainable development are usually presented: ecological, economic and social. But this ordering can also be seen as a reflection of its relative unimportance as perceived by biologists, traditional ecologists and health professionals. Although it is the last chapter, it is a candidate for becoming the "trouble maker" of the project. Social issues, after all, have been a source of unrest (at least) since the publication of *Capital* by Karl Marx, and more recently since 2011's Occupy movement. The word "social" has become a signifier of claims for justice and the fair distribution of resources. These manifestations were fuelled by the resistance to hyper-capitalism, which is at the foundation of the chosen methodology of the NEA – the ecosystem service assessment. Hence, our chapter is potentially "explosive" for the content as well as for the methodology. In light of these concerns, the confusion between services and contributions, or the overlap between the social and the cultural chapters, seem minor and solvable.

My essay will open with a brief description of the methodology of Ecosystem Services (ES). Then the tight relations between the social and the cultural aspects of sustainability are discussed and an attempt made to draw some distinctions between them. To complete the background, I will dive into the role of shared values and focus on one shared value that has interested me most over the last few years – happiness. After having introduced the links between happiness and well-being and I will then develop the notion of social happiness. My main effort will be to link social happiness with sustainability and show how they co-constitute each other. My contention is that happiness is not only a psychological goal and personality trait, but that it is a social construct that depends on a culturally rich environment.

Ecosystem service assessment

Sustainable development discourse typically centres on assessing the ecological impacts of human actions on the environment (it is important to stress here that most of these assessments leave society and culture in the fringes, a point I will return to later). Most assessment methods seek to quantify, and specifically to "monetarize", these impacts; that is, to express them in dollar or Euro terms. This practice has been expanded to earlier stages of projects' pre-evaluation and selection. In these stages, the needs to be answered by a project and the potential harms the project might create are "translated" into monetary terms, in order to sort and prioritize actions.

The ES approach emerged in the 1980s as a reaction to the then common practice of evaluating effects through measuring people's "willingness to pay" for environmental "commodities", in practice or in theory. The novelty of ES resides in the measurement of the services generated from the environment for the benefit of human beings (Norton 2012). Unlike "willingness to pay", ES raises awareness of human dependencies on nature. There are three major areas of services: first, provisioning (food and fibre); second, regulation (maintenance of energy flows and maintaining the resilience of systems); and, third, culture (such as the value of place). In 2005 the ES methodology was adopted by the World Health Organization for its *Millennium Assessment*, and became the basis for the National Ecosystem Assessments.

Between the social and the cultural

Whereas sustainable development defines the **social** as a pillar, the *Millennium Assessment* (2005) focuses on **cultural** services. The *Millennium Assessment* recognizes six categories of cultural Ecosystem Services:[2] heritage values; cultural identity; spiritual services (sacred, religious, or other forms of symbols and practices); inspiration (use of natural motifs or artefacts in art, folklore etc.); aesthetic appreciation of natural and cultivated landscapes; and, recreation and tourism. The social is also mentioned in the *Millennium Assessment*, as one constituent of well-being, along with health and security to name a few.[3] Yet, both the cultural and the social are considered relatively minor: In the diagram describing ES and well-being, the arrows to and from the social and cultural elements are the weakest – in colour and in width.

After the publication of the *Millennium Assessment*, three major practices emerged in the research literature for dealing with the cultural and the social aspects of sustainability. The first is simply ignoring the cultural or social aspects: The practice of ignoring cultural aspects is criticized by Plieninger et al. (2013), who urge researchers to study and map cultural Ecosystem Services. Yet these scholars ignore the social aspects, and mention only environmental and economic considerations.[4] The second practice involves referring to culture and society as synonyms: For example, de Groot et al. (2010: 262) mention them as a combined entity titled "socio-cultural values", as part of their analysis of cultural ES. Lastly, some researchers chose to assess the relations between the social pillar and cultural Ecosystem Services. One of the major works is a COST Action (2016; see also Dessein et al. 2015)[5] that explored the role of culture in sustainable development. In the final report, the Action's members suggest differentiating between culture and society while acknowledging that they co-shape and co-constitute each other. Yet the COST Action for cultural sustainability does not provide definitions for either. One of the important end results of this project was a set of concise models for the integration of culture into sustainability. Three models were developed: the "in" proposition which represents culture as a fourth pillar of sustainable development; the "for" proposition which represents culture as mediating between the three pillars; and the "as" proposition which represents culture as the foundation for sustainable development.

In the Israeli NEA, the distinction between social and cultural is expressed in the structure of the report, as each has a separate chapter written by a different team. This layout is a variation on COST's "for" option: the cultural serves as a class of ES while the social is located in the last part discussing contributions that serves as a mediating layer for the services chapters. I must admit that the differences in practice are difficult to maintain. Still, I am afraid the two chapters will share a similar destiny, of being considered of less importance compared to the biological-ecological and economic chapters.

Even today, we still need to overcome a major trend in ecological thinking, which addresses society and culture together, sometimes marginalized together and even ignored together. This trend is relevant to both the social and cultural aspects because they share a similar challenge – these services and constituents cannot be translated into money. Who can attach a fixed monetary valuation to a sense of place, to heritage or to landscapes (to name but a few)? Social and cultural services and constituents, as well as their impacts, cannot be "monetized".

Shared values (as substitution for monetization)

Even if we manage to find a magic formula to monetize social and cultural services and constituents, a deeper problem arises, which is inherent to evaluating the future. A Hebrew proverb says: Since the destruction of the Temple, prophecy has been given to fools. It is difficult, if not impossible, to assess – based on current parameters – an action that will affect future generations. Here is one example: In the nineteenth century, when the Ottoman Empire ruled Palestine, a rail project was initiated. It was a large project even by contemporary standards, and it included the laying of train tracks, the construction of bridges, stations and other facilities. During the First World War, the train became a strategic asset of the empire and hence a target of British attacks. At the same time, the supply of coal was disrupted, and an alternative source of energy was needed to move the locomotives. The quickest solution at-hand was to cut the local oak forests to supply wood to the engines. This raw material was also useful for the maintenance of the rails. The result was a destruction of almost all natural oak forests in Palestine. The interest of future generations in natural forests was overruled or overlooked. At best this decision falls under the category of prophecy after the destruction of the Temple. Today similar lessons can be applied to mining, fracking and water desalination to name a few. All these mega projects accentuate current human needs and tend to disregard the cultural and social effects that will deprive future generations of cultural environmental assets.

While the future cannot be predicted, defining shared values and taking them into account in the policy-making and planning phases can mitigate the effects of most projects, as these values call attention to the losses in cultural assets and the social costs that tend to be disregarded. Moreover, while monetization typically calculates only twenty-five years into the future, values can encompass longer periods of time. Shared values can also better represent the past through the notion of heritage. And in the present, they enable one to take into account experiences that cannot be translated into money, such as cultural ceremonies whose contributions to the sense of belonging, self-identity and subjective happiness are disproportionate to their costs of implementation. Values can represent competing vectors, just as monetization involves calculations that add and deduct. As an alternative to monetization, one can think of balancing values, similar to the practice of a court of justice when two competing values or rights are raised.

From well-being to happiness

Instead of attempting to "monetize" values, I suggest assessing them qualitatively and balancing between them. When considering this balance, a "meta value" emerges – well-being. Well-being is a central component of the *Millennium Assessment*, and it is presented very early in the 2005 report:

> "*The Millennium Assessment* examines how changes in ecosystem services influence human well-bein. Human well-being is assumed to have multiple

> constituents, including the basic material for a good life, . . . health, . . . good social relations, . . . security, . . . and freedom of choice and action.
> (2005: xiv)

In the *Millennium Assessment*, well-being is understood in the broadest sense and hence as having multiple aspects. It orients the *Millennium Assessment* to an anthropocentric approach to sustainability, in which human well-being is of top importance.

Interestingly, the *Millennium Assessment*'s definition of well-being does not refer specifically to the psychological aspects of happiness. The constituents related to happiness are "good social relations" and "freedom of choice and action" (2005: v). These conform to a certain definition of psychological well-being known as Self Determination Theory (SDT). According to SDT, long-term happiness arises when three basic needs are answered: relatedness, autonomy and competence. It is the last need that is left outside the *Millennium Assessment*'s definition for well-being. The two that were included are more socially oriented, as they reflect the inter-subjective character of happiness. Yet, SDT and the other psychological happiness theories focus on the individual, and do not develop a notion of social happiness.[6] In the next section, I will map the differences between psychological well-being and social happiness.

Social happiness

At the beginning of the nineteenth century Jeremy Bentham developed a utilitarian principle of happiness calling for the "greatest happiness for the greatest numbers" (Thin 2012: 4). His approach has dominated happiness scholarship across disciplines, especially since the 1990s. In most contemporary happiness research, the goal is to measure maximum levels of happiness in the largest number of people participating in self-reporting surveys. These surveys are conducted either in rich countries or among university students in poorer countries (Thin 2012: 6), but these limitations and other methodological obstacles do not slow down the adoption of such surveys as a key component in various indices, such as the OECD's Better Life Index, the Happy Planet Index or Bhutan's Gross National Happiness Index. They all measure the psychological subjective happiness through self-reports on life satisfaction. If more people report greater happiness, then the index goes up, thereby following Bentham's formula.

But what if happiness cannot be arithmetically accumulated? What if happiness is not totally subjective? What if it has a social dimension, which requires a happy socio-cultural environment and a well-functioning inter-subjective space? Let us take the US as an example. Each year we read that the number of millionaires in the US is growing, and this country is ranked fairly high in most happiness indices. Yet, the number of people who become addicted to anti-depressants like Prozac, pain killers, not to mention drugs, is on the rise.

Or take another example: Scandinavian countries occupy the top spots in most indices of happiness. Slavoj Žižek, the king of paradoxes, points to the fact that these countries also lead in the number of suicides (2009). Since the research of Emile Durkheim in the late nineteenth century, the number of suicides has been an indicator for societal health – that is, the health of a society as a whole. Measuring personal happiness, therefore, misses an important social dimension.

Happiness, I argue, is not the sum of subjectively happy people. It is not like money where the more you have the richer you are. Social happiness is not an accumulation of all (or the majority of) personal happiness-es. Social happiness does not automatically emerge when a certain statistical threshold is crossed. Social happiness takes into account another set of considerations.[7] It can be regarded as the result of a combination of factors such as good social relations, a rich cultural environment and fair institutions.[8] This is an additional dimension of happiness that resides in the inter-subjective level. This dimension allows us to overcome Bentham's maximization formula.

Within happiness studies, the notion of social happiness has emerged only recently. In his 2012 book *Social Happiness*, Neil Thin positions happiness as "a social and public good, not just a private and individual concern. To understand and promote it, we must look between and around persons, not just within persons" (2012: x). This type of happiness is just as important as psychological happiness. Thin suggests that public discourse on happiness should evolve from the subjective to the inter-subjective and social sphere: "It is not just about private self-interest, it is about love, empathy and engagement in the workplace and in communities, and about collective aspirations for a world that could be even better than it already is" (2012: xi). This direction coincides with sustainable development principles that likewise aims at promoting collective inspirations for a better world in terms of ecology, economy and society.

Thin considers the "social" in social happiness as an interim between micro-personal level and macro-political policy-oriented level. The in-between is composed of families, workplaces, communities etc. (2012: 15–17). He terms it the meso-level. When we consider sustainable development, all three levels of happiness are of relevance. That is why I refer to social happiness as including any inter-subjective relations, macro- and meso-level alike, leaving the micro-level to psychological investigations. In order to link social happiness with sustainable development, two parallel steps are required: One should position happiness (and specifically social happiness) as a legitimate consideration in the shaping of policy (remember that the NEA is intended for policy makers); the other step should map the complex relations between happiness and sustainability. Let us start with the policy issues.

Social happiness as a consideration for policy

Thin's research is aimed at including happiness considerations in public policies. He asserts: "To talk about values, or justice, or prosperity without reference to happiness . . . is perverse and potentially very damaging" (2012: ix). Yet, in spite of the importance of happiness, "[i]t is still rare to find substantial attention to happiness in research or policy texts on 'ethical' (or 'green', or 'sustainable', or 'eco-friendly', or 'fair trade') consumption or investment, or even in texts on corporate social responsibility (CSR)" (2012: 212).

Thin mentions few initiatives that attempt to deal with happiness at the policy level. One initiative is the 'Wellbeing Manifesto' promoted by the New Economics Foundation (NEF). In the manifesto, the NEF highlights eight key areas for action:

> national well-being accounts; ensuring access to meaningful work; a maximum thirty-five hour week, more flexibility in working hours, and more bank holidays; all schools promoting emotional, social, and physical well-being, with more emphasis on thriving and less on performance; a health service that promotes health; more support for early childhood and parenting; discouraging materialism and promoting authentic advertising; and strengthening active citizenship.
>
> (Thin 2012: 84)

As you can see, most of these considerations take into account the individual and do not attempt to capture the social dimensions of happiness like social justice or a rich cultural environment.

Another example is The Happy Planet Index, which is one of the few attempts to link happiness with sustainability. The index is simple: Happiness in a given country is calculated by the multiplication of the result of life satisfaction surveys by average life expectancy by inequality index, divided by the ecological footprint. The division by the ecological footprint gives an advantage to third world countries, and so at the top of the index you will not find "the usual suspects" – i.e. the Scandinavian countries – but rather those of South and Central America. The index limitedly refers to the social aspects of happiness through the inequality index. There are no references to culture. Moreover, Thin criticizes The Happy Planet Index for its over-simplicity. In his words, it is "frighteningly naïve" and "an insult to the intelligence" (2012: 85).

Sustainable development and happiness each invokes significant policy issues. Their intersection has been poorly explored. Social happiness and cultural sustainability have the potential to link sustainability and happiness policies: Cultural sustainability can serve as a shared domain whose values are elaborated in social happiness studies. I shall hence review in the next section the intersection between sustainability and happiness.

Happiness and cultural sustainability

Thin notes,

> the tide of interest in *environmental responsibility* and *sustainable development* is strongly linked with the projects of rethinking prosperity, transcending short-termist and selfish 'materialism', and development of more socially responsible self-projects and business approaches. Although very often

> associated with fear-mongering, pessimism, and ascetic values, the sustainable development movement makes little sense without a concept of good lives and good societies that would be worth sustaining.
>
> (Thin 2012: 19–20)[9]

The link between social happiness and sustainability seems natural, even a must. Yet, even psychological happiness is currently absent from many new societal policies, as well as from the majority of sustainable development policies. What can be found is research on psychological happiness in the sustainable context.

In a series of studies, psychologist and SDT scholar Tim Kasser explored the co-constitution of sustainability and personal happiness. In one research paper published in 2011, Kasser and colleagues experimented with US college students who were led to think about times when their nation had acted to support freedom, to build family values and to be generous to others, i.e. think of intrinsic values that signal subjective happiness (Sheldon et al. 2011). The researchers found that those students later endorsed more sustainable ecological policies, such as support for public transportation and smaller homes. The happiness explored in this research is not only personal but also social, by referring to freedom and empathy in the first phase. This experiment demonstrates how happiness that focuses on inter-subjective values leads people to think more sustainably.

A similar direction was taken by Catherine O'Brien who coined in 2005 the notion of "sustainable happiness" to denote the pursuit of happiness that does not exploit other people, the environment or future generations (2008). However, the relations between environment and happiness are bi-lateral, which is to say that not only do our moods influence the environment but also the environment can influence happiness: When the environment is polluted it lowers happiness, and when it is "good", it "facilitate(s) good lives and good experiences" (Thin 2012: 14). In other words, sustainability serves as a condition for happiness.[10] My last example relates to this connection, albeit on the psychological and personal value level: In 2014, Kasser et al. published a set of studies, the longest spanned over 12 years, showing that people's well-being improves as they place relatively less importance on materialistic goals and values. Put differently, those who orient toward less sustainable goals are more likely to experience a decrease in well-being over time. The researchers examined the effects of placing importance on attaining money and having many possessions. The results were that "when people oriented away from materialistic values and goals over time, their well-being improved, whereas when they increased the relative priority they placed on materialistic values and goals over time, their well-being declined" (Kasser 2014: 18). This research is of importance to my argument as it shows the links between sustainable culture and happiness.

Conclusion: happiness as a cultural value

My original intent was to integrate happiness as a value into the social contribution chapter of the Israeli NEA. I realized that happiness can be also in the cultural

ES chapter, because in the context of sustainable development the borderline between the cultural and the social is not clear-cut. I also found that happiness is missing from the definition of well-being in the *Millennium Assessment*, though shades of it can be identified, if one follows a specific theory like SDT. All this led me to define social happiness: The notion of social happiness differs from personal happiness due to its ingredients, including good social relations, rich cultural environment and fair institutions. Social happiness is a shared value that can assist decision-makers in assessing the impacts of a given action on cultural and ecological environments. Social happiness should be a consideration in policy-making.[11]

The relations between happiness and sustainability are complex: In a sustainable environment – ecologically, socially and culturally – happiness can flourish. It can flourish because the environment is clean, economically fair, socially balanced, culturally rich and diverse. But in a deeper sense, considering sustainable development leads us to re-think what makes us happy. Shopping, driving a sports car, building a swimming pool in the back yard – all these change their meanings once we examine them with "sustainability lenses".

As my last point I would like to invoke an ancient Greek distinction, which may help us in this rethinking process. The distinction between hedonism and Eudaimonia draws a timeline between two types of happiness, the short-term and the long-term. Sustainable development urges us to target the long-term eudemonic happiness and reject short-term hedonic pleasures.

Notes

1 One of the pioneers was the UK commencing such an assessment in 2009 and publishing the findings in 2011.

2 The *Millennium Assessment* states that the importance of cultural services and values is not currently recognized in landscape planning and management and that these fields could benefit from a better understanding of the way in which societies manipulate ecosystems and then relate that to cultural, spiritual and religious belief systems. The *Millennium Assessment* also states that the ecosystem approach implicitly recognizes the importance of a socio-ecological system approach, and that policy formulations should empower local people to participate in managing natural resources as part of a cultural landscape, integrating local knowledge and institutions.

3 On the first page of the *Millennium Assessment* the third paragraph states: "*The Millennium Assessment* examines how changes in ecosystem services influence human well-being. Human well-being is assumed to have multiple constituents, including the *basic material for a good life*, such as secure and adequate livelihoods, enough food at all times, shelter, clothing and access to goods; *health*, including feeling well and having a healthy physical environment, such as clean air and access to clean water; *good social relations*, including social cohesion, mutual respect, and the ability to help others and provide for children; *security*, including secure access to natural and other resources, personal safety, and security from natural and human-made disasters; and *freedom of choice and action*, including the opportunity to achieve what an individual values doing and being. Freedom of choice and action is influenced by other constituents of well-being (as well as by other factors, notably education) and is also a precondition for achieving other components of well-being, particularly with respect to equity and fairness" (2005: xiv).

4 Plieninger et al. (2013) mapped cultural services into aesthetic values, social relations and educational values.

5 COST Action is a European funding framework intended for bottom-up science and technology networks, which are open to researchers and stakeholders with a duration of four years. The action is active through a range of networking tools, such as workshops, conferences, training schools and dissemination activities. COST does not fund research itself (see www.cost.eu/COST_Actions). In this work, I refer to COST Action IS1007 *Investigating Cultural Sustainability*, which was active in the years 2011–2015.

6 On the link between happiness and ethics, see Wellner (2015).

7 For example, high self-esteem is perceived in Western culture as a positive trait for individuals, but society cannot be made up entirely of high-confidence people (Thin 2012: 19).

8 Thin roots social happiness with the theories of Hume and Adam Smith (Thin 2012: 4). "[T]o the extent that individual-level interventions are good for individuals, this doesn't necessarily amount to a social good in aggregate" (19).

9 Thin lists the disciplines that should take into consideration happiness (2012: 8–12): Psychology is the first, followed by sociology, economics and education, but the furthest Thin could get was gender studies, Law and business management. Ecology and sustainability are missing from his list. Happiness, according to Thin, has to do with development of poor countries, but he fails to see it as a value or goal for sustainability.

10 Brown and Kasser (2005) examined the links between subjective well-being (SWB) and ecologically responsible behaviour (ERB) and found compatibility between them: Higher subjective well-being was associated with higher ecologically responsible behaviour. The researchers concluded that sustainable way of life enhances both personal and collective well-being, although the studies are focused solely on the subjective one: "happier people were living in more ecologically sustainable ways" (2005: 360). The researchers point to a common ground to SWB and ERB – intrinsic value orientation and mindfulness. They explain: "a mindful consideration of one's inner states and behavior along with a set of values oriented more towards intrinsic than extrinsic aims appear to simultaneously benefit both individual and ecological well-being" (2005: 360). They add: "the focus on community that is a component of an intrinsic value orientation . . . might lead individuals try to decrease the ecological impacts of their behavior so as to benefit future human generations as well as other species" (2005: 361). Brown and Kasser built a model in which SWB, mindfulness and values affect ERB. An opposite link is absent, and so the model does not encompass the influences of ERB on WB.

11 The link between ES and values is parallel to that between cultural and social: Culture is assigned to ES, while values are dealt in the social sphere, elaborated through series of endless discussions that constantly shape the shared values of a given society.

References

Brown, K. W. and Kasser, T. (2005). "Are Psychological and Ecological Well-being Compatible? The Role of Values, Mindfulness, and Lifestyle". *Social Indicators Research*, 74(2), 349–368.

COST. "COST Actions". *European Cooperation in Science and Technology*. 26 September 2016. www.cost.eu/COST_Actions [Accessed 27 April 2017].

De Groot, R., Alkemade, R., Braat, L., Hein, L. and Willemen, L. (2010). "Challenges in Integrating the Concept of Ecosystem Services and Values in Landscape Planning, Management and Decision Making". *Ecological Complexity*, 7(3), 260–272.

Dessein, J., Soini, K., Faircough, G. and Horlings, L. (2015). *Culture in, for and as Sustainable Development: Conclusions from the COST Action IS1007 Investigating Cultural Sustainability*. Jyvaskyla: University of Jyvaskyla.

HaMaarag (2018). *Hamaarag – Israel's National Ecosystem Assessment Program.* www.hamaarag.org.il/en/content/inner/hamaarag-israel's-national-ecosystem-assessment-program [Accessed 13 January 2018].

Kasser, T. E. (2014). "Changes in Materialism, Changes in Psychological Well-being: Evidence from Three Longitudinal Studies and an Intervention Experiment". *Motivation and Emotion*, 38(1), 1–22.

Millennium Ecosystem Assessment (2005). *Ecosystems and Human Well-being: Synthesis. A Report of the Millennium Ecosystem Assessment.* Washington, DC: Island Press.

Norton, B. G. (2012). "Valuing Ecosystems". *Nature Education Knowledge*, 3(10), 2.

O'Brien, C. (2008). "Sustainable Happiness: How Happiness Studies Can Contribute to a More Sustainable Future". *Canadian Psychology*, 49(4), 289–295.

Plieninger, T., Dijks, S., Oteros-Rozasc, E. and Bielingd, C. (2013). "Assessing, Mapping, and Quantifying Cultural Ecosystem Services at Community Level". *Land Use Policy*, 33, 118–129.

Sheldon, K., Nichols, C., and Kasser, T. (2011). "Americans Recommend Smaller Ecological Footprints when Reminded of Intrinsic American Values of Self-expression, Family, and Generosity". *Ecopsychology*, 3, 97–104.

Thin, N. (2012). *Social Happiness: Theory into Policy and Practice.* Bristol: Policy Press.

Wellner, G. (2015). "Ethics in Times of Posthumanism". *Foundations of Science*, 1–4.

Žižek, S. (2009). *In Defense of Lost Causes.* London: Verso.

9 The challenges of cultural sustainability on an (inter) planetary scale

Andreas Losch

Introduction

"Planetary sustainability", understood here according to NASA's use of the term (NASA 2014), concerns the idea of sustainable development on a planetary scale, that is, not only regarding planet Earth, but also regarding the surrounding solar system (cf. Losch 2018). Recent space activities including space mining initiatives make it necessary to consider this dimension, at least economically, even though it may sound like science fiction to many. Taking this planetary or even interplanetary scale into account, this chapter discusses how economic and ecological dimensions of sustainability need to be extended to incorporate cultural ideas. The concept of planetary heritage sites might count as an example, but where else, on an (inter)planetary scale, does the cultural dimension of sustainability need to be brought into the discussion?

NASA is not the only institution currently propagating "planetary sustainability".[1] In fact the term was coined by UNESCO when describing a new relationship "between humankind and our habitat" (UNESCO 2003: 1). As the leading UN agency for sustainable development from 2005 to 2015, UNESCO was aware of the necessity of implementing a "renewed vision of culture and education", which could form the core of this new relationship in light of the emergence of the Information Age. (UNESCO 2003: 5).

Can culture be situated at the core of sustainability? It largely depends on the epistemological approach one likes to adopt. From a social constructivist point of view, one is already interpreting nature through the lenses of human culture by framing the situation of our planet as "problematic". While in this sense the problems of our planet are indeed culturally or "socially constructed" (Soini and Birkeland 2014: 215), one must immediately add that this evaluation of our situation is certainly *necessary* to safeguard not only our own survival on this planet. The constructivist approach does not intend to diminish existing problems, but to recall that it requires a certain degree of reflection and education to be aware of them. From this point of view, the UN is also right to put "culture and education" at the core of their vision. The constructivist perspective emphasizes that culture is no potential additional pillar of sustainability in addition to economic, ecological and social dimensions, but the *means* through which we encounter the topic, which permeates every aspect of sustainability.

The development of a holistic approach to sustainable development, such that it might be termed planetary sustainability, is only needed if we wish to ensure the ongoing survival of the biosphere. Whilst such an aim may itself be understood as culturally constructed – the result of an intra-species dialogue on human responsibility for the planet – it is nevertheless, very meaningful. The value judgement inherent to the assumption that sustainable development is positive is that, were the biosphere to be destroyed, the planet would not cease to exist but that it would continue to exist in a state less desirable, irrespective of whether the judgement of this state requires a human perspective. Planet Earth will continue to exist even if it is over-exploited, albeit most likely with an irrecoverably altered and vastly diminished biosphere. That we, as humankind, value it very highly as a living environment and want to preserve it in a shape close to the current one (hopefully with a diversity of species) is a *value issue*, focussed upon our *own* well-being. Most of us probably care about our planetary environment enough to want to safeguard our survival within it. Maybe we should also care about the planet for its own sake, yet this idea is not necessarily included within the current concept of sustainable development. Be that as it may, taking care of life on our planet (or not) is obviously our *decision*, informed by whichever sociocultural point of view we choose to take.[2]

In addition to this meta-role assigned to culture from a social constructivist perspective, I also want to explore culture as a *content* of sustainability. Here, we need to expand the concept of sustainability to embrace notions that might help safeguard our cultural heritage on Earth and beyond. While I share the sociocultural value-orientation expressed by UNESCO's use of the term "planetary sustainability", the extent of the term in this chapter will differ somewhat. I will attempt to focus less on our planet itself, but on its *surroundings*: our orbit, near Earth objects (NEO's), our Moon, and the planet Mars, insofar as they affect us. With this understanding of "planetary sustainability", I follow the NASA initiative with the same name, which not only envisions

- a world in which all people have access to abundant water, food and energy, as well as protection from severe storms and climate change impacts,
- healthy and sustainable worldwide economic growth from renewable products and resources.

but also

- a multi-planetary society, where the resources of the solar system are available to the people of the Earth (NASA 2014).

I want to make use of NASA's valuable idea that space activities be included within the consideration of planetary sustainability. Reflecting on this, one should perhaps call the deliberations of this chapter "*inter*planetary sustainability" but for the fact that all these reflections are still quite Earth-focused. For this reason I stick with the term "planetary sustainability", emphasizing, however, what we all know but often forget: Earth is a tiny planet in a vast tract of space, and our *space environment* is also important for our well-being.[3]

Recent space activities and their legal evaluation

One may ask if NASA's vision of "a multi-planetary society" is not more *science-fiction* than actual science. Without doubt the idea is futuristic. Yet the very concrete visions of space agencies and space companies are close to being realized. The UN is also active in these regards.

At the end of 2015, the American Congress passed the S.P.A.C.E. *(Spurring Private Aerospace Competitiveness and Entrepreneurship)* act, which opened to US citizens the opportunity "to engage in commercial exploration for and commercial recovery of space resources" (H.R.2262 2015: para. 51302 (3)), understood as abiotic resources including water and minerals (para. 51301).

Equally, Luxembourg started an initiative whose

> goal is to ensure that space resources explored under its jurisdiction serve a peaceful purpose, are gathered and used in a sustainable manner compatible with international law and for the benefit of humankind.
>
> (The Government of the Grand Duchy of Luxembourg 2017: n.p.)

Luxembourg's Chamber of Deputies passed the law on 13 July 2017 (Planetary Resources 2017).

Who is interested in such a legal framework? Private companies today not only invest in satellite launches and space tourism.[4] Some private enterprises (e.g. Elon Musk's *SpaceX)* even help to supply the International Space Station (ISS). The plans of companies such as *Planetary Resources* or *Deep Space Industries* to exploit near Earth objects (e.g. asteroids) show that the Space Age is on the verge of entering a new phase, a time which the European Space Agency (ESA) has framed as "Space 4.0":

> [S]pace is evolving from being the preserve of the governments of a few spacefaring nations to a situation in which there is the increased number of diverse space actors around the world, including the emergence of private companies, participation with academia, industry and citizens, digitalisation and global interaction.
>
> (ESA 2016: n.p.)

In response and as "a vision for global cooperation and Space 4.0", ESA is sketching its ambitious ideas for a "moon village", less a place to live, but "a community created when groups join forces without first sorting out every detail" (Woerner 2016: n.p.). The times have changed, and hence the aims are not purely scientific, although a radio telescope on the far side of the moon is part of the bundle of ideas. "You might see not only scientific and technological activities, but also activities based on exploiting resources or even tourism" (Woerner 2016: n.p.). There is actual money involved in exploring such visions. For their initiative, the Grand Duchy of Luxembourg announced a 25-million-euro investment cooperation with *Planetary Resources* (The Government of the Grand Duchy of

Luxembourg 2016). One can evaluate this critically, and I concur with Capova, who states that:

> All things considered, it seems reasonable to assume that the New Space Age is primarily an economically driven one; that is to say an effort to gain control over space resources and pursuit of business. The upcoming of this era is also ushered in changes in human relations to outer space; that is from something to be explored to something that could be exploited.
>
> (Capova 2016: 307)

Increased human activity starts to create problems. Besides the enhanced benefits with which a more diverse suite of satellites and launchers and an influx of commercial investment provide mankind, "there is also more orbital debris threatening the new activities in space" (Bonnal and McKnight 2017: 5). In low Earth orbit (LEO) alone, over 200 breakup events have created tens of thousands of objects too small to be sensed. We can only see objects above 10 cm in LEO, yet already "impacts from objects as small as 5 mm are likely to disrupt or terminate a satellite's operations" (Bonnal and McKnight 2017: 5).

These developments, including the extension of economy into space, make it important that we start to take better care of our space environment. The application of the concept of sustainability could be a decisive step and may help sustainable development on Earth as well. Indeed, the UN's *Committee on the Peaceful Uses of Outer Space* (UNCOPUOS) has established within its *Scientific and Technical Subcommittee* a *Working Group on the Long-term Sustainability of Outer Space Activities* which addressed thematic areas, "including sustainable space utilization supporting sustainable development on Earth" (UNOOSA 2017: n.p.). Parallel to the establishment of the Sustainable Development Goals (United Nations 2015), which still *lack* a consideration of the space surrounding Earth, the *Office for Outer Space Affairs* (UNOOSA) has launched a new framework strategy called Space 2030, advocating space as a driver for sustainable development.

> We want to make sure that space technology and applications are used to bring concrete benefits to all humankind, paying special attention to the future space-faring and developing countries while also carefully considering the long-term sustainability of outer space activities for current and future generations.
>
> (UNOOSA 2017: n.p.)

This Space 2030 framework will be built upon four pillars:

- Space accessibility: all communities using and benefiting from space technologies;
- Space diplomacy: building and strengthening international cooperation in space activities;

- Space economy: development of space-derived economic benefits;
- Space society: evolution of society and societal benefits stemming from space-related activities.

(Di Pippo 2017: n.p.)

The legal framework for international space activities is provided by the *Outer Space Treaty*. At this point it might be worth paying heed to article 1 of the *Treaty* from 1967, which is reflected by the UN office's activities:

> The exploration and use of outer space, including the moon and other celestial bodies, shall be carried out for the benefit and in the interests of all countries, irrespective of their degree of economic or scientific development, and shall be the province of all mankind.
>
> Outer space, including the moon and other celestial bodies, shall be free for exploration and use by all States without discrimination of any kind, on a basis of equality and in accordance with international law, and there shall be free access to all areas of celestial bodies.
>
> (United Nations 2002: 4)

The Luxembourgian initiative draws on this wording, too. But does the spirit of the *Outer Space Treaty* really sit well with the vision for space mining? With regard to this most space actors follow the provisions of the *International Institute of Space Law* (www.iislweb.org/), which interprets "use" as including the permission for "recovery". "In view of the absence of a clear prohibition of the taking of resources in the *Outer Space Treaty* one can conclude that the use of space resources is permitted" (International Institute of Space Law 2015: 3). By contrast, one could also argue that the economic exploitation of space was simply not considered in this treaty, which dealt with military threats and science. Private space actors are left out, because they hardly existed in 1967, and hence the IISL goes on to say that

> It is an open question whether this legal situation is satisfactory. Whether the United States' interpretation of Art. II of the Outer Space Treaty is followed by other states will be central to the future understanding and development of the non-appropriation principle. It can be a starting point for the development of international rules to be evaluated by means of an international dialogue in order to coordinate the free exploration and use of outer space, including resource extraction, for the benefit and in the interests of all countries.
>
> (International Institute of Space Law 2015: 3)

One can ask whether the plain declaration that the Luxembourgian initiative is "for the benefit and in the interests of all countries", as the country has done, is sufficient. Luxembourg's quasi-second legislative chamber, the Council of

State, had expressed several formal oppositions to the space mining project. The responsible official is even reported

> to have confirmed his objective . . . to ask for a revision of the question of property in the Outer Space Treaty. He wants the UN to create a legal framework which would allow companies worldwide to act in this domain.
> (Huberty 2017: n.p.; cf. Conseil d'etat du grand-duché de Luxembourg 2017; Allen & Overy 2017)

Cultural challenges and value issues

The current legal framework does not really seem to guarantee the rights to recover elements of our space environment. In truth, it seems it would need to be extended to take the private actors of "Space 4.0" into account. A fundamental question would be, however, whether private actors can actually act "for the benefit and in the interests of all countries" (United Nations 2002: 4). When economy extends into space, at least the principles of sustainable development should be applied to our space environment as an essential element of any legislative overhaul. This should certainly be for the benefit and in the interest of all countries.

A more bizarre development of space economy is the phenomenon of exo-burials, "making it possible to place a symbolic portion of cremated remains into orbit, onto the lunar surface, or be sent into deep space" (Capova 2016: 308). The Moon, it seems, already hosts a sort of a graveyard, and we have also seen that ESA activities involve this celestial body. Should not places like, for instance, the Apollo landing site or remarkable moon landscapes be protected against commercial use, or even against any further human intervention?

The 1979 Moon Agreement (Art. 7) "foresees the possibility of zones of special protection being established on celestial bodies" (Bohlmann 2011: 302). Article 4.1 even evokes the principle of intergenerational equity (fairness of distribution), which is part of the idea of sustainability. Unfortunately, the Moon Agreement has so far only been signed by thirteen State Parties (as opposed to the 100 State Parties to the *Outer Space Treaty*).

With regard to our neighbouring planet Mars, the idea of protected zones has been discussed in some detail:

> The most interesting and unique regions on Mars that might merit conservation and preservation are by definition the ones where we might wish to send robots and human explorers to explore and exploit. This creates the same paradox as we face on Earth with sites of scientific or natural beauty – how do we preserve such sites while at the same time allowing them to be explored and studied? On Earth the answer to this problem has come in the form of the National Park system, widely replicated in a number of countries around the world in various forms.
> (Cockell and Horneck 2004: 291)

It makes sense to discuss these questions and the related intrinsic and utilitarian reasons for protecting our natural and cultural heritage (Cockell and Horneck 2004: 294). Before I develop this further, let us consider the ethical fundament we stand on, and the basic ethical stances we cling to. *Anthropocentrism* attributes moral status (being the object of moral concern) only to *human* beings, while *sentientism* (also called "pathocentrism") assigns this to all (and only) *sentient* beings. *Biocentrism* extends this status to all (and only) *living* beings, while *ecocentrism* assumes whole ecosystems (cf. the Gaia theory) and sometimes even non-living nature to have moral status (Persson 2012). The latter would be especially important for cultural sustainability in the traditional "pillar" sense, where geological sites are regarded as cultural heritage as well.

The fundamental question to be evaluated here is, what is a living system? As extensions of this question, one might ask what and where are the fuzzy borders of live and non-live; are there any limits to the interconnectedness between organism and environment? And finally, how do we categorize feelings, and what precisely, is mind? Such questions lie behind these ethical distinctions and I fear it will be almost impossible to find an answer, in our time at least. For now, we need to allow intuition to dictate our stance, albeit not at the expense of remaining open to dialogue.

I personally think that *ratiocentrism* (a universal version of anthropocentrism: the idea that only rational beings morally matter) and the Gaia hypothesis on some deep level belong together, because I regard life, mind and consciousness as complexity phenomena. By way of example, even gaseous nebulae are more complex than previously thought (Benz 2016). To fully apprehend these phenomena would require a God's eye perspective. We can but hope to work with clues, although some "overview effect" of course helps (White and O'Neill 1987). As well have our own individual perspective, we need all of us and everyone else to get an overall idea.[5]

Now, while the Moon is regarded as lifeless, we do not know whether there is life on Mars. I am not talking here about little green men, but about potential *microbial* life. Even on Earth, so-called extremophiles have proven to be highly resilient in environments lethal for the human being, and actually, a crucial Viking lander experiment technically had a positive outcome concerning live-detection in the Martian soil (Losch 2017). Hence, we should be cautious at least and include the *possibility* of microbial life on Mars in our considerations.[6] Therefore, one also needs to include the dangers of forward or backward contamination, summarized under the label "planetary protection" (COSPAR 2011). Should we not care about this, we might face the problem that we only find on other planets what we ourselves brought there before. Or, worst case, we could even endanger the Earth's biosphere by importing extraterrestrial life.

This brings us back to a discussion of the values involved in space exploration. Much depends on the issue of whether we actually find extraterrestrial life out there. In considering why planetary parks might be created on Mars, we must distinguish between the reasons given in the absence of life and those given if microbial life were to be discovered. We could, for instance, protect regions of Mars purely for the enjoyment of future generations. This aesthetic reasoning would be an *instrumental* value, even with an anthropocentric stance provided, which could weigh in against the economic value of mining operations. Also, for scientific purposes, an epistemic value of unharmed geological figures should be considered. Some places, such as landing sites, could also be preserved for historical reasons. As on Earth, there might be those who argue beyond an instrumental value, in favour of some *intrinsic value* of Martian figurations, especially should there be areas with life on Mars (Cockell and Horneck 2004). This is where cultural sustainability comes into play. Microbial extraterrestrial life could have an instrumental value for our research as well, or receive an aesthetic or monetary evaluation, yet one also has to ask whether it deserves a *moral status of its own* (Persson 2017).

Conclusion

We have stated from the outset, that *any* human agenda is a value-laden cultural activity. Which foundational ethical principles should frame our collective space activities as we explore outer space (cf. Rummel et al. 2012)? It is a matter requiring a more in-depth discussion than can be done here. Yet the globally agreed-on concept of sustainability, a "development that meets the needs of the present without compromising the ability of future generations to meet their own needs" (World Commission on Environment and Development 1987, 16.41), is certainly one such ethical principle, in all its dimensions. In its cultural dimension, we face not only a legal challenge to combine international cooperation with the involvement of private investors in Space 4.0 and to regulate their permissions. In space, quite divergent national cultures meet with each other, and the ethics the nations may apply in the scenarios presented will depend on their respective cultural heritage. An *intercultural dialogue* is of the essence and appears as the greatest cultural challenge on a planetary and interplanetary scale. A constructivist approach may useful here, as it reminds us that our fundamental ethical assumptions are beliefs as opposed to knowledge; our divergent convictions and ethical rules are cultural constructions born out of an individual or communal perspective. Finally, if there really should be extraterrestrial life within reach of discovery, an important debate on its value and status would need to start.

In sum, *cultural* sustainability is of the essence to protect ourselves, our history and our living environment against hasty conclusions concerning how deeply connected we are with the biosphere and beyond, so we are able to refresh our memory of these insights in heritage parks at least. Besides, *everything we do* is a value-laden cultural activity. We are consciousness grown from this planet, hence we are responsible for caring for our home. In this regard, we cannot avoid evaluating, and hence the discussion needs to continue, with as many participants as possible.

Acknowledgements

I thank the editors of this volume, the native speakers and copy editors for helpful comments and corrections.

Notes

1 The status of the respective NASA initiative is unclear to me.
2 My own epistemological stance is a constructive-critical realism, which embraces constructivist thought, while emphasizing the importance of a realist approach, understood as a firm belief (Losch 2011, Ch. 10). For this chapter, however, it seems sufficient to focus on the social constructivist interpretation of epistemology.
3 That is why I call for an eighteenth UN Sustainable Development Goal, see Losch (2018).
4 For legal reflections on space tourism, see Masson-Zwaan (2010).
5 Theologically spoken (if I may), complete knowledge is an eschatological principle anyway.
6 The emerging discipline of astrobiology tries to take this possibility in our solar system into account. For some initial considerations on astrobiology and sustainability, see Arnould (2009).

References

Allen & Overy (2017). "Luxembourg Space Resources Act: Paving the Legal Road to Space". *Allen & Overy* (28 September 2017). www.allenovery.com/publications/en-gb/Pages/Luxembourg-Space-Resources-Act-Paving-the-legal-road-to-space.aspx [Accessed 4 October 2017].

Arnould, J. (2009). "Astrobiology, Sustainability and Ethical Perspectives". *Sustainability*, 1(4), 1323–1330.

Benz, A. D. (2016). *Astrophysics and Creation: Perceiving the Universe Through Science and Participation*. New York: The Crossroad Publishing Company.

Bohlmann, U. (2011). "Connecting the Principles of International Environmental Law to Space Activities". *Proceedings of the International Institute of Space Law*, 4, 301–310.

Bonnal, C. and McKnight, D. S. (2017). *IAA Situation Report on Space Debris – 2016*. Paris: International Academy of Astronautics.

Capova, K. A. (2016). "The New Space Age in the Making: Emergence of Exo-mining, Exo-burials and Exo-marketing". *International Journal of Astrobiology*, 15(4), 307–310.

Cockell, C. and Horneck, G. (2004). "A Planetary Park System for Mars". *Space Policy*, 20(4), 291–295.

Conseil d'etat du grand-duché de Luxembourg (2017). *Avis 51.987 du 7 avril 2017: Projet de loi sur l'exploration et l'utilisation des ressources de l'espace*. http://conseil-etat.public.lu/content/dam/conseil_etat/fr/avis/2017/07042017/51987.pdf [Accessed 4 October 2017].

COSPAR (2011). "Planetary Protection Policy: with Explanatory Annotations". https://planetaryprotection.nasa.gov/file_download/65/IAA_COSPARPPPolicy%28v32411%29human_w_ExplAnnotations.pdf [Accessed 15 October 2017].

Di Pippo, S. (2017). "Space Weather: Developments Within the United Nations. United Nations/United States of America Workshop on the International Space Weather Initiative". Unispace +50. *UNOOSA* (31 July 2017). www.unoosa.org/documents/pdf/psa/activities/2017/ISWI%20Boston/ISWIBostonDay1/01.pdf [Accessed 22 January 2018].

ESA (2016). "What is Space 4.0?". *Ministerial Council 2016*. www.esa.int/About_Us/Ministerial_Council_2016/What_is_space_4.0 [Accessed 28 July 2017].

The Government of the Grand Duchy of Luxembourg (2016). "The Luxembourg Government Becomes a Key Shareholder of Planetary Resources, Inc., the U.S.-based Asteroid Mining Company". *Space Resources*. www.spaceresources.public.lu/en/press/press-release.html [Accessed 11 October 2017].

The Government of the Grand Duchy of Luxembourg (2017). "Luxembourg Aims to Contribute to the Peaceful Exploration and Sustainable Utilization of Space Resources for the Benefit of Humankind". *Space Resources*. www.spaceresources.public.lu/en.html [Accessed 28 July 2017].

H.R.2262–114th Congress (2015). "U.S. Commercial Space Launch Competitiveness Act". Public Law No: 114–190. 129 STAT. 704 (25 November 2015). www.congress.gov/114/plaws/publ90/PLAW-114publ90.pdf [Accessed 22 January 2018].

Huberty, M. (2017). "Doubts over Space Mining Legal Status". *DELANO* (13 April 2017). http://delano.lu/d/detail/news/doubts-over-space-mining-legal-status/142567 [Accessed 23 May 2017].International Institute of Space Law (2015). "Position Paper on Space Resource Mining". *IISL* (20 December 2015). http://iislweb.org/iisl-position-paper-on-space-resource-mining/ [Accessed 15 October 2017].

Losch, A. (2011). *Jenseits der Konflikte: Eine konstruktiv-kritische Auseinandersetzung von Theologie und Naturwissenschaft*. Göttingen: Vandenhoeck Ruprecht.

Losch, A. (ed.) (2017). *What is Life? On Earth and Beyond*. Cambridge: Cambridge University Press.

Losch, A. (2018). "The Need of an Ethics of Planetary Sustainability". *International Journal of Astrobiology*, 1–8.

Masson-Zwaan, T. (2010). "Regulation of Sub-orbital Space Tourism in Europe: A Role for EU/EASA?". *Air and Space Law*, 35(3), 263–272.

NASA (2014). "Our Vision for Planetary Sustainability". *NASA* (4 November 2014). www.nasa.gov/content/planetary-sustainability-our-vision/#.WBgtmiTBZsl [Accessed 1 November 2016].

Persson, E. (2012). "The Moral Status of Extraterrestrial Life". *Astrobiology*, 12(10), 976–984.

Persson, E. (2017). "Ethics and the Potential Conflicts Between Astrobiology, Planetary Protection, and Commercial Use of Space". *Challenges*, 8(1), 12.

Planetary Resources (2017). "Space Mining Law Passes in Luxembourg". *Planetary Resources* (13 July 2017). www.planetaryresources.com/2017/07/space-mining-law-passes-in-luxembourg/ [Accessed 28 July 2017].

Rummel, J. D., Race, M. S. and Horneck, G. (2012). "Ethical Considerations for Planetary Protection in Space Exploration: A Workshop". *Astrobiology*, 12(11), 1017–1023.

Soini, K. and Birkeland, I. (2014). "Exploring the Scientific Discourse on Cultural Sustainability". *Geoforum*, 51, 213–223.UNESCO (2003). "Planetary Sustainability in the Age of the Information and Knowledge Society for a Sustainable World and Future: Working Toward 2015 for a Sustainable World and Future". www.un-documents.net/psaiks.pdf [Accessed 11 December 2017].

United Nations (2002). "United Nations Treaties and Principles on Outer Space". *United Nations*. No. E.02.1.20 ST/SPACE/11. www.unoosa.org/pdf/publications/STSPACE11E.pdf [Accessed 11 December 2017].

United Nations (2015). "Sustainable Development Goals: 17 Goals to Transform our World". *United Nations*. www.un.org/sustainabledevelopment/sustainable-development-goals/ [Accessed 21 October 2017].

UNOOSA (2017). "Working Groups of the Committee and its Subcommittees: Scientific and Technical Subcommittee". *United Nations Office for Outer Space Affairs*. www.unoosa.org/oosa/en/ourwork/copuos/working-groups.html [Accessed 23 May 2017].

White, F. and O'Neill, G. K. (1987). *The Overview Effect: Space Exploration and Human Evolution*. Boston: Mifflin.

Woerner, J. (2016). "Moon Village: A Vision for Global Cooperation and Space 4.0". *ESA*. www.esa.int/About_Us/Ministerial_Council_2016/Moon_Village [Accessed 15 October 2017].

World Commission on Environment and Development (1987). *Our Common Future*. Oxford: Oxford University Press.

10 Sustainability's promise of salvation?

A Kuhnian reconstruction of sustainability from resource management to contingency management

Marius Christen, Peter Seele, Lucas Zapf

Introduction: the sustainability debate from the perspective of the Kuhnian paradigm shift theory

Thomas Kuhn examined the "structure of scientific revolutions" and determined that the theory of paradigm shifts is applicable not only to the concept of science and scientific disciplines but also to cultural history and nature in general (Kuhn 1972: 122). In this context, we refer to the prevailing description of a paradigm shift in the sustainability discussion and reconstruct it with the help of Kuhnian terminology. In so doing, we understand the idea of sustainability as a 'paradigm candidate'. Kuhn explains that "a crisis may end with the emergence of a new candidate for paradigm and with the ensuing battle over its acceptance" (Kuhn 1972: 84). We differentiated sustainability as new paradigm candidate into sub-candidates providing different answers to different sub-problems: Sustainability, when implemented, would therefore be a new paradigm. According to the mainstream discourse, the current paradigm of the 'Anthropocene' (Ehlers and Krafft 2005) in its irreversibility of human- and company-induced climate change and underlying system understanding of unlimited economic growth is incompatible with potential future paradigms of sustainability (Redclift 2005). In Kuhn's terminology they are 'incommensurable'. What is crucial here is the transition in which a new paradigm replaces the prevailing paradigm after a "revolutionary phase" and then reverts back to a "normal science" (Kuhn 1972: 10). Currently we are far from sustainability as being the paradigm of human development. Our paradigm could be considered more as unsustainability. Therefore, the candidates we discuss here are to be seen as ideal or potential future paradigms. In the sustainability debate different paradigm candidates, which interpret the idea of sustainability controversially, compete with one another and with the prevailing paradigm. In this regard, we also aim to reconstruct the sustainability debate as ongoing and sustainability as an interpretative concept subject to discourse.

In this chapter, we examine the predominant sustainability paradigm candidates and differentiate these various types of sub-paradigm candidates of the term sustainability. In Kuhn's terminology we analyse the "Diffusion of the new paradigm" (Kuhn 1972: 169) along historically defined classifications since the idea arose in the eighteenth century.

Building on this we pose our research question: *How has the idea of sustainability, as used and formed by individuals, corporations, NGOs and other stakeholders, changed since its inception in the search to find answers to different and time-dependent demands?* Accordingly, our sustainability research subject is a diachronic system classification. We introduce a typological schematization of the sustainability concept in its chronological context. In so doing – and based on Redclift's historiographical study of the sustainability concept epochs (2005) – we understand the idea of sustainability, in its function as a paradigm candidate (its thematic content structure notwithstanding), to be a focused and future-oriented solution strategy. As such it proposes answers to practical problems pertaining to the integrative relationship between societal and/or economic development and its natural conditions (Christen and Schmidt 2012). Consequently, we are assuming provisional sustainability conceptions, i.e. different interpretations of the concept (cf. Grunwald 2013: 40; Christen 2013: 34), respective to the specific form of the problem at hand. Based on the literature we classify them into three existing paradigm candidates. Candidate one, 'sustainability as resource management', started with the forestry debate in the eighteenth century. Candidate two arose in the mid-twentieth century and culminated with the 'Club of Rome'-debate on 'Sustainability as environmental management'. With the publications by the Brundtland Commission (WCED 1987) and the UN-Rio Conference (UN 1992) we can speak of candidate three as 'sustainability as social management'. Furthermore, we have recently seen indications that not only social management is in focus, but increasingly also a connection emerges regarding the individual and society's sense-making and contingency – areas that are traditionally part and parcel of the 'offerings' of mainstream religions. Hence, we propose a fourth paradigm candidate under the heading 'sustainability as contingency management' addressing individual salvation mediated through a sustainable way of living.

In Table 10.1 the four paradigm candidates of sustainability are summarized. We will explain and describe them in further detail in the next section. It should be noted here that the four candidates are presented diachronically in order of their emergence. However, the candidates do not present consecutive epochs of paradigm candidates, but rather conceptions with a distinct emergence and period of validity. Therefore, in the overview we present the four candidates not only as *diachronic* but also as *synchronic*, and also illustrate the exponents of the respective candidates in terms of their relevance today (for more historical approaches, see e.g. Grober 2012; Heinrichs et al. 2016: Ch. 2).

Three existing paradigm candidates of the idea of sustainability

As a first step we conceptually develop three paradigm candidates along the following criteria: What is the problem, the specific candidate provides an answer

Table 10.1 Overview of the sustainability conceptions as paradigm candidates

Sustainability as . . .	*Diachronic*	*Synchronic*	*Nature of the problem*	*Guiding principle*	*Normative core*
Resource management	Since the eighteenth century	Carlowitz as historical reference	Resource crisis	Economy	Living from renewable production and not from the substance
Environmental management	Since mid-twentieth century	Club of Rome	Environmental crisis	Conservation	Intrinsic value of nature
Social management	Since 1987, 1992	Brundtland Commission Declaration Rio	System crisis	Quality of life today and tomorrow	The good life for everyone today and tomorrow
Contingency management	Twenty-first century	Sustainability as religion	Multi-crisis; appearance of an ultimate crisis	Contingency management	Ending the multi-crisis; attaining world healing and salvation

Source: Marius Christen, Lucas Zapf, Peter Seele

for? What is the intrinsic guiding principle? And what is the normative core idea of the candidate?

Paradigm candidate one: sustainability as resource management

Fundamental problem: resource crisis

In the sixteenth century Europe began to suffer from the so-called 'timber shortage', a scarcity of one of the main resources of that time (Sieferle 1982). Practical solutions were needed to meet the challenge of this resource crisis. It is not surprising that the first written evidence of the idea of sustainability appeared in these circumstances at the beginning of the eighteenth century. In 1713 Hans Carl von Carlowitz's solution to the timber shortage was the 'sustainable use' of the forests. He proposed that only so much wood should be cut within a certain phase as would grow again.[1] The understanding of 'sustainable use' in that way strives for a balance between availability and the use of a renewable resource in a local regenerative ecosystem, i.e. the 'maximum sustainable yield'. It espouses that the management of a forest, for example, is sustainable when the current use of resources does not endanger the future availability of that resource. Of course this 'management principle of renewable resources' is still valid today. The Forest Steward Council (FSC) Standard's Principle 5.6.1 e.g. requires that: "The systematic use of wood does not surpass the potential for sustainable exploitation" (FSC Deutschland 2012). According to this paradigm candidate, sustainability is understood to be a guideline for the management of renewable resources. This resource approach is very prominent and an essential part of the following paradigm candidates. However, it is nowadays nowhere found as an independent, stand-alone paradigm candidate.

Guiding principle: economy

The fundamental idea behind this early understanding of sustainability is distinctly economic. The management principle is not about protection of life in the forest. The understanding is much more about securing the continuous potential use of a resource. Human intervention should not be prohibited, but regulated. This interpretation of sustainability has – as have all the following interpretations – an inherent future prospect as well as a provision for the human-nature relationship. Contrary to subsequent emerging conceptions, this first paradigm candidate focuses exclusively on guaranteeing the management and exploitation of renewable resources.

Normative core idea: live from the interest, but not from the capital

This economic guiding principle is, as every interpretation of sustainability, ethically substantiated. The first paradigm candidate refers on an assumption of fairness, namely that subsequent beneficiaries should receive at least as much as do the current beneficiaries. Respectively, a pertinent mot has entrenched itself as a management policy in the sustainability discussion: 'Live from the interest, but not from the capital'. Therefore, this first paradigm candidate is based on the logic of a specific type of resource, i.e. renewable resources, and on an intergenerational principle of justice.

Paradigm candidate two: sustainability as environmental management

Fundamental problem: environmental crisis

With continuing population growth and an enormous increase in economic performance after World War II, the human-nature relationship changed dramatically. On one hand, thanks to cheap primary energy there was an unprecedented increase in prosperity in broad levels of the population unparalleled in human history. On the other hand, these epochal dynamics led to an increasing exploitation of non-renewable resources as well as to an immense additional burden on natural systems through emissions (Steffen et al. 2004).

In contrast to the timber shortage as resource crisis this development has intensified since the middle of the twentieth century to reach a wide-ranging ecological crisis. Not only did it include the danger of a shortage in renewable resources, more importantly it brought the threat of a complete resource scarcity – if resources are exploited that do not regenerate – as well as an overload in absorption capacity in natural systems. Debates on worldwide loss of biodiversity (see e.g. Carson's Silent Spring 1962), on the endangerment of the ozone layer through CFC emissions (starting in the mid-1970s), anthropogenic climate change (cf. the first report of the Intergovernmental Panel on Climate Change in 1990) are some examples of the global, and to some extent irreversible, impacts of human involvement in environmental systems.

This change in situation demands for a change in solution. Consequently, 'sustainability' was not only interpreted as an answer to a local resource crisis but as a response to the global ecological crisis (for example Meadows et al. 1972). Correspondingly, the solution entails more than just the management of renewable use of resources. Sustainability is conceptualized as comprehensive environmental management policy. In addition to the above-described resource management there appeared new rules for the use of non-renewable resources, output of substances in the environment, stress speeds and handling dangers and risks.

According to this second paradigm candidate, development is sustainable when there is no danger to the present use of and stress on the environment, and to the potential for its future local and global use. Sustainability is systematically conceptualized as an environmental conservation principle (IUCN 1980) and claims to regulate the relationship between society and nature at an encompassing and global level.

Guiding principle: environmental protection

While continuous use of renewable resources is the main goal of the paradigm candidate 'resource management', the environmental system in all its complexity comes to the fore in the paradigm candidate 'environmental management'. This reflects the fact that increasing attention has been paid to the scientific examination of natural systems as well as to the impact of humans on their environment. Terms like 'ecosystem services' or 'resilience' dramatically shaped the debate on sustainability. From this perspective, it comes as no surprise that 'sustainability' is often pushed into the 'green corner'.

Normative core idea: intrinsic value of nature

The normative core of this paradigm candidate is the integrity of the relationship between humans and nature. Sustainability is about protecting the environment from human interference and misuse. Nature's utility function moves into the background or is only of secondary importance. Contrary, arguments highlighting the intrinsic value of nature and its integrity gain significance (cf. influential positions in early environmental ethics by e.g. Aldo Leopold, Arne Naess, James

Lovelock, Holmes Rolston, J. Baird Callicot etc.). At the turn of the millennium, the *Earth Charter*, for example, formulated an entire programme of sustainable development that is based on this environment-centred understanding that the integrity of nature without reference to its (direct) use for humans constitutes the ethical foundation for the demands for protection (Earth Charter 2000).

Paradigm candidate three: sustainability as social management

Fundamental problem: system crisis

In the 1980s a growing number of people perceived the environmental crisis not only as a threat to nature, but also a social danger. With this understanding, the problem receives the status of a system crisis. Emanating from two challenges that were previously considered completely separate, the elimination of poverty on the one hand and the overexploitation of nature on the other, the *World Commission on Environment and Development* (WCED) developed an integrated stance on the matter (cf. Di Giulio 2004). The pressure on nature and the creation of human wealth were seen as inextricably interwoven, following that the environmental and poverty crises should not be regarded independent of one another. The WCED warned that an improvement in the quality of life following the model of the rich nations would endanger the "basic integrity" of the global ecosystem. Conversely it also applies that the generation of prosperity of those living in the rich nations has an increasingly critical impact on global ecosystems.

According to the WCED, both challenges for societal development and the environmental pressure can only be mutually solved. That means that, on the one hand, long-term societal development is only possible when natural conditions are taken into consideration, and on the other, that the environmental issue can only be alleviated when societies develop in a certain way.

The sustainability idea then is interpreted as a guiding principle for societies as a whole. It not only regulates the relationship between society and nature but beyond that also makes provisions for development in the sub-system 'society', whereby everyone can satisfy hers or his "essential needs" (WCED 1987: 43). This paradigm candidate affects all relevant development aspects of society and thus its progress as a whole. With the WCED-report the perspective pivoted from environmental management, i.e. from the focus on the natural conditions of social development, to the subject of enabling good life while considering the available natural preconditions.

Guiding principle: focus on society as a whole for the quality of life for all

With this broadening of the perspective, to guarantee the option of fulfilling (basic) needs became a central factor in the discussion of sustainability. Sustainability is no longer exclusively about regulating the exploitation and the pressure on nature. Moreover, the formula for sustainability has become so generalized that no human behaviour at all should endanger the basis for a decent life for present and future human beings.

The new guiding principle then is a good quality of life for all human beings living at present and in the future. This transcends the two previous paradigm candidates in two aspects: First, in addition to the intertemporal and global perspective sustainability obtains an ethical-universal entitlement. Second, it considers the interpretation of the intact functioning of the ecosystems as necessary, but not as sufficient – since the use of nature could be organized in such a manner that it ensures its future functionality, but at the same time drastically limits other basic options for good life. According to this paradigm candidate, progress in the direction of sustainability, thus, is incompatible with the establishment of an eco-dictatorship, which regulates the relationship of society to nature in an environmentally sustainable way but at the same time restricts the human quality of life. Aspects such as the guarantee of health, education, political participation, economic production capability etc. became essential parts of the sustainability topic. This broad understanding is also reflected, for example, in the FSC standard. The FSC standard takes into account the well-being of all people affected by forestry management, in addition to the regulation of the use of nature as Carlowitz's original idea of 'sustainable exploitation' had in mind.

The guarantee of a good quality of life must be in accord with the natural preconditions. This interactivity of the developmental and environmental perspectives is just as essential to this understanding of sustainability as its forward-looking character. It reflects the dominating interpretation of sustainability in the political, social and economic discussions of our days.

Normative core idea: a good life for all today and tomorrow

The normative core idea behind this paradigm candidate is the universal right to a decent life: "all human beings – those here and those who are to come – have the right to life, and to a decent life" (WCED 1987: 41). It follows that all human beings have the right to resource consumption. But the right to a good life encompasses even more. Which other conditions additionally have to be considered essential elements of sustainable development is one of the main topics of the (theoretical) debate on sustainability. According to paradigm candidate three, sustainability is an anthropocentric concept. From this standpoint, sustainability formulates the most comprehensive and therefore the highest imaginable ethical norm.

A new paradigm candidate? Sustainability as contingency management

The connections between traditional religion and sustainability are diverse. For one, religion can promote sustainable action. Johnston convincingly described (2010, 2013) sustainability as advocacy for a mindful and respectful dealing with godly creation that can serve as a guideline for secular, political advocacy. In other research, religious care for sustainability has been elaborated regarding the degree and hierarchical value of sustainability within different religions, arriving at a characterization of "dark green religions" (Taylor 2010: 10) placing nature higher than humans. A trait that is not common in Christianity but sometimes found as a political claim when taking action towards a more sustainable future. Building on this idea, there are increasing – and critical – references made in literature where sustainability is linked to religious traits, up to the point of being considered a 'sustainability religion'. There is a hidden motif in these attributions that we would like to explore further by developing the fourth paradigm candidate, mindful of the three paradigm candidates of 'sustainability' presented: sustainability as contingency management, as sense-making and as a way out of the multi-crisis. The multi-crisis thereby is connected to the system crisis of candidate three and described by some as the 'ultimate' crisis, ending life on earth as we know it. In view of this threat the well-being of the individual, and following along the religious theme, the salvation of the individual in the sense of a positive solution of the apocalyptic outlook is in danger. With the conception of sustainability as contingency management, a solution-oriented approach will be found that either improves or restores world healing and salvation of the individual.

Similar to the three previously described paradigm candidates, in the following we will describe the possibility of such a fourth candidate along the three established categories: *Fundamental problem*, *Guiding principle* and *Normative core idea*.

Fundamental problem: multiple crises and contingency

When a cluster of existential problems threatens the present and the future of the humankind, we experience a multi-crisis. Therefore, the problem to which the sustainability idea should react has intensified. It is increasingly considered to be existential, with the extent and diversity of the problems threatening the existence of the humankind as a whole. The result is a fundamental uncertainty, a lack of clarity about where and how one is positioned in this danger.

The multi-crisis is reflected in economic, financial, climate, food and biodiversity crises. In light of this assessment, the future seems to hold nothing short of the collapse of the overall system (Metz et al. 2007).

The multi-crisis is unlike other perceptions of the world's last days. Sustainability holds out the prospect of an anthropocentric solution, because the crisis is the result of human action. To avoid the threatening doom, the economy, the interaction with nature and society as well as our individual behaviour must change. The multi-crisis necessitates reflection, adjustment and renewal in all spheres of life. Herein, sustainability is put forward as the guiding principle of solution. Thus, the idea of sustainability is seen as a way out of the threatening situation by 'managing' the contingency emerging from the multi-crisis.

The comprehensive change in all spheres of life calls the *status quo* into question, thereby endangering the familiar (cf. Blumenberg 1981: 157–158). The multi-crisis thus both generates and amplifies contingency. Contingency entails that life experiences are fundamentally open-ended and uncertain. Situations can arise as expected, but they very well can develop differently, too. Contingency is the "cask of all conceivable alternatives to the case" (Hörisch 2013: 148). Contingency arises through the personal perspective: "events can be contingent because of restrictions in the particular lenses" (Bruun et al. 2002: 108). Another source of contingency is the lack of expectation value: "events can be contingent because they fall outside any detectable trend" (108). Contingency therefore is rooted in the person's expectations of outcomes or developments while depending on the variability of nature: Everything that is neither necessary nor impossible is contingent.

'Religion' in this research is understood as concept that provides a strategy to cope with contingency by providing guidelines for individual action and thereby connecting the individual to a super-individual framework (or narrative). This is a functional understanding of religion with a focus on the believing individual and the effects thereon (cf. McCleary 2007: 50; Lübbe 1986: 219). 'Religion' in this sense provides contingency with a structure and an objective. Contingency management is considered a unique selling point of religion: "The characteristic of religion through the function of contingency management refers religion to that function in which nothing else can be a substitute" (Lübbe 1986: 227).

This refers both to concepts that are traditionally described as 'religious' (e.g. Christianity) as well as concepts that offer the same functional framework, yet are not traditionally viewed as religious concepts (cf. Zapf 2014: 256) – e.g. as we will see, sustainability that serves as an individual guideline for action and the contingency evoked through the apocalyptic narratives connected to it (cf. Taylor 2010).

Religion provides a narrative with which human can position themselves in a contingent world. The religious story contains stability and points to a transcendent world. This transcendent world is not affected by inner worldly coincidences and chaos, and therefore not by contingency (Troeltsch 1913: 771–772). By contributing to contingency management, sustainability questions the religious monopoly to contingency management.

Guiding principle: contingency management and the fear of the collapse in the overall system

The message of the multi-crisis communicates: 'If the behaviour of humans continues as it is, the overall system will collapse in the foreseeable future'. Presupposing human's instinct of self-preservation, this assessment warrants the compulsory challenge to take countermeasures. Only through immediate, planned and collective action the multi-crisis, and with it the existential threat, can be overcome. This guiding function is given to the idea of sustainability. Sustainability offers a way out of the multi-crisis, overcomes the associated contingency and provides guidance.

The perception of sustainability as a solution for apocalyptic conditions is already apparent at the very start of the sustainability discussion. Fear of the human-made destruction of livelihoods is stated in the work of Carl v.

Carlowitz. He interprets the resource crisis as a punishment for human misconduct willed by God. Humans and their sins, his ingratitude, laziness and carelessness are the reason for the God-sent infertility. Carlowitz calls for a reversal: Greater care and the duty to preserve creation are the sustainable solution for this divine punishment and subsequent vegetative neglect (Carlowitz and Rohr 1732).

Today the sustainability discussion no longer needs to refer back to religion to cope with contingency. Humans carry problems and solutions equally within themselves and a metaphysical entity is no longer considered necessary.

Normative core idea: attaining world healing and salvation

The argumentative relation between sustainability and religion offers a new perspective in the exploration of the idea of sustainability.

For the first three paradigm candidates, sustainability refers to concrete challenges in resources, environmental and social management. It provides answers to specific, current and tendentious perceptible or measurable problems.

Sustainability as contingency management goes further. It functions against the backdrop of the multi-crisis whose threats are in part virtual and often indirect. The disastrous consequences of the multi-crisis are subtle and vague, but present. They often have the effect of a threat and nightmare scenario that awaits us in future. The sustainability discourse responds to this indirect and rather virtual threat with equally abstract analysis of the situation through scientific justification, mathematical models and theoretical studies. At the same time it offers visions and views of the future that provide information on the passage into a new sustainable world. In Christian theology this form of contingency management is called eschatology: an elaborate narrative that influences human's way of life by correlating their actions to an ultimate aim (cf. Edsman 1962).

Religious contingency management is related to the existence of risks in life and existential insecurity. In western societies these challenges so far only exist to a limited extent, because those dramatic risks and uncertainties are minimized by progress and prosperity. As religion no longer assumes this function, a gap opens for new forms of contingency and its management.

With respect to the contingency generated from the multi-crisis the absence of religion indicates: With its contingency management, sustainability approximates religion in a functional way.

This transfer of contingency management from religion to (apparent) areas that have little or nothing to do with religion is not new. This was done especially with reference to the economy. Economics offer scientific strategies to manage decisions in uncertainty, that is, in contingent situations. There are some useful findings that, as Hörisch (2013) explains it, can be actually seen as baked goods to reduce hunger. In view of the multi-crisis, sustainability provides a similar function. In the face of the impending world destruction it indicates the way with specific actions toward a future without multi-crisis.

In summary, sustainability has a dual function in contingency management from which its strength as a paradigm candidate shapes a new world culture:

1 *The social function*: The end to the multi-crisis and the deliverance of humans from the impact of threats to their existence. Sustainable actions demand restrictions in the here and now: the *Pay-off* of these restrictions, if not directly, then in every case perceptible in this world. While eschatology of classical religions – e.g. Christian eschatology – promises redemption from life's hardships in a carefree afterlife, the eschatology of sustainability puts this redemption into the temporal, inner worldly future (cf. Priddat 2010: 31 for the similarities of this process with regard to the economy).
2 *The individual function*: The individual function of sustainability complements the social function. Sustainability offers individual guidance for action on three levels. On a cognitive and perception-forming level, when existential threats like climate change and pollution are felt through personal experience and sustainability is accepted as an escape. At a linguistic level when the sustainability discourse on these experiences provides respective content, expression and concepts. And on the intentional level when out of this conflict situation the conviction of one's own effectiveness arises to avert the problems. Individual human actions and what is beneficial for the environment are present in this level of sustainability. Its meaning transcends the individual and entails the securing of world healing and salvation. The rapidly increasing consumption of so-called sustainable products paradigmatically displays this function.

With these reflections describing sustainability as an eschatologically charged strategy for contingency management, we portray a new paradigm candidate of sustainability. Through the individual's sustainable thinking and actions arises a paradigm that looks to be capable of supplanting unsustainable life styles because individual understanding and actions are viewed as embedded in a life-threatening context.

Conclusion

Conceptualizing sustainability from a Kuhnian perspective allows for a new understanding of sustainability as interpretative concept. This implies several conclusions, but before we present the theoretical and practical conclusions in detail, we present a summary of the conceptual contribution of the four discussed paradigm candidates as well as the overall perspective from a Kuhnian paradigm shift.

Incommensurability: It is important to note that sustainability here is seen as approach of 'revolutionary science' seeking for an answer of the current paradigm of the Anthropocene characterized as unsustainable. The Kuhnian approach here is particularly instructive as the concept of incommensurability between the old paradigm (Anthropocene) and the (possible) new paradigm (full sustainability)

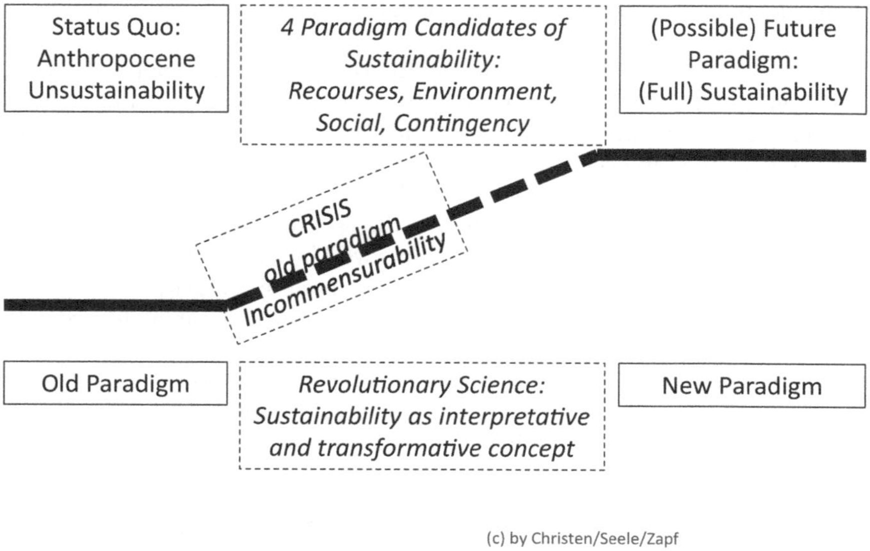

Figure 10.1 Overview of the sustainability conceptions as paradigm candidates

Source: Marius Christen, Lucas Zapf, Peter Seele

reveals that there is no coexistence, but a more distinct either-or. The 'revolutionary science' aspect of the paradigm crisis and the emergence of new paradigm candidates leads to the more existential and fundamental crisis, as discussed here in religious term regarding salvation and contingency. Figure 10.1 summarizes this conceptual model bearing theoretical (4.1) and practical consequences (4.2):

Theoretical implications

Kuhn's conception of the diffusion of paradigm candidates right through to the breakthrough of a new paradigm offers a novel version of how to understand the transformative character of sustainability in conflict with its critical nature.

Therefore, our contribution consists of three points.

First, we have moved the sustainability discourse into the transitory logic of a paradigm shift. This enables an understanding of sustainability as an answer to existing challenges and crises.

Second, we differentiate between various types of sustainability paradigm candidates.

Third, building on this methodology and the three existing interpretations allows us to add a fourth candidate that we call contingency management, and which contains the social as well as individual function of sense-making and improvement/salvation through sustainability.

Practical implications

The historical examination of the sustainability discourse along the lines of the paradigm shift theory of Kuhn also has consequences for the practical implementation of sustainability: All four candidates serve as responses to problems and gaps that are considered unsolved today.

For example, the four candidates offer companies the potential to direct their sustainability engagement according to the four solution approaches. In sustainability reports that follow the standards of the Global Reporting Initiative (GRI) fifty-six Key Performance Indicators (GRI 3.1) are enlisted to report on environmental, social, ethical and economic issues (Knebel and Seele 2015). Here the four candidates are called upon to address the various dimensions according to the nature of the problem.

For *resource management*, it is worth reporting on the economic and environmental aspects of corporate activities. For *environmental management* on the other hand, pure environmental topics are important, in particular the preservation of and benefit to nature. However, for *social management* it would be good to cite social indicators to confront the system crisis.

As Gayá and Philips show, the focus on fundamental concerns like 'a new era' and the exposure to problems of contingency can achieve on an organizational business-level "a meta-level counter-story which emphasizes their paradigm-shifting and future-shaping potential" (Gayá and Phillips 2015: 17). *Contingency management* therefore concerns a dimension of corporate activity that over the last few years has played an increasingly important role in product design and product responsibility. Sustainability presents a component that enables consumers to connect to issues of contingency via the product – thereby connecting religious and socio-ecological issues.

Note

1 Carlowitz also proposed other measures to combat the resource crisis: more efficient utilization, increase in resources and the search for substitutes.

References

Blumenberg, H. (1981). "Paradigma, grammatisch". In *Wirklichkeiten in denen wir leben*. Stuttgart: Philipp Reclam jun., 157–163.

Bruun, H., Hukkinen, J. and Eklund, E. (2002). "Scenarios for Coping with Contingency: The Case of Aquaculture in the Finnish Archipelago Sea". *Technological Forecasting and Social Change*, 69(2), 107–127.Carson, R. (1962). *The Silent Spring*. New York: Mariner Book.

Christen, M. (2013). *Die Idee der Nachhaltigkeit: Eine werttheoretische Fundierung*. Marburg: Metropolis-Verlag.

Christen, M. and Schmidt, S. (2012). "A Formal Framework for Conceptions of Sustainability – a Theoretical Contribution to the Discourse in Sustainable Development". *Sustainable Development*, 20(6), 400–410.

Di Giulio, A. (2004). *Die Idee der Nachhaltigkeit im Verständnis der Vereinten Nationen: Anspruch, Bedeutung und Schwierigkeiten*. Münster: LIT.

Earth Charter (2000): Earth Charter. www.earthcharter.org.

Edsman, C. M. (1962). "Eschatologie". In K. Galling, H. Frhr. v. Campenhausen and E. Dinkler (eds.). *Die Religion in Geschichte und Gegenwart: Handwörterbuch für Theologie und Religionswissenschaft, Band 2*. Tübingen: J.C.B. Mohr, 651–655.

Ehlers, E. and Krafft, T. (2005). *Earth System Science in the Anthropocene: Emerging Issues and Problems*. New York: Springer Netherlands.

FSC Deutschland (2012). *Deutscher FSC-Standard*. Deutsche übersetzte Fassung, Version *2.3 vom 01.07.2012*. Freiburg.

Gayá, P. and Phillips, M. (2015). "Imagining a Sustainable Future: Eschatology, Bateson's Ecology of Mind and Arts-based Practice". *Organization*, 23(6), 1–22.

Grober, U. (2012). *Sustainability: A Cultural History*. Dartington Hall: Green Books.

Grunwald, A. (2013). "Wie viel und welche Theorie benötigt nachhaltige Entwicklung? Konzeptionelle Überlegungen". In J. C. Enders and M. Remig (eds.). *Perspektiven nachhaltiger Entwicklung: Theorien am Scheideweg*. Marburg: Metropolis-Verlag, 27–46.

Hamerman, E. J. and Johar, G. (2013). "Conditioned Superstition: Desire for Control and Consumer Brand Preferences". *Journal of Consumer Research*, 40(3), 428–443.

Heinrichs, H., Pim, M., Michelsen, G. and Wiek, A. (eds.) (2016). *Sustainability Science: An Introduction*. Dordrecht: Springer Netherlands.

Hörisch, J. (2013). "Man muss dran glauben – ein ökonomisch-theologischer Traktat". In P. Seele and G. Pfleiderer (eds.). *Kapitalismus – eine Religion in der Krise I. Grundprobleme von Risiko, Vertrauen, Schuld*. Baden-Baden: Nomos, 137–222.

IUCN (1980). *World Conservation Strategy: Living Resource Conservation for Sustainable Development*. Gland: IUCN.

Johnston, L. (2010). "The Religious Dimensions of Sustainability: Institutional Religions, Civil Society, and International Politics since the Turn of the Twentieth Century". *Religion Compass*, 4(3), 176–189.

Johnston, L. F. (2013). *Religion and Sustainability: Social Movements and the Politics of the Environment*. Sheffield-Bristol, CT: Equinox.

Knebel, S. and Seele, P. (2015). "Quo Vadis GRI? A (critical) Assessment of GRI 3.1 A+ Non-financial Reports and Implications for Credibility and Standardization". *Corporate Communications: An International Journal*, 20(2), 196–212.

Kuhn, T. (1972). *The Structure of Scientific Revolutions* [1960]. Chicago: University of Chicago Press.

Lübbe, H. (1986). *Religion nach der Aufklärung*. Graz: Styria.

McCleary, R. (2007). "Salvation, Damnation, and Economic Incentive". *Journal of Contemporary Religion*, 22(1), 49–74.

Meadows, D. H. et al. (1972). *The Limits of Growth: A Report for the Club of Rome's Project of the Predicament of Mankind*. London: Earth Island.

Metz, B., Davidson, O. R. and Bosch, R. (2007). *Climate Change 2007: Mitigation of Climate Change*. Cambridge: Cambridge University Press.

Priddat, B. P. (2010). "Ökonomie des Glaubens?". *Zeitschrift für Wirtschafts- und Unternehmensethik*, 11(1), 25–34.

Redclift, M. (2005). "Sustainable Development (1987–2005): An Oxymoron Comes of Age". *Sustainable Development*, 13, 212–227.

Sieferle, R. P. (1982). *Der unterirdische Wald: Energiekrise und industrielle Revolution*. München: Beck.

Steffen, W. et al. (2004). *Global Change and the Earth System: A Planet under Pressure*. Berlin: Springer Netherlands.

Taylor, B. R. (2010). *Dark Green Religion: Nature Spirituality and the Planetary Future*. Berkeley: University of California Press.
Troeltsch, E. (1913). *Gesammelte Schriften Bd. 2: Zur religiösen Lage, Religionsphilosophie und Ethik*. Tübingen: J.C.B. Mohr and Paul Siebeck.
UN (1992). *Agenda 21, Resolution from United Nations Conference on Environment and Development*, Rio de Janeiro.
von Carlowitz, H. C. and von Rohr, J. B. (1732). *Sylvicultura oeconomica, oder haußwirthliche Nachricht und Naturmäßige Anweisung zur wilden Baum-Zucht*. Leipzig: Johann Friedrich Brauns.
WCED (1987). *Our Common Future*. Oxford: Oxford University Press.
Zapf, Chr. L. (2014). *Die religiöse Arbeit der Marktwirtschaft: Ein religionsökonomischer Vergleich*. Nomos: Baden-Baden.

Part III

The arts, literature and cultural sustainability

Theoretical, historical and contemporary perspectives

11 Culture and the arts in sustainable development

Rethinking sustainability research

Sacha Kagan

Introduction

In policy discourses concerned with sustainability/sustainable development (besides any policies specific to the "cultural sector"), such as policies for sustainable urban development, culture is considered only marginally. However, there is a growing awareness of some of the functions attributed to culture: When culture is understood as a prerequisite for social change, the functions attributed are those of value-system, mode of place-making and identity-building (Barthel-Bouchier 2012; UNESCO 2013a; Lehmann 2010). When it is seen as a motor for transformation, its functions are to enable creativity and engagement (James 2015; UN Habitat 2013; UNESCO 2013a, 2013b). When it is perceived as a social challenge, its functions relate to cultural diversity and multi-ethnicity (DifU 2011; Meuleman 2013). Further attributed functions of culture relate to "well-being", "happiness" and sustainable ways of life away from consumer culture (UNESCO 2013a; Davies 2015).

The groundwork for this growing attention to culture came from an increasing number of discourses about culture and sustainability, carried out by professional, academic and policy-related actors. In the years that followed the Brundtland Report and the 1992 Rio UN Summit, a minoritarian – albeit growing – discursive arena emerged that gave space to arguments for a greater attention to culture in sustainable development policies, discourses, practices and research. Unsurprisingly, actors from the cultural sector and cultural policy were involved in this development, and one UN agency that supported this trend from the start was UNESCO. The UNESCO Summit on Culture and Sustainable Development in 1998 was the result of a process that lasted several years. Under the title "The Power of Culture", it proclaimed in general terms the interdependence of culture and sustainable (mainly economic) development. Over the following decades, several discourses around sustainable development and sustainability emerged that introduced a cultural component, in a variety of ways (see section as follows).[1] Furthermore, an imaginative and arts-informed approach to sustainability has also emerged, which has gathered a growing attention and bears promising potentials for the rethinking of sustainability (see second section).

Four expressions of the relation between culture and sustainability

One type of discourse that champions the integration of a cultural component into sustainability does so around the notion of culture as a "fourth pillar of sustainability". The expression was explicitly foregrounded as the title of a publication by Australian cultural policy analyst and community arts advocate Jon Hawkes in 2001 (*The Fourth Pillar of Sustainability: culture's essential role in public planning*). His four pillar model of sustainability stressed cultural vitality as an important dimension, pointing at the inherent value of cultural diversity and of a vibrant cultural life amongst human communities, necessary to quality of life (Hawkes 2001).

Hawkes' discourse is connected to the policy field of "cultural planning", which emerged in the 1980's in the US in urban and regional policy, and was further articulated in Europe from the 1990's onwards as a trans-sectoral approach to policy-making, in particular by Franco Bianchini (1993). It is also connected to "community cultural development", which is a North American term for policies supporting community arts and the cultural practices of communities under a logic of "cultural democracy" and human development (see Adams and Goldbard 2001). Both policy fields are based on a rather wide (anthropological) definition of culture as way of life and of cultural practices as rooted in the everyday lives of local communities.

The fourth pillar discourse was taken up by other protagonists, including "United Cities and Local Governments" (UCLG), an international organization federating local governments, which since the mid-2000's has been advocating (with its "Agenda 21 for Culture") for the strategic positioning of culture (and culture-related policies) as a "fourth pillar" for sustainable development in its own right, i.e. not only as an instrument for achieving goals related to the ecological/environmental, social and economic "pillars" but also as an own domain, acknowledging cultural goals for human development (UCLG 2004). Indeed, the development of a rich, historically grounded, creative and lively culture is as important to the sustainable development of human societies as a rich and evolving biosphere, an economy that meets human needs, and a society that is just and equitable for diverse social groups.

Another discursive development emerging from the late 1990's onwards – and eventually finding some common ground with the fourth pillar discourse – took the shape of discourses about "cultural sustainability", i.e. investigation/research into, and advocacy for aspects of culture that are deemed worthy of consideration from the perspective of sustainability, as well as research on the ways in which cultural processes and cultural institutions are sustaining human communities and societies, as well as landscapes and other natural-cultural ensembles. The term "cultural sustainability" is also sometimes used for investigations into "sustainable" processes and effects located within "the cultural" field – which may lead back to a deleterious fragmentation and re-disciplining process, defeating the transdisciplinary ambitions of sustainability research. Meanwhile, some authors have also used the German equivalent of the expression "cultural sustainability", i.e. "kulturelle Nachhaltigkeit" (see e.g. Krainer and Trattnigg 2007), but in a way that conceptually bears more resemblance to the discourses of "culture(s) of sustainability" that I discuss later.

These sets of discourses stimulated the emergence of a research field on "cultural sustainability", hosting a diversity of themes, topics and approaches (from memory studies to speculative fiction, making links to pre-existing fields such as ecocriticism, geography and various branches of humanities and social sciences). As a result, the expression "cultural sustainability" has thus been used in varied ways over the past decade, and although no consensus emerged on one definitional framework, some international research networks have attempted to sort out the multiple understandings of cultural sustainability, such as e.g. the COST Action network "Investigating cultural sustainability" (Dessein et al. 2015).

In the meantime, while continuing to pragmatically use the term "pillar", Hawkes noted the difficulties that come with the use of such "a limiting and misleading metaphor" as that of pillars of sustainability/sustainable development, reckoning that "lens, framework, dimension or perspective offer a much clearer impression of the proposed applications of this conception" (Hawkes 2010: n.p.). A comparable direction, rejecting the sturdy symbolism of "pillars", had already been taken up in the late 1990's in Germany by several authors associated with the "integrative" definition of sustainable development, who explicitly rejected a "pillarized" discourse on sustainability and called instead for a recognition of a "cultural dimension of sustainable development" (Stoltenberg and Michelsen 1999; Holz and Muraca 2010; Holz and Stoltenberg 2011; Holz 2016). This discourse is related to the field of "Education for Sustainable Development", whereby the importance of cultural education and of a transdisciplinary, integrative approach to education, are heralded by the advocates of "the cultural dimension". According to this conception, one can never understand the integrative challenge of sustainability/sustainable development as long as one focuses solely on single dimensions taken separately from one another (be they environmental/ecological, economic, social, cultural or other dimensions). Indeed, the very question of sustainable development is one of a complex integration of different dimensions. Thus, any non-integrative conceptualization of sustainability that merely perpetuates the traditional modernist juxtaposition of sectorally conceived policies is a basic misunderstanding of the very question at hand.

In parallel to these developments, discourses also emerged that advocated and described elements, whether historical, contemporary or foreseen, of a culture/cultures of sustainability (Rowson 1997; Worts 2006; Nadarajah 2007; Brocchi 2008; Kagan 2010). Another term used by a comparable strand of discourse is "ecocultures" (Slack and Whitt 1992; Ivakhiv 1997; Böhm et al. 2014). The "culture(s) of sustainability" approach aims to identify the characteristics of cultures that are able to evolve and sustain human development in challenging environments, and especially able to learn from crises and transform themselves accordingly (Kagan 2012).

Discourses surrounding "cultures of sustainability" understand "culture" in the sense of a shared set of world views and value systems, and a shared symbolic universe. They argue that the move toward culture(s) of sustainability is comprised of several ethical and normative elements:

> a fundamental shift in contemporary culture away from a hyper-consumption oriented, hyper-industrialised, hyper-modern culture and towards a culture infused with an understanding of and a respect for life in all its complexity; a culture empowering people to change their lives in order to re-invent another, more sustainable "good life" that is inclusive of human groups until now oppressed or disadvantaged.
>
> (Kagan and Kirchberg 2016: 1490)

They also identify an epistemological quality to such a cultural change, which involves developing a set of richer, more diversified and integrated skills, competences, capabilities, reflexivities and ways of knowing reality as a basis for re-inventing possible futures in an inter- and transdisciplinary search process (Dieleman 2008; Kagan 2011; Eernstman et al. 2012; Shrivastava 2012; Eernstman and Wals 2013). Social creativity (i.e. combining individual and collective creativity – cf. Montuori and Purser 1999), serendipity (Kagan 2012) and emergent cultures are central to the cultural change process advocated for in this discourse because "we need to learn to deal with complexity and uncertainty rather than learning a predetermined 'sustainable' set of values and behaviors" (Sandri 2013: 767).

Discourses on culture and sustainability have also been differentiated according to the normative frameworks guiding them. One attempt in this direction was made by Soini and Birkeland (2014) who conducted a systematic discourse analysis of peer-reviewed articles including the term "cultural sustainability" in the period 1997–2011 (i.e. focusing only on one of the key expressions I discussed earlier), and identified four divergent frameworks (conservative, neoliberal, communitarian and environmentalist). However, the discourses referred to previously do not necessarily exclude each other and often exist in hybrid forms, whereby the different expressions are often used interchangeably (also causing confusion). Common across most of the discourses on culture and sustainability is the thesis that sustainability transformation, across and beyond all its "dimensions" or "pillars", requires wide-ranging cultural transformations (see e.g. Hawkes 2001; Brocchi 2008; Kagan 2011; Holz and Stoltenberg 2011). Furthermore, these discourses generally share the view (clearly articulated in Hawkes 2001) that sustainability goals should include the vitality of cultural and artistic expression in their definition of diversity, allowing for a rich cultural life and guarding against cultural homogenization (as could unfortunately happen under the goals of economic or ecological "efficiency").

Culture, sustainability and the arts

All of the four directions introduced earlier (and most of the authors cited above) do involve the arts in their discussion of the relations of culture and sustainability, albeit in partly different ways and with different foci. The fourth pillar discourses usually advocate for the importance of renewed and extended support for the arts through strategically central cultural policies that would allow professional arts organizations as well as community arts initiatives to contribute to community cultural development. The cultural sustainability discourses often add a specific focus on the preservation and safeguarding of cultural heritage (also in relation to landscape preservation) and on cultural memory.

Discourses on the cultural dimension of sustainability, for their part, often stress the benefits of cultural education, including artistic, as part of an integrated approach to education for sustainable development, as well as the contributions of the arts to transversal and transdisciplinary thinking, ethical-reflection and research for sustainability. In a similar vein, the discourses advocating cultures of sustainability tend to focus on artistic research, arts-based research and artful knowing rooted in a variety of arts practices, as well as the role aesthetics plays as a form of experience (after John Dewey) and site of reflexivity (Kagan 2017). They are thereby striving toward new transdisciplinary qualities in the search for viable and desirable futures (Dieleman 2017). In this respect, some discourses on cultures of sustainability have focused their attention on a range of creative cultural practices: artistic practices rooted in the socio-ecologically transformative ambitions of "social sculpture" (initiated by Joseph Beuys and further developed by others such as Shelley Sacks at Oxford Brookes University); "ecological art", with its complex aesthetic, cultural and research-like qualities (Kagan 2011, 2014); the cultural-ecological reflexive practice of ecocriticism in relation to literature and cinema (cf. Zapf 2016a, 2016b); and the political potential of recent "social practice" arts.

With regard to the part that I played in these discourses on the "cultural dimension" and on "cultures of sustainability", I have focused on the need to cultivate an "aesthetics of complexity" (Kagan 2011) and a culture of qualitative complexity as a core element for (inter)personal, social and civilizational developments. In my analysis, any culturally meaningful approach to sustainability should work with a "procedural" definition of sustainability (Miller 2011) where "sustainability is the emergent property of a discussion about desired futures" (Robinson in Miller 2011: 31), not a Brundtlandian "universalist" definition that aims to merely determine standards that can be agreed upon at the level of international organizations. A procedural definition recognizes and works with the unavoidable and necessary conditions of emergence, unpredictability, uncertainty and "situated knowledges" (Haraway 1988). Unlike universalist sustainability, it does not level out cultural difference but instead engages in a process of articulation, (re)interpretation and negotiation of cultural difference with respect to questions posed by global sustainability challenges (cf. Bhabha 1994). This then implies a departure from the illusion of direct control over the complex processes at play. (This analysis of mine is, however, not necessarily shared across the whole range of discourses relating culture and sustainability that I discussed earlier.)

In its procedural definition, sustainability is about re-inventing worlds, and is thus primarily a cultural project. The importance of imagination becomes salient, as well as the interface of memories and futures in the production of social imaginaries and in the constant (re)negotiation of diverse goals and priorities. Nevertheless, sustainability goals, even under a procedural definition, are not completely free-floating, nor reducible to a "culturalist" dimension. Rather they are bound to (1) consider all the dimensions of sustainable development and especially the interrelations (and contradictions) between them, and (2) acknowledge certain planetary conditions about which a wide scientific consensus exists, such as climate change and the massive extinction of species/biodiversity loss.

Consequently, I hold that the procedural definition of sustainability is best associated with complexity-based "culture(s) of sustainability". The "culture(s) of sustainability" approach offers a distinct way to deal with a crucial dilemma in the relationship between culture and sustainability, which is the relation between normative prioritization and cultures. Normative prioritization is a general question facing the diverse discourses of sustainability/sustainable development: On the one hand, the infrastructural dependence of humanity on the planetary biosphere demands a top-prioritization of the ecological dimension of sustainability, whereas the economy should be considered as a means to achieve ecological, cultural and social sustainability before considering economic ends in themselves. Hence Niko Paech's argumentation in favour of a prioritization whereby the ecological dimension is granted highest priority and the economic dimension lowest priority, i.e. the exact opposite of contemporary effective prioritization in international political practice (Paech 2006). On the other hand, this prioritization should not annul the importance of manifold normative goals and the need to strike balanced compromises between conflicting goals. As pointed out by Robert Hauser and Gerhard Banse,

> if values such as life, infant mortality, nutrition, or the ability to adapt to changing environmental conditions are used as indicators of sustainability, the fictitious people living close to nature probably would not compare well with Western cultures.
>
> (Hauser and Banse 2011: 45–46)

Furthermore, the very normative value-construction of sustainability (with its principles of justice(s) and of responsibilities) is culturally determined (Hauser 2011), as are all interpretive frameworks by which we come to know reality. Therefore, in order to be able to conceptualise the mutual grounding of the ecological and cultural dimensions in one another, a meta-level must be reached with a higher degree of complexity, where concepts no longer suffice, but "macro-concepts" (Morin 1977) come into play that allow a "dia-logic" (in Morin's sense, not "dialogic" in Bakhtin's sense) intellectual culture to emerge beyond the limitations of both logical positivism/critical rationalism and dialectics, and beyond the opposition of naturalism, culturalism and radical social constructivism. While there exist no "sustainable cultures" per se, as Hauser and Banse (and many others) argue, a complex, transdisciplinary (in the sense employed by Basarab Nicolescu) and uniplural culture(s) *of* sustainability is called forward that would allow both a planetary naturecultural co-piloting of the biosphere and a historically (and humanities- and social-scientifically) informed approach to cultural complexity.

The imaginative yet scientifically informed process that is called forward, one that is cultural yet recognizant of the non-human, entails a challenging and often uncomfortable search. It is a search which cannot succeed without the development of an acute sensitivity toward the manifold transdisciplinary "patterns that connect" (Kagan 2011) different dimensions of our planetary interdependencies. Training the capacity to perceive these patterns and to embrace a complex world without fear (and without the subsequent simplification that would result in ideological impasses) is what I have been pointing towards with the notion of an "aesthetics of complexity" (Kagan 2011, 2015a), which draws from John Dewey's and Gregory Bateson's previous works on aesthetics together with Edgar Morin's paradigm of complexity. Deweyian aesthetics points at personal affectivity in everyday life experience and at a human being's overall interrelationship with his/her environments. I associate it with Batesonian aesthetics as sensibility to the "pattern that connects", i.e. a pre-ethical yet already transdisciplinary perceptivity for the linking patterns across different levels of reality, which can be further informed by insights from the arts and sciences. An aesthetics of complexity, as a co-evolving movement of the self in contact with the movements of the living world (cf. Merleau-Ponty 1969; Ingold 2011; Weber 2013), forms a basis for ethics, knowing and action through dynamic connections with a complex world that involve the body-mind through multisensorial (Pink 2015), emotional and cognitive processes in integrated ways. Complex interdependencies, tensions and interpenetrations between culture and nature are thereby perceived, involving cooperation, competition, antagonism and unity through mutual constitution rather than a simple dichotomic opposition (or a likewise simplistic, "holistic" unification of nature and culture).

From the perspective of an aesthetics of complexity, the arts and especially art as a form of enquiry and (re)search process into the qualitative nuances of the human experience of the world, bear special qualities. Arts-based approaches to enquiry/research elicit unusual ways of thinking about social and natural phenomena, through the stimulation of uncertainty, risk-taking, and confrontation beyond superficial and taken-for-granted understandings and meanings, "broadening and deepening conversations" (Savin-Baden and Wimpenny 2014: 79). An arts-based approach opens up new ways of asking questions and uncovers new questions to be asked (Leavy 2009: 12). It aims to make questions and enquiry more interesting, to "stimulate problem formulation" (Barone and Eisner 2012: 171), rather than to directly and unequivocally answer research questions and offer some "definitive" meanings, as it "revisits the world from a different direction, seeing it through fresh eyes" (Barone and Eisner 2012: 16).

Abandoning the claim to produce universal knowledge, arts-based research generates multiple perspectives on the questions it poses, which are rooted in multiple "attentions" (Mersch 2009: 37–38). Those attentions address complex and subtle interactions and they make them noticeable in the first place. This deepens our understanding of issues and makes it more (qualitatively) complex

(Barone and Eisner 2012: 3). An "artful sustainability research" (Kagan 2017), integrating the qualities of artistic and arts-based research briefly evoked here, thus forms a desirable horizon for sustainability research and sustainability-oriented activism. Examples of the manifold insights and resulting innovative perspectives that can result from the involvement of the arts in the search process of sustainability already abound, for example in the works of ecological artists over the past four decades (cf. Blanc and Ramos 2010; Kagan 2011, 2012, 2014; Weintraub 2012; Blanc and Benish 2016; Mayer Harrison and Harrison 2016; Demos 2016).

Procedural sustainability as a cultural project affects both the professional worlds of culture and the arts, and the worlds of sustainability research and advocacy/activism. It integrates arts-based enquiry into an artful sustainability research agenda while re-inventing worlds and rooting itself in an aesthetics of complexity. This results in potential opportunities and responsibilities for the professional "cultural sector" (i.e. the fields of cultural expression and arts production). Organizations working in this sector are potential bearers of "spaces of possibility" where alternatives can be imagined, experimented with, designed and directly experienced (Kagan 2016).

> This is not just about professional artists (who of course can be very inspiring initiators), or about artists in social practice and communities (also playing essential roles), but it is also about sharing response-ability for more diffused artful doing and learning by local communities in spaces of challenging experience, imagination and experimentation.
>
> (Kagan 2015b: 29)

Arts organizations deal directly with meaning-making and structures of meaning, with world views and values, and with the histories and lives of cultural artefacts and intangibles (practices, traditions, non-written transmissions, etc.). Thereby, they contribute to memories and changes in the symbolic universe that we build and inhabit, as well as to place-making in the locations in which we dwell. They thus hold a share of responsibility regarding the ways in which individuals, communities and societies (do not) relate to their environments, from local to planetary scales.

Procedural sustainability as a cultural project poses a challenge to the business-as-usual of sustainability research (including the neo-discipline of sustainability science) and of sustainability advocacy, which both need to place social imaginaries at the centre of their attention and imagination at the core of their practice. Sustainability research and advocacy then need to integrate scientific and arts-based forms of research, learning and expression. Sustainability research requires a major qualitative shift in its transdisciplinary gravity point, away from its current claim to form a "sustainability science" as a kind of neo-discipline and towards the formation of an integrated research practice of artful sustainability.

Only a few of the leading researchers in this field have already realized this need and acted upon this realization. One of them is John Robinson, who was an early proponent of procedural sustainability as "a kind of discursive playing field in which [conflicting views] can be debated" and as "the emergent property of a conversation about what kind of world we collectively want to live in now and in the future" (Robinson 2004: 382). As Provost for Sustainability of the University of British Columbia, in Vancouver, Robinson launched in 2014 the interdisciplinary project "Sustainability in an Imaginary World" that developed an arts-based platform to collectively explore visions of possible futures with the hope of stimulating "deeper ontological and epistemological questions" on worlds-making, ideally "challenging ontological presuppositions" (Bendor et al. 2015: 54–55).

Another current project integrating artful sustainability is the "City as Space of Possibility" project at the Leuphana University Lüneburg (led by Volker Kirchberg, Ute Stoltenberg, Ursula Weisenfeld and myself). Our project investigates creative and innovative approaches to sustainable urban development in the city of Hanover, Germany. We look for physical, social and mental spaces where potentially sustainable futures are already taking some shape, emerging and experienced in today's urban society – involving networks or/and coalitions of various urban actors including artists and other cultural professionals and creative practitioners (and 'life-artists'), civil society and social movements, educationalists, local entrepreneurs, city-government administrators and policy-makers (Kagan et al. 2018). Other sustainability research projects incorporating an artful sustainability are being developed around the world (e.g. those by Hans Dieleman in Mexico, Karen O'Brien in Oslo etc.). Early-career researchers pursuing this approach are starting to organize themselves (e.g. at the international Workshop-Conference "Realizing Potentials: Conversations and Experiments at the Frontier of Art-based Sustainability" held in Barcelona in November 2016).

Conclusion

Over the last two to three decades, a variety of discourses around sustainable development and sustainability have introduced and discussed culture in multiple ways. Beyond the differences, contradictions and confusions between different discourses (which I clustered in this text into four approaches: culture as a fourth pillar of sustainability, cultural sustainability, the cultural dimension of sustainability/sustainable development, and culture(s) of sustainability), certain features know a near-consensus among many (if not most) researchers: the importance of a rich and lively cultural heritage and cultural development for the vitality of human societies, and the need for wide-ranging cultural transformations in contemporary societies and polities that are excessively ruled by a narrowly conceived mainstream economic rationality. Most of these discourses also intersect by or converge with approaches relating the arts and aesthetics to theoretical and practical aspects of sustainability.

Some researchers stressed in particular the importance of social imaginaries (as pools of cultural resources), and of the human imagination (as a generative principle) as engines for potential social transformations. This direction can be engaged with through procedural sustainability, understood as a cultural project.

Furthermore, to deal with the complexity of sustainability's multiple challenges, a cultural continuous-learning process rooted in an aesthetics of complexity allows an embrace of qualitative complexity, rather than a simplifying response strategy reinforcing illusions of control, as do certain discourses of the so-called "anthropocene" and certain solutionist sustainability discourses.[2] Within this process, artistic and arts-based practices of (re)search offer opportunities to both develop imaginative approaches and deepen the qualitative enquiry into complexity. Cultural-academic and artistic-discursive reflections and their critiques of social developments, including critiques of sustainability and sustainable development, may sometimes allow a more challenging and thus more diverse and deeper reflective process than do the usual academic fora of sustainability research. However, a more detailed discussion of this aspect would fall outside the space available for this text.

In this sense, the discourses and research investigating the relation between culture, the arts and sustainability/sustainable development not only constitute a valuable sub-theme for specialized researchers and for cultural practitioners interested in this intersection. They also offer a crucial opportunity for sustainability research to realize the potential of a deeper epistemological programme of transdisciplinarity, in order to help re-orient human development with more inventive, imaginative and challenging reflective approaches and practices, beyond the limitations of an engineering of sustainability-solutions into which sustainability science may otherwise trap itself.

Notes

1 For an overview of the development of policy discourses on culture and sustainability over the 1990's and 2000's, see Duxbury and Jeannotte (2010).

2 A discussion and critique of these complexity-obscuring trends lies beyond the scope of this text.

References

Adams, D. and Goldbard, A. (2001). *Creative Community: The Art of Cultural Development*. New York: Rockefeller Foundation.

Barone, T. and Eisner, E. W. (2012). *Arts Based Research*. Thousand Oaks: Sage.

Barthel-Bouchier, D. (2012). *Cultural Heritage and the Challenge of Sustainability*. Walnut Creek, CA: Left Coast Press.

Bendor, R. et al. (2015). "Sustainability in an Imaginary World". *Interactions*, XXII.5 (September–October 2015), 54–57. http://interactions.acm.org/archive/view/september-october-2015/sustainability-in-an-imaginary-world [Accessed 15 September 2016]

Bhabha, H. K. (1994). *The Location of Culture*. London: Routledge.

Bianchini, F. (1993). *Urban Cultural Policy in Britain and Europe: Towards Cultural Planning*. Brisbane, QLD: Institute for Cultural Policy Studies, Griffith University.

Blanc, N. and Benish, B. L. (2016). *Form, Art, and Environment: Engaging in Sustainability*. London: Routledge.

Blanc, N. and Ramos, J. (2010). *Écoplastie: art et environnement*. Paris: Manuella.

Böhm, S., Pervez Bharucha, Z., and Pretty, J. (eds.) (2014). *Ecocultures: Blueprints for Sustainable Communities*. London: Routledge.

Brocchi, D. (2008). "The Cultural Dimension of Sustainability". In S. Kagan and V. Kirchberg (eds.). *Sustainability: A New Frontier for the Arts and Cultures*. Frankfurt am Main: VAS – Verlag für akademische Schriften, 26–58.

Davies, W. K. D. (2015). "Background to Sustainable Cities". In W. K. D. Davies (ed.). *Theme Cities: Solutions for Urban Problems*. Dordrecht: Springer Netherlands, 151–205.

Demos, T. J. (2016). *Decolonizing Nature: Contemporary Art and the Politics of Ecology*. Berlin: Sternberg Press.

Dessein, J. et al. (2015). *Culture in, for and as Sustainable Development. Conclusions from the COST Action IS1007 Investigating Cultural Sustainability*. Jyväskylä: University of Jyväskylä. www.culturalsustainability.eu/conclusions.pdf [Accessed 26 July 2017].

Dieleman, H. (2008). "Sustainability, Art and Reflexivity: Why Artists and Designers May Become Key Change Agents in Sustainability". In S. Kagan and V. Kirchberg (eds.). *Sustainability: A New Frontier for the Arts and Cultures*. Frankfurt am Main: VAS – Verlag für akademische Schriften, 108–146.

Dieleman, H. (2017). "Transdisciplinary Hermeneutics: A Symbiosis of Science, Art, Philosophy, Reflective Practice, and Subjective Experience". *Issues in Interdisciplinary Studies*, 35, 172–201.

DifU (Deutsches Institut für Urbanistik) (2011). *Städte für ein nachhaltiges Deutschland. Gemeinsam mit Bund und Ländern für eine zukunftsfähige Entwicklung*. Berlin: DifU.

Duxbury, N. and Jeannotte, M. S. (2010). *Culture, Sustainability, and Communities: Exploring the Myths*. Working paper no. 353. Coimbra: Oficina do CES, Universidade de Coimbra.

Eernstman, N. et al. (2012). "Inviting the Unforeseen: A Dialogue about Art, Learning and Sustainability". In A. E. J. Wals and P. B. Corcoran (eds.). *Learning for Sustainability in Times of Accelerating Change*. Wageningen: Wageningen Academic, 201–212.

Eernstman, N. and Wals, A. E. J. (2013). "Locative Meaning-making: An Arts-based Approach to Learning for Sustainable Development". *Sustainability*, 5 (4), 1645–1660.

Haraway, D. (1988). "Situated Knowledges: The Science Question in Feminism and the Privilege of Partial Perspective". *Feminist Studies*, 14 (3), 575–99.

Hauser, R. (2011). "Implicit Normative Settings in Concepts of Sustainability and Cultural Diversity: Solving the Antagonism". In G. Banse, G. L. Nelson, and O. Parodi (eds.). *Sustainable Development – The Cultural Perspective: Concepts – Aspects – Examples*. Berlin: edition sigma, 159–166.

Hauser, R. and Banse, G. (2011). "Culture and Culturality: Approaching a Multi-faceted Concept". In G. Banse, G. L. Nelson, and O. Parodi (eds.). *Sustainable Development – The Cultural Perspective: Concepts – Aspects – Examples*. Berlin: edition sigma, 31–50.

Hawkes, J. (2001). *The Fourth Pillar of Sustainability: Culture's Essential Role in Public Planning*. Common Ground Publishing, Champaign, Illinois.

Hawkes, J. (2010). *Framework of 4.1*. http://community.culturaldevelopment.net.au//CHART FrameworkOf4.1.html [Accessed 15 March 2017].

Holz, V. (2016). *Bildung für eine nachhaltige Entwicklung: Kulturwissenschaftliche Forschungsperspektiven*. Opladen: Verlag Barbara Budrich.

Holz, V. and Muraca, B. (2010). „'Kultur' als Herausforderung für die Nachhaltigkeitsforschung". Bericht vom 7. Workshop des Forums on Sustainable Technological Development in a Globalising World "Sustainability 2010: The Cultural Dimension". In Institut für Technikfolgenabschätzung und Systemanalyse (eds.). *Technikfolgenabschätzung, Theorie und Praxis* 19(3). Karlsruhe: ITAS, 120–124.

Holz, V. and Stoltenberg, U. (2011). „Mit Dem Kulturellen Blick Auf Den Weg Zu Einer Nachhaltigen Entwicklung". In G. Sorgo (ed.). *Die Unsichtbare Dimension. Bildung*

Für Nachhaltige Entwicklung Im Kulturellen Prozess. Wien: Forum Umweltbildung im Umweltdachverband, 15–34.

Ingold, T. (2011). *Being Alive: Essays on Movement, Knowledge and Description*. London: Routledge.

Ivakhiv, A. (1997). "Ecocultural Critical Theory and Ecocultural Studies: Contexts and Research Directions". In *Cultures and Environments: On Cultural Environmental Studies* [online conference], American Studies Program, Washington State University, 20–22 June 1997. www.uvm.edu/~aivakhiv/eco_cult.htm [Accessed 20 October 2015].

James, P. (2015). *Urban sustainability in theory and practice: Circles of sustainability*. Abingdon: Routledge.

Kagan, S. (2010). "Cultures of Sustainability and the Aesthetics of the Pattern that Connects". *Futures: The Journal of Policy, Planning and Futures Studies*, 42(10), 1094–1101.

Kagan, S. (2011). *Art and Sustainability: Connecting Patterns for a Culture of Complexity*. Bielefeld: transcript Verlag.

Kagan, S. (2012). *Toward Global (Environ)mental Change: Transformative Art and Cultures of Sustainability*. Berlin: Heinrich Böll Stiftung.

Kagan, S. (2014). "The Practice of Ecological Art". *Plastik*, 4. http://plastik.univ-paris1.fr/document.php?id=866 [Accessed 26 July 2017].

Kagan, S. (2015a). "Complexity as Experience: The Contribution of Aesthetics to Cultures of Sustainability". In C. Crouch, N. Kaye and J. Crouch (eds.). *An Introduction to Sustainability and Aesthetics: the Arts and Design for the Environment*. Boca Raton: Brown-Walker Press, 25–32.

Kagan, S. (2015b). "Prefiguring Sustainability: Response-Ability and Spaces of Possibility". In M. Le Sourd and H. Van Den Bergh (eds.). *Art for the Planet's Sake: Arts and Environment* (Fresh Perspectives, vol. 4). Brussels: IETM, 29–32. www.ietm.org/en/system/files/publications/ietm-art-for-the-planets-sake_jan2016_0.pdf [Accessed 15 April 2017]

Kagan, S. (2016). "The Emergence of Creative Sustainable Cities". In M. Ziehl, C. Rabe and T. Haupt (eds.). *City Linkage: Art and Culture Fostering Urban Futures*. Berlin: Jovis Verlag, 21–29.

Kagan, S. (2017). "Artful Sustainability: Queer-Convivialist Life-Art and the Artistic Turn in Sustainability Research". *Transdisciplinary Journal of Engineering & Science*, 8: 151–168.

Kagan, S. and Kirchberg, V. (2016). "Music and Sustainability: Organizational Cultures towards Creative Resilience – A Review". *Journal of Cleaner Production*, 135: 1487–1502.

Kagan, S., *et al.* (2018). "Culture in Sustainable Urban Development: Practices and Policies for Spaces of Possibility and Institutional Innovations". *City, Culture and Society*, 13: 32–45.

Krainer, L. and Trattnigg, R. (eds.). (2007). *Kulturelle Nachhaltigkeit: Konzepte, Perspektiven, Positionen*. Munich: Oekom Verlag.

Leavy, P. (2009). *Method Meets Art: Arts-based Research Practice*. New York: Guilford Press.

Lehmann, S. (2010). "Green Urbanism: Formulating a Series of Holistic Principles". *S.A.P.I.EN.S*, 3(2). https://sapiens.revues.org/1057 [Accessed March 15 2017].

Mayer Harrison, H. and Harrison, N. (2016). *The Time of the Force Majeure: After 45 Years Counterforce is on the Horizon*. Munich: Prestel.

Merleau-Ponty, M. (1969). *La prose du monde*. Paris: Gallimard.

Mersch, D. (2009). "Kunst als epistemische Praxis". In E. Bippus (ed.). *Kunst des Forschens: Praxis eines ästhetischen Denkens*. Zurich: Diaphanes, 27–47.

Meuleman, L. (2013). "Cultural Diversity and Sustainability Metagovernance". In L. Meuleman (ed.). *Transgovernance*. Berlin: Springer Netherlands, 37–81.

Miller, T. R. (2011). *Constructing Sustainability. A Study of Emerging Scientific Research Trajectories*. PhD Thesis. Phoenix: Arizona State University.
Montuori, A. and Purser, R. E. (1999). *Social Creativity*. New York: Hampton Press.
Morin, E. (1977). *La méthode, vol. 1: la nature de la nature*. Paris: Seuil.
Nadarajah, M. (2007). "Culture of Sustainability: Multicultural Reality and Sustainable Localism – A Case Study of Penang (George Town), Malaysia". In M. Nadarajah and A. T. Yamamoto (eds.). *Urban Crisis: Culture and the Sustainability of Cities*. Tokyo: United Nations University Press, 107–133.
Paech, N. (2006). "Nachhaltigkeitsprinzipien jenseits des Drei-Säulen-Paradigmas". *Natur und Kultur* 7, 42–62.
Pink, S. (2015). *Doing Sensory Ethnography*. 2nd ed. London: Sage.
Robinson, J. (2004). "Squaring the Circle? Some Thoughts on the Idea of Sustainable Development". *Ecological Economics*, 48, 369–384.
Rowson, J. M. (1997). *Community-based Sustainability and the construction of Difference on Galiano Island, British Colombia*. Vancouver: University of British Columbia.
Sandri, O. J. (2013). "Exploring the Role and Value of Creativity in Education for Sustainability". *Environmental Education Research*, 19 (6), 765–778.
Savin-Baden, M. and Wimpenny, K. (2014). *A Practical Guide to Arts-related Research*. Rotterdam: SensePublishers.
Shrivastava, P. (2012). "Enterprise Sustainability 2.0: Aesthetics of Sustainability". In P. Bansal, A. J. Hoffman (eds.). *The Oxford Handbook of Business and the Natural Environment*. Oxford: Oxford University Press, 630–638.
Slack, J. D. and Whitt, L. A. (1992). "Ethics and Cultural Studies". In L. Grossberg (ed.). *Cultural Studies*. New York: Routledge, 571–590.
Soini, K. and Birkeland, I. (2014). "Exploring the Scientific Discourse on Cultural Sustainability". *Geoforum*, 51, 213–223.
Stoltenberg, U. and Michelsen, G. (1999). „Lernen nach der Agenda 21: Überlegungen Zu Einem Bildungskonzept Für Eine Nachhaltige Entwicklung". In U. Stoltenberg, G. Michelsen and J. Schreiner (eds.). *Umweltbildung – Den Möglichkeitssinn wecken*. Schneverdingen: Alfred-Toepfer-Akademie für Naturschutz, 45–54.
UCLG (United Cities and Local Governments) (2004). *Agenda 21 for Culture*. Barcelona: UCLG. www.agenda21culture.net/ [Accessed 20 October 2015].
UN Habitat (2013). *Planning and Design for Sustainable Urban Mobility: Global Report on Human Settlements 2013 (Full Report)*. Nairobi: UN Habitat. http://unhabitat.org/books/planning-and-design-for-sustainable-urban-mobility-global-report-on-human-settlements-2013/ [Accessed 20 October 2015].
UNESCO (2013a). *Creative Economy Report 2013*. Paris: UNESCO. http://unesco.de/kultur/2013/creative-economy-report-2013.html [Accessed 27 September 2016].
UNESCO (2013b). *Create Cities Network: Mission Statement*. Paris: UNESCO. https://en.unesco.org/creative-cities/sites/creative-cities/files/Mission_Statement_UNESCO_Creative_Cities_Network.pdf [Accessed 27 September 2016].
Weber, A. (2013). *Enlivenment: Towards a Fundamental Shift in the Concepts of Nature, Culture and Politics*. Berlin: Heinrich Boell Stiftung.
Weintraub, L. (2012). *To life! eco art in pursuit of a sustainable planet*. Berkeley: University of California Press.
Worts, D. (2006). "Fostering a Culture of Sustainability". *Museums & Social Issues*, 1(2), 151–172.
Zapf, H. (2016a). *Literature as Cultural Ecology*. London: Bloomsbury.
Zapf, H. (ed.) (2016b). *Handbook of Ecocriticism and Cultural Ecology*. Berlin: De Gruyter.

12 Cultural ecology and the sustainability of literature

Hubert Zapf

Introduction: literature as cultural ecology

Let me start by outlining some of the main assumptions of a cultural ecology of literature.[1] Between an anthropocentric cultural studies perspective, in which nature is dematerialised into a discursive human construct, and an ecocentric naturalism, in which cultural processes are simply subsumed under naturalist premises, cultural ecology looks at the interaction and living interrelationship between culture and nature, without reducing one to the other. Literature is seen as a cultural form in which this living interrelationship is expressed and explored in specifically productive ways, providing a site of critical self-reflection of modern civilisation as well as a source of creative cultural self-renewal.

One of the primary theoretical references is Gregory Bateson's *Steps to an Ecology of Mind* (1973), which explores what he calls 'connecting patterns' between mind and life, expanding the mind of the individual ego toward an ecology of vital interrelations with other minds and with their material as well as historical-cultural environments. These patterns are shared across different epistemic and cultural domains and do not designate fixed properties of given realities but non-linear, emergent processes characterised by interacting networks and recursive feedback relations. Another key reference is Peter Finke's notion of 'cultural ecosystems' (2006) that he develops from Bateson's ecology of mind and from Jakob von Uexkuell's distinction between *Umwelten* and *Innenwelten*, between external environments and internal worlds. Finke points out that the characteristic environments of human beings are not just external but internal environments, the inner worlds and landscapes of the mind, the psyche and the cultural imagination which follow their own metabolic processes of energy transformation and make up the habitats of humans as much as their external natural and material environments. Literature, art and other forms of cultural creativity are, in this view, essential to increase the richness, complexity and diversity of those cultural ecosystems and to ensure their continuing evolutionary potential for self-correction and self-renewal.

In this more specific sense, I am arguing that literature can itself be described as the symbolic medium of a particularly potent form of 'cultural ecology.' This involves the assumption that literature is not only a preferred site for complex representations of the culture-nature relationship but that, in its aesthetic transformation of experience, it acts like an ecological force in the larger cultural system. Literary texts have staged and explored the complex interactions between culture and nature in ever new scenarios, and have derived their specific power of innovation and cultural self-renewal from the creative exploration of this boundary. Since the beginnings of literature, cultural narratives of personified natural forces, human-animal symbioses, hybrid trickster figures, and nature-culture metamorphoses have shaped the literary imagination, and have supplied body- and nature-connoted alternatives to instrumental reason and to an increasingly self-referential anthropocentric civilisation. As Louise Westling (2006) among others has pointed out, the 'human animal dance,' i.e. the performative staging of a close symbiogenetic interaction between human and nonhuman life forms, has shaped literary narratives from archaic to modern times, from the Gilgamesh epic to Virginia Woolf. The liminal space between the wild and the domesticated became a generative principle, inscribing the fundamental ecological interrelationship between culture and nature into texts. The dance of life was translated into a dance of words, and in this expressive symbiosis of wildness and language, ecosemiotic communication in literary narratives evolved as a cultural analogue to communicational energies and autopoetic processes of life itself, as Wendy Wheeler (2016) has pointed out so persuasively in her recent work.

This attention to the life-sustaining significance of the mind/body and culture/nature interaction became especially prominent in the era of romanticism, but continues to be characteristic of literary stagings of human experience up to the present. The aesthetic mode of textuality involves an overcoming of the mind-body dualism by bringing together conceptual and perceptual dimensions, ideas and sensory experiences, reflective consciousness and the performative staging of complex dynamical life processes. Literature as a medium of cultural ecology thus specifically focuses on this interactivity of mind and life as a liminal phenomenon on the boundary between culture and nature, self and other, autopoetic and ecopoetic processes (Zapf 2002, 2016a, 2016b, 2017). Literary texts provide a transformative site of cultural self-reflection and cultural self-exploration, in which the historically marginalised and excluded is semiotically empowered and activated as a source of artistic creativity, and is thus reconnected to the larger cultural system in both deconstructive and reconstructive ways. As the medium of a potentially radical civilisational critique, literature simultaneously provides a sustainable generative matrix for the continuous self-renewal of the cultural ecosystem.

Sustainability and literature

The relation of sustainability to literature has only recently begun to be explicitly addressed. Apart from occasional contributions in various journals, a special issue of *American Literary History* has been devoted to the subject, in which the concept of sustainability is examined in a spectrum of contributions addressing topics such as the relation between the humanities and the sciences (D'Arcy Wood

2012); environmental justice aspects of sustainability implied in the 'indigenous cosmopolitics' at play in literary works by indigenous authors (Adamson 2012); and more general epistemological questions on the potential contribution of the humanities to the sustainability discourse – in terms of a critical and self-reflexive metadiscourse on the categories that define it (Philippon 2012). All of these aspects of the relation between literature and sustainability reveal, as Slovic rightly points out, that sustainability is no fixed concept, content or programme, but "a moving target, a distant goal, not a permanently achievable plateau of being" (2012: 187). In a 2012 issue of *PMLA*, the topic was addressed from different critical angles, ranging from a radical critique of the concept as an instrument of corporate greenwashing, to the postulate formulated by Stephanie LeMenager and Stephanie Foote that "we scholars of literary and cultural studies need to claim our stake in sustainability" (2012: 577), and to Lynn Keller's contention that "the arts and the human imagination deployed by the arts have significant roles to play" if, as she puts it, "popular ideas of sustainability are to be reclaimed from the blurry, feel-good realm of corporate advertising and given meaningful, hard edges" (2012: 581). In this sense, it is important, rather than only considering thematic issues and environmental contents, to explore the ways in which art and literary aesthetics itself can be conceived of as a site and medium of cultural sustainability.

This adoption of the term for ecocritical literary studies takes into account that, as Stacy Alaimo observes, the dominance of a "sanitized term of sustainability" that does not "in any way question capitalist ideas of unfettered expansion" creates a problem especially for the humanities and for literary studies, because its adherence to the "gospel of efficiency" involves a preference for disciplines such as "engineering, the sciences, and maybe architecture," at the expense of "philosophical questions, social and political analyses, historical reflections, or literary musings," which seem a waste of time in the face of pressing environmental realities and responsibilities and thus "irrelevant for the serious business of sustainability" (Alaimo 2012: 560). In its adaptation to cultural and humanist studies, then, the discourse of sustainability needs critical questioning and epistemic extension from such a narrow 'techno-scientific focus' towards an inclusion of "issues of human choice involved in putting sustainability into effect and . . . the socio-cultural practices, behaviours, and structures such choice involves" (Goeminne 2011: 20). It requires recognition that "one's very self is substantially interconnected with the world" (Alaimo 2012: 561), and that this entails a "regrounding of the subject in a materially embedded sense of responsibility for the environments he or she inhabits" (Braidotti 2006: 137). Braidotti helpfully differentiates the concept further by adapting Deleuze's processual thought to the idea of "sustainable becoming," relating it to everyday practices of personal life as "transformative micro-practices" (Alaimo 2012: 561), which should be complemented on a collective level by practices of environmental justice and environmental-health activism (561). Sustainability in this cultural sense would then involve the transformation of a "technocratic, anthropocentric perspective" towards "more complex epistemological, ontological, ethical, and political perspectives," which counter the tendency "to externalize and objectify the world" and instead broaden this revised concept of sustainability by incorporating the "lively relationalities of becoming of which we are part" (Braidotti 2006: 393). It is my contention that literature and art provide one medium of cultural representation and communication in which this more complex, self-reflexive and ethically responsive concept of sustainability or "sustainable becoming" (393) is part of its generative potential and transformative function within the larger discursive system of cultural knowledge and semiotic practices.

A provisional definition of the notion of literature as a form of 'sustainable text' would include the following aspects:

(1) a *long-term* perspective of culture-nature co-evolution vs. short-term concerns;
(2) a double orientation on continuity *and* change, on past *and* future, cultural memory *and* cultural creativity;
(3) a sensitivity to the *multi-layered forms of relationality* between self and other, mind and life, humans and the nonhuman world, encompassing perceptual, sensory, emotional, cognitive, spiritual, communicational and creative dimensions;
(4) an attention both to life-sustaining *diversities* and to patterns of *connectivity* across the boundaries of categories, discourses and life forms;
(5) a transactive and participatory concept of sustainability that describes not an objectively given set of properties, but rather a *potentiality* of texts that only comes alive through its ever new actualisations by ever new generations of readers, within always changing historical, social and individual conditions of living across different periods and cultures.

Thinking sustainability is thus by no means opposed to innovation and creativity as such. While sustainability contradicts and endeavours to overcome the currently prevalent short-term, instrumental and profit-driven form of economic and scientific innovation, it necessarily presupposes an alternative, more complex notion of cultural creativity oriented on the long-term survival of cultural and natural ecosystems as interdependent realities enabling the continuation of life on the planet. It must constantly negotiate between the poles of continuity and change, in the imaginative anticipation of possible future scenarios in a newly conceived, democratically legitimated form of cultural evolution, which combines the awareness of its natural-historical conditions of emergence with the consideration of the ethical-ecological consequences of its future agendas. Thinking sustainability in this sense involves "no dogma, no fixed set of formulas and rules that tell you you have to do this and that in order to be sustainable" (Dürr in Grober 2002: 175, my trans.). Rather, it provides "a playing field where one can experiment" (175, my trans.). And it seems to me that art and literature represent a form of cultural creativity that provides important playing fields for such experiments in the interest of long-term cultural evolution.

Extending the frame of sustainability: 'Wildness' and the regenerative energy of literature

To indicate some of the ways in which literature is such an experimental site of sustainable cultural practices, I am addressing in the following the topic of 'wildness' as a non-anthropocentric signifier of creative energy which has pervaded the ecocritical as well as the literary discourse in connection with processes of both natural and cultural creativity.

As an ecological force in culture, literature recurrently draws on narratives and images of the wild, as an area untamed by human regulation and control, in order to activate the critical and creative energies that it circulates in culture. In the poiesis of literary texts, it is a source of radical disruption and defamiliarisation, but also of imaginative counterdiscourses that reconnect the cultural system with its excluded other, which nevertheless turns out to be its deeper condition of survival. Thus wildness, as evoked in the language, metaphors and imaginative scenarios of texts, signifies an ecosemiotic communication that opens existing categories of thought and agency to the living processes of human-nonhuman interaction, which have been subjected to an increasing economic and techno-scientific control in the Anthropocene but are brought out in their own vital dynamics in the aesthetic process.

In American culture, which is my special scholarly field, wildness has always been a prominent topic. In its connection with the more concrete and essentialist concept of wilderness, it has been one of the key terms in classical American Studies, which has been rightly criticised for its ideological implications in terms of gender, race, colonialism and national exceptionalism. It helped to construct an American master narrative that, on the one hand, demonised the wilderness and its human and nonhuman inhabitants as a realm of darkness and evil that had to be eliminated in the march of the nation across the continent; and on the other hand, it ennobled wild nature as the true origin of American democracy and timeless justification of existing power structures.

In American literature, however, these inherent contradictions and ambiguities were themselves becoming the topic of poetic and narrative explorations. In Henry David Thoreau's writings, wilderness and the wild gained shifting meanings, ranging from deep ecology to political ecology, and covering the spectrum from spiritual-empathetic to social-critical connotations of wildness, from wildness as personal freedom to wildness as ecological community of multispecies coexistence. In this spectrum, the environmental imagination in American nature writing has developed as part of the larger cultural discourse on civilisation and nature in the U.S., and as such has certainly served, beyond its ideological shortcomings, as a significant inspiration of environmental awareness and environmental policy.

Textual examples I: American novels

I would like to focus here, however, not on this nonfictional tradition of nature writing but on some of the major fictional works of American literature, in which wildness is used in less conspicuous but nevertheless substantive ways as a source of critical, counterdiscursive and regenerative energy in the narrative process. I am specifically referring to Nathaniel Hawthorne's *The Scarlet Letter* (1986/1850), Herman Melville's *Moby-Dick* (1956/1851), Kate Chopin's *The Awakening* (1993/1899) and Toni Morrison's *Beloved* (1991/1987).

In Hawthorne's *The Scarlet Letter*, 'wild' as an attribute first appears in the "wild rose bush" that grows by the prison door, that "black flower of civilized society" (1986/1850: 76) from which Hester Prynne emerges at the novel's beginning to be presented with her illegitimate child to the assembled community of Puritans, the scarlet letter A on the breast of her gown. In the course of the novel, Hester is herself associated with that wildness in her publicly stigmatised erotic transgression and her life on the border between town and forest, as is her daughter Pearl, who dances and communicates with wild animals and is explicitly associated with the wild rose bush, while also personifying the scarlet letter as the cultural sign of her mother's stigmatisation. This sign, however, gains itself a dimension of 'wildness' in that it exceeds, both in erotic-artistic and in gothic-demonic ways, the original meaning ascribed to it, turning into an ecocultural energy source that drives and transforms the relationships between the characters and their environments. This is underlined in the element of fire, which is associated with the creative force of the letter A. Already in the metafictional frame narrative of "The Custom-House," a fiery energy is emanating from this material signifier as the narrator finds the faded piece of cloth in the archive and drops it to the floor because of its "burning heat . . . as if the letter were not of red cloth, but of red-hot iron" (62). In the perception of the townspeople it assumes a demonic force: "They averred that the symbol was not mere scarlet cloth, tinged in earthly dye-pot, but was red-hot with infernal fire, and could be seen glowing all alight, whenever Hester Prynne walked about in the night-time" (112).

In Melville's *Moby-Dick*, the narrative voice of Ishmael builds up an alternative, contrastive view to Ahab's biophobic demonisation of the white whale, arising from the recognition of the human species' "Siamese connexion with a plurality of other mortals" (1956/1851: 254). Through Ishmael's perspective, the white whale as an embodied signifier of wildness becomes the central agency of an imaginative counterdiscourse that undermines and overwrites Ahab's civilisational will-to-power over creation. Moby-Dick's irreducible co-agency with the human actors in the narrative, which is conveyed in Ishmael's account of the events, undermines Ahab's civilisational hubris and turns the whale into a medium of trans-species connectivity. This imaginative counterdiscourse already emerges early in the novel when the narrator projects his half real, half dreamlike sea journey as a journey towards his deeper self. This self, however, is not a separate entity but co-exists with the whale in the shared medium of that "ungraspable phantom of life" (24) which connects all beings in a continuous process of metamorphic becoming that is translated into the imaginative process of the text:

> The great flood-gates of the wonder-world swung open, and in the wild conceits that swayed me to my purpose, two and two there floated into my

> innermost soul, endless processions of the whale, and, mid most of them all, one grand hooded phantom, like a snow hill in the air.
>
> (26–27)

The innermost centre of the narrative self is expressed in the imagery of whales. The 'essence' of the self is defined by its relation to a nonhuman other, whose irresistible presence overflows the boundaries between self and other, outside and inside. The figuration of the interior world of the narrator's mind fuses human and nonhuman domains, chaos and order, solid and fluid, wildness and the sacred in such a way that the white whale emerges as its unavailable ground and highest manifestation. The self's encounter with the whale implies the dissolution of the anthropocentric narrator-subject towards an 'intra-action' (Barad) between inseparable internal and external, material and mental, cultural and natural forces in the medium of a fluid imagination, in which the whale signifies an ecosemiotic reality that is co-emergent with the deeper reality of the human self, and which always also conceals itself in the forms of its manifold revelations ("one grand hooded phantom, like a snow hill in the air"). The mythopoetic language of the imagination is employed here in such a way that it envisions the process of the novel as an aesthetic transformation of the same forces that underlie the shapes and metamorphoses of life itself.

In Chopin's *The Awakening*, Edna Pontellier's resistance to imprisoning conventions and role patterns takes a twofold form: a conscious, intellectual form in her withdrawal from the imposed rules of social institutions, as evidenced in her growing defiance toward marriage, monogamy and the economic rationalism of her husband's stock market mentality; and an unconscious, intuitive form in her opening to the influence of the sea. As a compositional leitmotif of the text, the 'voice of the sea' becomes the source of an ever-intensifying rhythm of emotional, bodily and erotic awakenings that creates a wave-like, 'oceanic' (Den Tandt 1997) form of discourse and undulating flow of the narrative. In an ekphrastic way, this fusion of the elemental rhythms of the sea and the transfiguring effect of art is illustrated in a performance of Chopin's music (doubtlessly an ironic self-reference to the novel's author), to which Edna reacts in a deeply passionate way that is conveyed in images of intense bodily experience and immersion in the waves:

> The very first chords . . . sent a keen tremor down Mrs. Pontellier's spinal column. . . . [T]he very passions themselves were aroused within her soul, swaying it, lashing it, as the waves daily beat upon her splendid body. She trembled, she was choking, and the tears blinded her.
>
> (Chopin 1993: 44–45)

Again, an analogy is established here between the experience of art as transformative medium of cultural self-expression and self-exploration, and the experience of passion and 'wild' nature, which is metonymically associated with the sea. In the artificial order of musical signs and sounds, an original chaos becomes audible, in which the control of self and world threatens to be lost, but from which both art and life can gain the energy for creating ever new patterns of emergent order, thus ensuring their continued vitality. This ekphrastic scene is related to the narrative process of the novel as a whole, which aims at the 'dionysian' reconnection of life and art, culture and nature, in the paradoxical tension between linguistic articulation and prelinguistic experience, which Edna encounters in the medial translation of Chopin's music.

In Morrison's *Beloved*, the imaginative counterdiscourse to the system of nineteenth-century American slavery that forms the cultural-historical context of the novel is personified in the ghost of Beloved, the murdered child who returns into the present as an incarnation of the repressed past, initiating a multi-voiced process of 'rememory' that shapes the nonlinear dynamics of the narrative. Beloved makes possible the confrontation and overcoming of the trauma of slavery in the polyphonic story-telling which is sparked off by her reappearance, and in which Morrison combines modern stream-of-consciousness techniques with forms of African-American folklore and jazz, as indicated in the name of the "Blues[-]tone Road" on which Sethe's haunted house is located. Beloved represents the countless anonymous victims of slavery, but also the return of the power of feelings, of 'loving' and 'being loved,' which had been symbolically destroyed in Sethe's killing of her own child as her desperate, self-destructive act of resistance. Beloved reintroduces a sphere of tenderness, longing and excessive desire as a world of magic and reenchantment, and as a wild force of strange but irresistible attraction, which counteracts the violence of racial, cultural and personal separations. Beyond her role as victim, Beloved becomes a powerful agency, a catalyst of radical change and metamorphosis, resembling in some ways the trickster figure of African American folk tales, which transgresses cultural taboos in order to liberate the subliminal fears and desires of human beings. Beloved is a medium of metamorphic contact and transformation, which emerges from water as a spectral hybrid being on the boundary of culture and nature ("A fully dressed woman walked out of the water." [1991/1987: 63]), and in the end returns into her element of water as a pregnant naked woman, "with fish for hair" (328). Human and nonhuman agency converge as water, trees and the regenerative cycles of more-than-human life contribute to an ecosemiotic counterdiscourse that is also expressed in another central recurrent signifier of the narrative, the deep scar on her mother Sethe's back, which is the brutal mark of her violent slavery past that, in the course of time, assumes the shape of a blooming tree. The metamorphic blending of the bodily trace of her trauma with a signifier of possible regeneration is a transformative process which characterises the imaginative process of the novel as a whole (cf. Bonnet 1997).

Wildness in these major works of American fiction from different periods and cultures of writing becomes the source of imaginative counterdiscourses to a biophobic civilisation, activating critical and regenerative energies in ecopoetic transformations that envision more sustainable communities both among the living beings depicted in the narratives and between the texts and their readers.

Textual examples II: American poetry

In American poetry, Walt Whitman is certainly a prime example of this poiesis of regenerative energy when he invokes "Nature without check with original energy" as the source of his *Leaves of Grass* (Whitman 2002: 27). He famously adopts the "barbaric yawp" (77) of the spotted hawk as a poetological motto for his "Song of Myself," in which the experimental openness of his democratic free verse gains its avant-garde modernism precisely through its translation of the creative energies of wild nature into the flow of his dionysian verse. Grass as central symbol becomes an inspirational source of both infinite diversity and infinite connectivity, which in the end includes the poet himself, who metamorphs in a very material sense into the regenerative cycle of nature: "I bequeath myself to the dirt to grow from the grass I love,/If you want me again look for me under your boot-soles" (77) – a metamorphosis of text into life which autopoetically reflects back to the ongoing regenerative energy transaction between text and reader.

Whitman's influence on American poetry has of course been immense and has shaped its forms of writings through modernists like William Carlos Williams and Wallace Stevens, Beat Poets like Allen Ginsberg and Gary Snyder, and into the present. This also involves a critical assessment of the ways in which the expanding civilisation threatens to undermine itself by the increasing subjugation and colonisation of the wild, as Wallace Stevens' well-known poem "Anecdote of the Jar" (1979: 1154) illustrates. The poem is clearly structured in its sequence of three four-line stanzas. The flow of its four-stressed iambic rhythm is however fractured in several places, such as at the close of the first and second stanzas, before it regains dominance at the end with the triumph of the "Jar" over nonhuman nature. The poem deals with the relationship between culture and nature, with the autonomisation of a civilisational order that defines the natural ecosystem not in terms of living interconnectedness but as a mere external environment ("it made the slovenly wilderness surround that hill," 1979: 1154). The first person speaker is the symbolic agent of this anthropocentric civilisation, which realises its goal of superior order through the total control of nature. At the time of the New Criticism and beyond, the "Jar" was seen as a symbol of the artwork, which achieves an imaginative control over an otherwise chaotic reality. On closer inspection, however, the aesthetic process of the text consists in undermining this claim of dominance and authorial omnipotence ("it took dominion everywhere," 1979: 1154), and in conveying the apparently harmless domestication of the wilderness as an all-pervading paralysis of life, in which the uniformity and monotonous circularity of the civilisational system goes hand in hand with the paralysis of the poetic imagination ("the jar was gray and bare," 1979: 1154). The death of the wilderness is the death of the imagination, and the living interrelationship with nonhuman nature appears not only as a vital condition of human life but of poetic creativity. This signifies a counterdiscursive force in the text whose subliminal presence is marked by the fractures in the rhythm and the blind spots of the anthropocentric perspective, but also in the grotesque and fantastic mode of representation. In its polysemic meaning, the word "jar" (connoting not just a "container made of glass or pottery, especially one used for storing food," but also something that is "incongruous in a striking or shocking way," *OED*) introduces a cognitive dissonance in the text implying tension, conflict and incongruity in the relation between cultural object and natural environment. As a signifier of the silent noise contained in expanding circles of exclusion, the jar, which grows in its eerily magnified relation of size to its surroundings, furthermore contains a socioeconomic meaning connected with its commercial use, which has been linked to a real brand of marmalade from the time named Dominion Wide Mouth Jar (see Pearce 1977: 65), thus pointing to an industrial use of nature in a mass consumer society. In its mixture between sublime artefact and monstrous ready-made, the jar simultaneously satirises and pathologises the pressures of cultural normality and normativity. Stevens' jar could also be understood as what Timothy Morton calls a 'hyperobject' of civilisation; evidence of the Anthropocene, whose surreal scale reshapes and deforms the precivilisational world of nature (see Morton 2013).

Something similar applies to other admirers of Whitman like Christopher Morley, whose 1921 poem "Elegy Written in a Country Coal-Bin" critically rewrites the pastoral vision of an iconic poem of early romanticism, Thomas Gray's "Elegy written in a Country Churchyard," in the light of the crisis of coal energy in his time and region. I have this example from the very useful volume co-edited by Scott Slovic, James E. Bishop and Kyhl Lyndgaard, *Currents of the Universal Being*, which provides highly illustrative texts, films and other cultural forms in connection with sustainability and the energy of literature (Slovic, Bishop and Lyndgaard 2015). In another example from this volume, John Updike in his poem "Energy: A Villanelle" (1993), employs the pastoral form of the villanelle to convey, within its insistently repetitive verse and rhyme structure, the rising ecological costs of the expanding use of fossil energy. The poem consists of variations of this theme, contrasting the self-sustaining energy cycles of the earth with the massive costs of our petroculture, which destroys huge amounts of "microorganisms" (Updike 2015: 166) for very small amounts of petroleum and fuels a machinery of capitalist exploitation that pervades our everyday motorised lives ("every tire is by the fiery heavens lightly spun," Updike 2015: 167) and threatens the eventual eclipse of the planetary ecosystem ("So guzzle gas, the sunless night draws nigher," Updike 2015: 167). A purely economic thinking in terms of costs and profits is overwritten here by a holistic-ecological thinking that ironically exposes the long-term, self-destructive consequences of that anthropocentric utilitarianism. Literature as cultural ecology, as exemplified in these poems, is thus both a critical reflector of unsustainable cultural practices and the resulting ecological crisis, and a medium of translating this crisis into poetic form while inscribing its sustainable energies into the discourses of culture. The 'literature of energy,' as Slovic calls it, transforms into the energy of literature, which manifests itself in manifold ways in contemporary poetry.

Of the numerous examples I would just like to refer here to the recently published volume *Lyrik im Anthropozän* (*Poetry in the Anthropocene*) (2016), edited by Anja Bayer and Daniela Seel in collaboration with the Rachel Carson Institute and the German Museum Munich, in which manifold poetic approaches taken to the Anthropocene in the German-speaking context are anthologised. These approaches radically confront the global ecological crisis but also practise an experimental poetics that turns language, texts, images, the self and human-non-human relationships into high-energy fields of regenerative creativity even in the face of threatening disaster.

Conclusion

As ecological metanarratives of their culture, literary texts follow an underlying dynamic of transgression and polysemic excess which enables processes of revitalisation and symbolic regeneration – even if such regeneration is almost never fully achieved on the level of mimesis and plot but only in an aesthetic reflexivity of discourse that is to be realised in the cognitive-emotional participation of the reader. The 'life' to which literary texts relate includes, but also goes beyond, individual life. 'Life' as a precarious process of constant semiosis, interpretation and adaptation between self and other, autopoiesis and ecopoiesis, is also a crucial dimension of the way in which imaginative texts, *as* texts, are acting as an ecological force in culture. In their performative and communicative dimensions, they are primary examples of what has been distinguished as "*in vivo* knowledge" from the "*in vitro* knowledge" of the objectifying natural sciences (Nicolescu 2008: 3). They enact their aesthetic processes in force-fields between the poles of culture and nature, mind and matter, self and other, chaos and order, traumatic disruption and regenerative transformation. From the microstructure of tropes to the macrostructure of narratives, the imaginative processes of texts are made up by such tensions and connecting patterns, which transgress the separations of hegemonic discourses and release creative energies that can be activated for the continuous self-criticism and self-renewal of the cultural ecosystem.

Note

1 In my essay, I will in part be drawing from my recent book, *Literature as Cultural Ecology: Sustainable Texts* (2016a), London: Bloomsbury.

References

Adamson, J. (2012). "Indigenous Literatures, Multinaturalism, and *Avatar*: The Emergence of Indigenous Cosmopolitics". *American Literary History*, 24(1), 143–162.

Alaimo, S. (2012). "Sustainable This, Sustainable That: New Materialisms, Posthumanism, and Unknown Futures". *PMLA*, 127(3), 558–564.

Bateson, G. (1973). *Steps to an Ecology of Mind*. London: Paladin.

Bayer, A. and Seel, D. (eds.). (2016). *Lyrik im Anthropozän (Poetry in the Anthropocene)*. Berlin: Kookbooks.

Bonnet, M. (1997). "'To Take the Sin out of Slicing Trees . . .' The Law of the Tree in *Beloved*". *African-American Review*, 31(1), 41–53.

Braidotti, R. (2006). *Transpositions*. Malden: Polity.

Chopin, K. (1993). *The Awakening: Complete, Authoritative Text with Biographical and Historical Contexts, Critical History, and Essays from Five Contemporary Critical Perspectives*. Ed. N. A. Walker. Boston: St. Martin's.

Den Tandt, C. (1997). "Oceanic Discourse, Empowerment and Social Accommodation in Kate Chopin's *The Awakening* and Henrik Ibsen's *The Lady from the Sea*". In G. Debusscher, and M. Maufort (eds.). *Union in Partition: Essays in Honour of Jeanne Delbaere*. Liege: L 3 – Liege Language and Literature, 71–79.

Finke, P. (2006). "Die Evolutionäre Kulturökologie: Hintergründe, Prinzipien und Perspektiven einer neuen Theorie der Kultur". *Anglia: Journal of English Philology*, 124(1), 175–217.

Goeminne, G. (2011). "Once upon a Time I Was a Nuclear Physicist: What the Politics of Sustainability Can Learn from the Nuclear Laboratory". *Perspectives on Science*, 19(1), 1–20.

Grober, U. (2002). "Modewort mit tiefen Wurzeln: Kleine Begriffsgeschichte von 'sustainability' und 'Nachhaltigkeit' ". In *Jahrbuch Ökologie 2003*. München: Beck, 167–175.

Hawthorne, N. (1986/1850). *The Scarlet Letter and Selected Tales*. Harmondsworth: Penguin.

Keller, L. (2012). "Imagining Beyond, Beyond Imagining". *PMLA*, 127(3), 579–585.

LeMenager, S. and Foot, S. (2012). "The Sustainable Humanities". *PMLA*, 127(3), 512–578.

Melville, H. (1956/1851). *Moby-Dick, or, The Whale*. Ed. A. Kazin. Boston: Riverside.

Morrison, T. (1991/1987). *Beloved*. Harmondsworth: Penguin and Signet.

Morton, T. (2013). *Hyperobjects: Philosophy and Ecology after the End of the World*. Minneapolis: Minnesota University Press.

Nicolescu, B. (ed.) (2008). *Transdisciplinarity: Theory and Practice*. Cresskill: Hampton Press.

Pearce, R.H. (1977). "'Anecdote ot the Jar'. An Iconological Note." *Wallace Stevens Journal*, 1(2), 65.

Philippon, D. J. (2012). "Sustainability and the Humanities: An Extensive Pleasure". *American Literary History*, 24(1), 163–179.

Slovic, S. (2012). "Commentary". *American Literary History*, 24(1), 180–188.

Slovic, S., Bishop, J. E. and Lyndgaard, K. (eds.) (2015). *Currents of the Universal Being: Explorations in the Literature of Energy*. Lubbock: Texas Tech University.

Stevens, W. (1979). "Anecdote of the Jar". In R. Gotesmann et al. (eds.). *Norton Anthology of American Literature*. New York, London: Norton, 1154.

Updike, J. (2015). "Energy: A Villanelle" [1993]. In Scott Slovic, James E. Bishop and Kyhl Lyndgaard. (eds.). *Currents of the Universal Being. Explorations in the Literature of Energy*. Lubbock: Texas Tech University Press, 166–167.

Westling, L. (2006). "Darwin in Arcadia: Brute Being and the Human Animal Dance from Gilgamesh to Virginia Woolf". *Anglia: Journal of English Philology*, 124(1), 11–43.

Wheeler, W. (2016). *Expecting the Earth: Life, Culture, Biosemiotics*. London: Lawrence & Wishart.

Whitman, W. (2002). "Song of Myself". In M. Moon (ed.). *Leaves of Grass and Other Writings: Authoritative Texts, Other Poetry and Prose, Criticism*. New York: Norton, 26–78.

Wood, G. D'Arcy (2012). "What Is Sustainability Studies?". *American Literary History*, 24(1), 1–15.

Zapf, H. (2002). *Literatur als kulturelle Ökologie: Zur kulturellen Funktion imaginativer Texte an Beispielen des amerikanischen Romans*. Tübingen: Niemeyer.

Zapf, H. (2016a). *Literature as Cultural Ecology: Sustainable Texts*. London: Bloomsbury.
Zapf, H. (ed.) (2016b). *Handbook of Ecocriticism and Cultural Ecology*. Berlin: DeGruyter.
Zapf, H. (2017). "Ecological Thought in Europe and Germany". In L. Westling, and J. Parham (eds.). *A Global History of Literature and the Environment*. New York: Cambridge University Press, 269–285.

13 An entangled history of environmental and cultural sustainability

Satirical reflections on the German forest and the German oak as resources of cultural energy

Evi Zemanek

Introduction

In Sustainability Studies the modern concept of sustainability as the capacity of ecological, economic, sociocultural, or political systems to remain infinitely productive is commonly traced back to its roots in forestry. The German term 'Nachhaltigkeit' was introduced by Hans Carl von Carlowitz in his groundbreaking treatise on forestry *Sylvicultura oeconomica, oder haußwirthliche Nachricht und Naturmäßige Anweisung zur wilden Baum-Zucht* (1713), a text written in reaction to the scarcity of timber after large-scale clear cuttings. Von Carlowitz promoted a careful, farsighted forest management, what he described as "the conservation and cultivation of wood in a way that guarantees a continuous, durable and sustainable utilization, because it [the wood] is an essential thing, indispensable for the essence/character of the country" (von Carlowitz 1713: 105–106).[1]

Significantly, von Carlowitz does not primarily argue with the economic value of the wood, its indispensability for industry, house building and the military, but he asserts its ethic and aesthetic values for the people, and he does so even in this central passage of his treatise. In other words, von Carlowitz emphasized the importance of sylvan nature for a culture's sustainability, and others adopted his point of view. In fact, early forestry's narrow focus on yield increase has been broadened since by the acknowledgement of the multiple functions of the forest, not only as a source of raw material and as a protective barrier, but also as an important factor in the ecological system (in view of photosynthesis, air purity, water balance and climate altogether), as an aesthetic factor in landscaping, and, last but not least, as a space for recreation (cf. Steinsiek 2012: 94) – and even for self-fulfilment and creativity, as I want to add. Consequently, the present-day German law for the preservation of the forest (*Bundeswaldgesetz*) equally provides for the sustainment of its economic, ecological, aesthetic and recreational functions, while balancing the interests of the public and the forest owners.

Thus, the planting of trees for future generations is one of the oldest forms of provision and sustainable action that has been documented in sources and conveyed in pictures (cf. Steinsiek 2012: 92–93). To repair the already visible damage, von Carlowitz recommended reforestation and tree nurseries: Their German term 'Baumschulen' emphasizes the cultural formation of nature. The future-oriented symbolic meaning of reforestation was again taken up after the Second World War when in 1948 the Deutsche Mark was introduced and the new 50-pfennig coins showed a woman planting a tree – an oak tree, precisely, which has a long career as an outstanding symbol of German national identity.

The association of the German (or Germanic) people with their forests, and in particular with their oaks, can be traced back to the Romans. The idea has been taken up at several crucial moments in German cultural history. Going much further than von Carlowitz in the semantization of the forest, the cultural historian Wilhelm Heinrich Riehl promoted his conviction of "the close bond between landscapes and peoples" (Riehl 1854: 25) in his widely read *Natural History of the People* (*Die Naturgeschichte des Volkes als Grundlage einer deutschen Socialpolitik*, 1854–1869). He relates the different characters of peoples to their natural environments, on the one hand meaning that they were shaped by their environments – a basic idea of climate theories as proposed by Aristotle and Montesquieu – and on the other hand that their own shaping of their environments, especially their handling of the forest, not just through landscaping but also through artistic representation, reveals their characters. Delineating a collective German identity in contrast to its European rivals, his 'ethnography' begins with the assertion that only Germany still has a "real forest" with a "social meaning" (Riehl 1854: 25). He argues that the wild forest, as all untouched nature, is a source of strength for humans, thus it needs to be protected for the sake of the culture's sustainability.[2] Famous is his plea: "We must preserve the forest, not just to keep the oven warm in winter, but also to make sure that the people's pulse keeps beating warm and merrily, and that Germany remains German" (Riehl 1854: 32). Riehl's last words do not have a racial connotation, as could be assumed when the quote is isolated from its context. They refer to his ideal of a 'natural' lifestyle, which can only be realized in "personal freedom, untouched by police surveillance" and "free of moral pressure" (Riehl 1854: 34). Thereby he means a freedom of movement and opinion he terms "Germanic Waldfreiheit" in contrast to other nations' restrictive policies (34).

According to Peter Finke, who promoted the renewal of cultural studies as "Cultural Ecology", important criteria to measure the sustainability of a culture are, besides its resource management and generally its interaction with the natural environment, its diversity and its creative potential (cf. Finke 2003: 267). He links creativity to diversity by assuming that, just as biodiversity offers various survival strategies, cultural heterogeneity offers a whole range of lifestyles that foster creativity (cf. Finke 2003: 264). In this perspective, which rests on a comparison between cultural systems and ecological systems, languages are considered the most important sources of cultural energy, since they create and transform the structures of civilizations or cultural ecosystems (cf. Finke 2003: 271). Language, in turn, is the medium of literature and many art forms, "whose task is the constant critical examination, imaginative exploration, and creative self-renewal of these cultural sign systems" (Zapf 2016a: 65), as Hubert Zapf phrases it. Zapf's theory of literature as a powerful form of cultural ecology departs from the observation that literature and other arts act like ecological forces within the cultural system.[3]

These recent approaches are relevant for my essay in a twofold way: By presenting satirical texts and drawings from German magazines of the late nineteenth and early twentieth century that display concepts of sustainability, this essay analyses discourses of environmental and cultural sustainability with the aim of showing their historical entanglement. However, looking at satirical works that mock unsustainable uses of natural resources, unsustainable cultural politics and even certain concepts of sustainability is not the same as studying writings that seriously promote sustainable practices. Some studies trace the conceptual history of sustainability in various cultures and disciplines, and some of them may criticize earlier concepts of sustainability from today's perspective, but this essay tries to see these historical concepts through the eyes of contemporary satirists and cartoonists. It does not present a selection of caricatures in order to prove or illustrate a history of ideas that has already been written, but it considers these artworks for their own sake: It looks at the cultural function of these verbo-visual artworks, which are themselves a form of cultural energy to be taken into account within an ecology of knowledge. Furthermore, it reveals the enormous role mass media has played (and still plays) in the development and popularization of sustainability concepts, and it shows how deeply they are rooted in the German culture's very own history of ideas. In the late nineteenth century, different discourses emerged – such as landscape preservation and animal protection – that articulate proto-ecological thought without ever mentioning the proper term "Nachhaltigkeit". The latter is the keyword, however, in all discourses on the forest, and here the entanglement of environmental and cultural sustainability is more evident than in any other field.

The artworks I discuss in this essay require some preliminary remarks on the media history of caricatures and satirical magazines.[4] In newspapers and magazines from the second half of the nineteenth century, the dominant discourse consists of success stories that quite uncritically describe humans' conquest of nature. In newspapers like the *Illustrirte Zeitung* (1843–1944) or the entertaining and educational bourgeois journal *Die Gartenlaube* (1853–1944), which appeared weekly from mid-century, readers were being informed continually about technical innovations as well as the expansion of industrialization and infrastructure. Only after the readers had been made familiar with all these aspects of sweeping modernization could the 'achievements' be called into question through a demonstration of their downsides. The papers mentioned above, however, usually did not intend to alarm their readers. This goal was instead pursued by satirical magazines like *Fliegende Blätter* (1845–1928), *Kladderadatsch* (1848–1944), *Kikeriki* (1861–1933), *Der wahre Jacob* (1879–1933) and *Simplicissimus* (1896–1944). Besides poems, short stories and advertisements, these magazines featured countless caricatures. Unfortunately, historians often treat them as mere illustrations of events instead of unique verbo-visual forms of cultural criticism, sometimes being the only possible form of a critique otherwise too dangerous to be expressed publicly.

Since satire, and especially caricature, aims at making its readers or beholders realize that something is wrong, it deforms and exaggerates its subject, often to a degree that makes it seem ridiculous, although not all satires provoke laughter. Even if it uses wit to attract attention, its main goal is to make people think about the problem it displays. Its other common techniques are parody, irony and sarcasm, and they rely on allegation and provocation. They cannot be understood without knowing their horizon of reference, the sociopolitical status quo and the ideal they thwart. Both must be known in order to recognize them as dissent. It has to be kept in mind, however, that magazines had to respect censorship regulations for mass media.[5]

Leafing through satirical magazines of the second half of the nineteenth century, it becomes clear that satire primarily targeted the lifestyles of bourgeoisie and gentry, class conflicts, patriarchy, women's emancipation, children's education and other explosive sociocultural topics. During the German Empire much critique was directed against the monarchy and imperialism, the politics of the Kaiser and chancellor Bismarck, the military and warfare. Generally, caricature's favourite topics rest on human mistakes which do not have disastrous consequences and thus can be laughed at. Of the far fewer caricatures that depict industrialization, modernization and urbanization, some call attention to the consequences for human life and the natural world. Among those, only few address visible environmental problems of the time, such as air or water pollution, while more can be found that focus on the fate of the woods.

As is well-known, Germany's pristine woodlands had been shrunk and transformed, notably through being turned into timber, throughout the eighteenth and early nineteenth centuries. Environmental historians have asserted that an early consciousness of nature's value can be discerned in complaints of a wood shortage ("Holznot"), complaints that had been widespread in central Europe since the eighteenth century.[6] Whether these complaints were justified or not, they were motivated mainly by economic calculations and existential fear, rather than by the impulse to protect nature for its own sake. Some early environmentalists nevertheless did consider the excessive clear-cutting as historically the first step towards humanity's destruction of nature (cf. Schultze-Naumburg 1916/1917: 12). However, I would like to defend the thesis that forest clearance receives attention mainly because the destruction of the German forest is regarded as an existential threat to both nature *and* culture.

German forest and German oak: nature as a cultural symbol

The 'German forest', as described in poetry and fairy tales, hailed in folk songs, imagined in paintings and referred to in political discourse, has always been more than a natural landscape and even more than a cultural landscape. Going back to ancient times, it has become a national myth and a symbol of German culture; its popularity grew enormously between 1800 and the Nazi Regime, while its connotations varied over time.[7]

The earliest descriptions of the German forest date back to the Romans (e.g. Caesar's *De bello Gallico*, Pliny the Elder's *Naturalis historiae*, and Tacitus' *Germania*) and deserve to be briefly mentioned because these accounts, notably from a Mediterranean perspective, already prefigure its striking mythification. The forests of Germania, most often called the Hercynian forests, were described as pristine and endless, scary and mysterious, consisting of giant oaks (Pliny), inhabited by strange animals such as unicorns (Caesar), and populated by brave Germanic tribes, prototypical forest people, which were considered humble and hospitable by Caesar, and coarse and barbaric by Tacitus.[8] The latter mentions the decisive Battle of the Teutoburg Forest (or Hermannsschlacht) in 9 CE, that is, the victory of the Germanic tribes headed by the Cheruscan Arminius (Hermann the German) against Rome's legions led by Publius Quinctilius Varus. Here already, Hermann is idolized as "the liberator of Germania",[9] who was well suited to serve as a heroic model for the emerging German nationalism in the nineteenth century. In this view, Hermann stood for freedom from foreign rule and for the united 'nation' yearned for in the decades before 1871, but also for a genuine forest people (Waldvolk, Waldgemeinschaft). The Germans' self-understanding as descendants of the Germanic people naturally implies their self-perception as a forest people, now defined by exclusively positive attributes.

In the German Renaissance, the humanist Conrad Celtes, who edited Tacitus' *Germania* and complemented it with his own *Germania generalis* (around 1500), uncritically revived the idealized image of a vast Germanic forest full of giant oaks. In fact, according to forest historians, at least two thirds of this central European territory was densely wooded until the Middle Ages, during which period deciduous and mixed forests were much more prevalent than they are today, as now the forests are mostly conifer. Beginning in the middle ages there were several phases of extensive deforestation, which were already perceived as a problem in the sixteenth century, and complaints of a lack of wood did not stop before the mid-nineteenth century. It is commonly agreed that between 1750 and 1850 the Central European forest was in its worst condition, before coniferous forests were created in large-scale reforestation projects. Today, approximately one third of Germany is covered by forests, roughly the same extent as in the Renaissance.

During the following centuries, various influential poets (e.g. Casper von Lohenstein, Klopstock, Kleist and Grabbe) studied Tacitus and chose Hermann as a hero for fictions imagining the origins of the Germanic or German people. Klopstock especially took up the adoration of the oak tree, which symbolizes the Cheruscans and the fatherland in his drama (cf. Klopstock 2009: 30, 32, 34, 71, 80). In his odes, he turns it into a symbol for "genuinely German poetry", considering oak leaves more appropriate to crown a German poet than the classical laurel wreath.[10] Thus, by 'germanizing' the classical images of Greek poetry, Klopstock ties poetic inspiration, that is, cultural creativity and ultimately German culture, to the natural surroundings of a typically 'German' forest.

The increasingly differentiated symbolic spectrum ascribed to the oak tree and the forest becomes evident when Klopstock compares the German language of his earliest ancestors to a forest wilderness and states that Martin Luther, with his Bible translation, has transformed it into a grove (cf. Klopstock 1975: 157). He was not the only one in his times to compare language – as the prerequisite for culture – and its different stages of development with forests cultivated to various degrees. But while some, like Klopstock, favoured a highly cultivated language, others idealized its most 'natural' state, like the brothers Jacob and Wilhelm Grimm: they edited the most famous German fairy tale collection (*Kinder- und Hausmärchen/Grimm's Fairy Tales*, first published in 1812), promoted the literary tradition in their journal *Altdeutsche Wälder* (1813–1816) and initiated the great German Dictionary (*Deutsches Wörterbuch*, published since 1854) in order to preserve cultural and linguistic heritage for future generations. In view of one satire discussed later it is worth mentioning that the Grimms metaphorically describe both folk poetry and the development of the German language as plants that need to grow naturally and undisturbed. Jacob speaks of his native language as a 'giant tree', which, as he explains in his German Grammar (*Deutsche Grammatik*, 1819), is threatened by 'weeds', meaning foreign loanwords (cf. Grimm 1884: 557; Grimm 1881: 519–520).[11] His ideal of a 'pure' language was shared by many of his contemporaries; significantly, the journal published by the "Association for the Promotion of the Pure German Language" (*Verein zur Beförderung der deutschen Reinsprache*, founded in 1848), which promoted language purism much more uncompromisingly, was named "The German Oak; Journal for the Promotion of German Sense, German Civilization, and Pure German Language through Instruction and Entertainment" (*Die deutsche Eiche. Zeitschrift zur Förderung deutschen Sinnes, deutscher Gesittung und deutscher Reinsprache durch Belehrung und Unterhaltung*).[12]

Besides their predilection for sylvan metaphors, the Grimm brothers played a key role in mythifying the forest through their widely read fairy tales. Not only are many of the collected tales set in the woods, but in the editorial process the Grimms also amplified the fictional forests and emphasized their essential meaning as an exceptional sphere in which people and nature still exist in organic unity.[13] This corresponds to their 'Zeitgeist': Since deforestation and monocultural plantations had become a visible concern, verbal and visual imaginations of a healthy primaeval forest had become ever more popular. A few examples should suffice to demonstrate this: The political connotation of the German forest becomes visible in paintings from the French period in which German volunteer units (like the Lützowsche Freikorps) seek shelter in an oak grove as in Georg Friedrich Kersting's painting *Auf Vorposten/On an Outpost* (1815) or in which the French enemy is getting lost in the dark German forest as in Caspar David Friedrich's *Der Chasseur im Walde/The Soldier in the Forest* (1814).[14] Additionally, historical paintings depicting scenes from the Battle of the Teutoburg Forest popularized the German forest. Of even greater importance for the context of this essay is the pictorial association of the forest with core areas of culture such as religion (e.g. Adrian Ludwig Richter's *Genoveva in der Waldeinsamkeit/Genoveva in the Forest*

Solitude, 1841), architecture (e.g. August von Kreling's *Erwin von Steinbach im Waldesdom/E. v. S. in the Sylvan Dome*, 1849), painting (see Carl Spitzweg's *Der Maler auf einer Waldlichtung unter einem Schirm liegend/The painter lying in a forest clearing under an umbrella*, ca. 1850), music (e.g. Moritz von Schwind's *Des Knaben Wunderhorn/The boy's magic horn*, 1848) and poetry (e.g. Carl Spitzweg's *Der Lieblingsplatz/The Favorite Place*, 1849). When we look at these paintings of the forest as a sphere of the sacred and the sublime, of art and leisure, it becomes obvious why German Romanticism is often specified as "Forest Romanticism" ("Waldromantik"). Besides these paintings, it is very often Romantic poetry that serves as a point of reference for the verbo-visual satires discussed in this essay.

When asked about the most popular forest poems, literary scholars usually first of all name Joseph von Eichendorff, who often described an idyllic sylvan nature that stands in stark contrast to the urbanizing world and the political turmoil of his times (see, for example, *Klage/Lament*, 1809). The Catholic Eichendorff spent his childhood on a Silesian estate amidst vast forests before he experienced the drastic sociocultural transformations of the nineteenth century. Surprisingly, scholarship has largely neglected the fact that he witnessed nothing less than the destruction of the forest due to war, fire and overexploitation.[15] Extensive clear cuttings, which he explicitly criticized in personal notes, did not even spare oak forests (cf. Frühwald 1994: 19). Even his own mother sacrificed a forest belonging to the family to pay her debts (cf. Frühwald 1994: 20). Of course, these grim realities were no adequate subject for Romantic poetry. Eichendorff instead concealed the loss in his forest imaginations, which have often been called "landscapes of yearning" ("Sehnsuchtslandschaften") and should, more precisely, be seen as "endangered landscapes" and the "aesthetic revival of the lost nature experience" (Frühwald 1994: 17–18).

This is not the place to offer a comprehensive overview of Eichendorff's forest imaginations. In this essay, it is most important to note the forest's function as a place of both ethical and artistic inspiration: Firstly, it is a model of morality and a shelter for those who want to live a faithful, upright life which includes respect for the achievements of their predecessors (see *Der Jäger Abschied/The Hunters' Farewell*, written 1810/published 1837, and *Abschied/Farewell*, 1815). Secondly, Eichendorff naturally locates the creation and recitation of song and poetry in the forest, implicitly arguing that only a beautiful environment offers inspiration (see also *Die Zauberin im Walde/The Fairy in the Forest*, 1837), and this inspiration is needed to fulfil the poet's duty: to communicate his homeland's beauty. However, both cultural and natural heritage are in danger and need to be preserved: In his poem *Trost/Consolation* (1837) he mourns the death of great singers/poets, but at the same time he optimistically reminds his audience that their heritage can be sustained as long as the poems are being remembered and the forest greens every spring. Thus, the singer/poet prays for the forest, most prominently in the final line of *Der Jäger Abschied/The Hunters' Farewell*: "Schirm dich Gott, du schöner Wald!"/ "May God protect you, beautiful forest!". Here, hunters leaving the forest to fulfil their political duties praise it for being their spiritual home, which inspires them to loyalty and national feeling – they associate the forest with the

German banner. It is one of Eichendorff's best-known poems, which, along with *Abschied/Farewell*, had been set to music by Felix Mendelssohn-Bartholdy; it belonged and still belongs to the standard repertoire of choirs.[16] The poem's first lines ("Wer hat dich, du schöner Wald/Aufgebaut so hoch da droben?"/ "Who has built you, beautiful forest, so high up there?") and its refrain ("Lebe wohl,/ Lebe wohl, du schöner Wald!"/ "Farewell, farewell – you beautiful forest!") has been quoted many times in various popular contexts and taken up repeatedly in parodies modifying the poem's original content – or rather bringing its concealed subtext to the surface.

The fight over the forest in parody and satire

As early as in 1873, the hunters' personal farewell was reinterpreted as environmental lament: The satirical journal *Fliegende Blätter* dedicated a full page to a poem framing two illustrations under the heading *Elegie/Elegy* (Figure 13.1). The parodistic poem and the two realistic pictures – as a whole signed by Crassus [Sigmund Krassberger] – contrast past and present, a once healthy, beautiful forest versus an ugly field of tree stumps. Alluding to the first lines of Eichendorff's original, the question "Who has built you . . .?" is replaced with "When will [the forest] be cut down?" (rhyming: "auferbaut" – "niederg'haut"; as a fitting English translation one could think of an equally forced rhyme: "By whom were you built?" – "When will you be killed?"). The parody ends with the grandchildren blaming the stock company for the destruction of the forest.

This was only the beginning of the parodistic echo of Eichendorff's poem that could be heard well into the twentieth century, and sometimes even today. More than twenty years later, the same journal, *Fliegende Blätter*, featured another verbo-visual parody, titled *Abschied vom Walde/ Farewell to the Forest* (1895)[17] just like the poem's musical version by Mendelssohn. Its single picture by E. Wagner shows a bald hill, sawing woodworkers, and a wagoner handling the horses that carry away the dead trunks. Its text, written by Friedrich Detjens, is even closer to Eichendorff's original in its structural and metrical composition. Quite similar to the parodistic poem by Crassus, it alters the famous opening apostrophe to the forest by replacing "built" ("aufgebaut") with "cut" ("abgeholzt"), and its final lines also explain that sellout of the forest is a result of greed for money. The later parody from 1895, however, is a stronger plea for environmental and cultural sustainability. Explicitly criticizing that the sounds of the hunter's horns and the poets' songs inspired by nature have been replaced by the noises of tree-sawing and horse-whipping, it demands that the younger generation protect what their ancestors had protected: both nature and art.

Once the potential of Eichendorff's poem for environmental critique had been discovered, its matrix was taken up at various occasions during the following decades, for example to call attention to the destruction of the Grunewald Forest. Before coming to this particular case, it has to be mentioned that in the second half of the nineteenth century, against the backdrop of accelerating industrialization and urbanization, the forest had become more and more important as

Elegie.

Auf steilem Fels emporgeschossen
Ragt eine Tann' in stolzer Ruh',
Sie ist die Letzte der Genossen,
Und wär' nicht da — könnt' man ihr zu.

Steht sie wohl hier um zu verkünden,
Was keinem Auge mehr sich malt?:
Auf diesen Höh'n, in diesen Gründen
Rauscht' einstens frischer, grüner Wald;

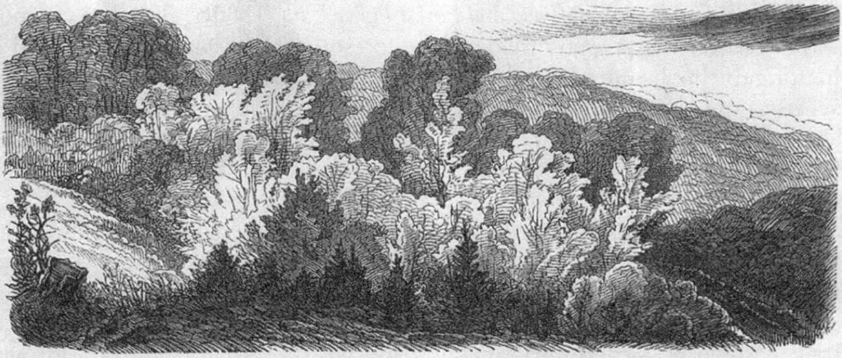

Einst pulste hier das regste Leben,
Hier kreischte wild des Adlers Schrei,
Hier trug der Hirsch mit stolzem Heben
Des Haupt's das zackige Geweih';
Hier sang der Urhahn Liebesweisen,
Das Reh sprang auf der blumigen Hald',
Das ist vorbei, seitdem das Eisen
Der Axt hinschmetterte den Wald;

So stieß der grause Allbezwinger
Tod, die Natur vom Thron' herab,
Nun reckt die Tanne sie als Finger
Zum Himmel dräuend aus dem Grab'.
„Du stolzer Wald," klang einst die Frage,
„Wer hat so hehr dich auferbaut?"
Wen kümmerte das heutzutage,
Jetzt heißt's: „Wann wird er niederg'haut?"

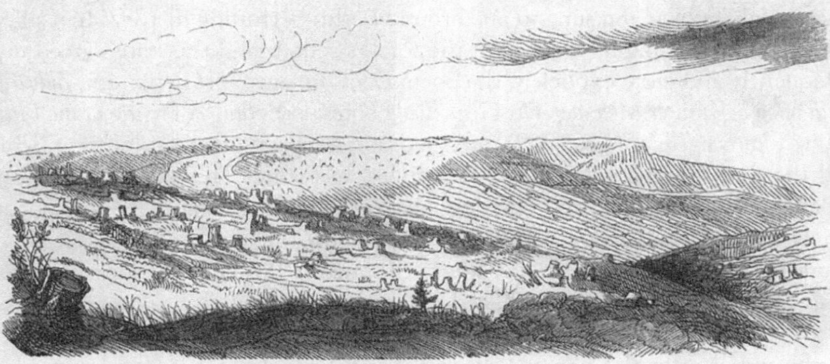

Wenn uns're Berg' in künft'gen Tagen
Nur mehr ein Haufe nackt Gestein,
Wird unser Enkel schaudernd sagen:
„Das that ein Aktien-Verein."

Crassus.

Figure 13.1 Crassus (1873). "Elegie". *Fliegende Blätter*, 58(1452), 159 http://digi.ub.uni-heidelberg.de/diglit/fb58/0163 [Accessed 1 August 2017].

a recreation area for the bourgeois and, a little bit later, also for working-class urbanites. While the Romantics around Eichendorff had famously described their wanderings in the forests, only several decades later, hiking became a fashionable leisure activity for the masses. In the last third of the century, growing conflicts over the question of to whom the forest belongs and whose needs should obtain priority (be it the urbanites and tourists or the landlords or the foresters) led to a series of controversial public and parliamentary debates, and finally to the *Feld- und Forstpolizeigesetz/Field and Forest Police Law* (1880).[18] This law's purpose was to protect the property rights of the forest owners against theft of wood by limiting the public use of the forest. Interpreted as an attempt to deny the people access to the forests and thus denying them nature, fresh air, freedom and leisure altogether, the law provoked an unprecedented wave of protest, which manifested itself in satire and caricature.[19] In the end, this law did not keep the masses out of the forest.

The Grunewald, located in the western part of Berlin, had been mainly used as an aristocratic hunting ground, until the rapidly growing city's dwellers conquered it in the 1880s in search of 'nature'. Reacting to public will, Wilhelm II finally stopped his hunting activities in the Grunewald around 1904 and announced that it would be turned into a public park, but the Prussian ministry for agriculture, led by Victor von Podbielski, nevertheless planned to sell it as land for building.[20] Already, with the expansion of the city, parts of the Grunewald had been turned into building land, for example the "Villenkolonie Grunewald", an upper-class residential area developed in 1889. The forest clearance carried out for this purpose was publicly criticized in the satirical popular song *Die Holzauktion/The Wood Auction* ("Im Grunewald, im Grunewald ist Holzauktion"), which was widely known and sung in and around Berlin beginning in 1892. In a similar vein, but addressing its critique of the forest's sellout and the state's greed more eloquently and more explicitly, the poem *Der Grunewald bei Berlin. Ein Zukunftsbild nach bekannter Melodie/ The Grunewald Forest in Berlin. A Picture of the Future in well-known Tunes* (J.S. 1909) offers a dystopic vision of a near future in which all the forest, along with its sales revenues, has gone. This poem, which once again takes up Eichendorff's opening lines and refrain, and modifies both accordingly, appeared in the journal *Der Wahre Jacob* in 1909, in the May issue. In the same month, the newly built racecourse "Rennbahn Grunewald", for which again part of the forest had been sacrificed, was to be inaugurated.[21]

In 1908, when the construction of the racecourse was well under way, another verbo-visual artwork appeared in the satirical magazine *Kladderadatsch* which emphasized the interdependence of nature and culture, or, respectively, the entanglement of nature's destruction and culture's decline. The picture (Fig. 13.2) signed by Arthur Johnson fills an entire page and bears the title *Das Schweigen im Grune-Wald (frei nach Böcklin)/The silence of the Grunewald forest (after Böcklin)*. Below the picture it says "Hier malte Walter Leistikow seine schönsten Bilder"/ "Here, Walter Leistikow has painted his most beautiful pictures".

Only readers familiar with art history and contemporary painting might have been able to decipher the meaning of the double reference to two painters.

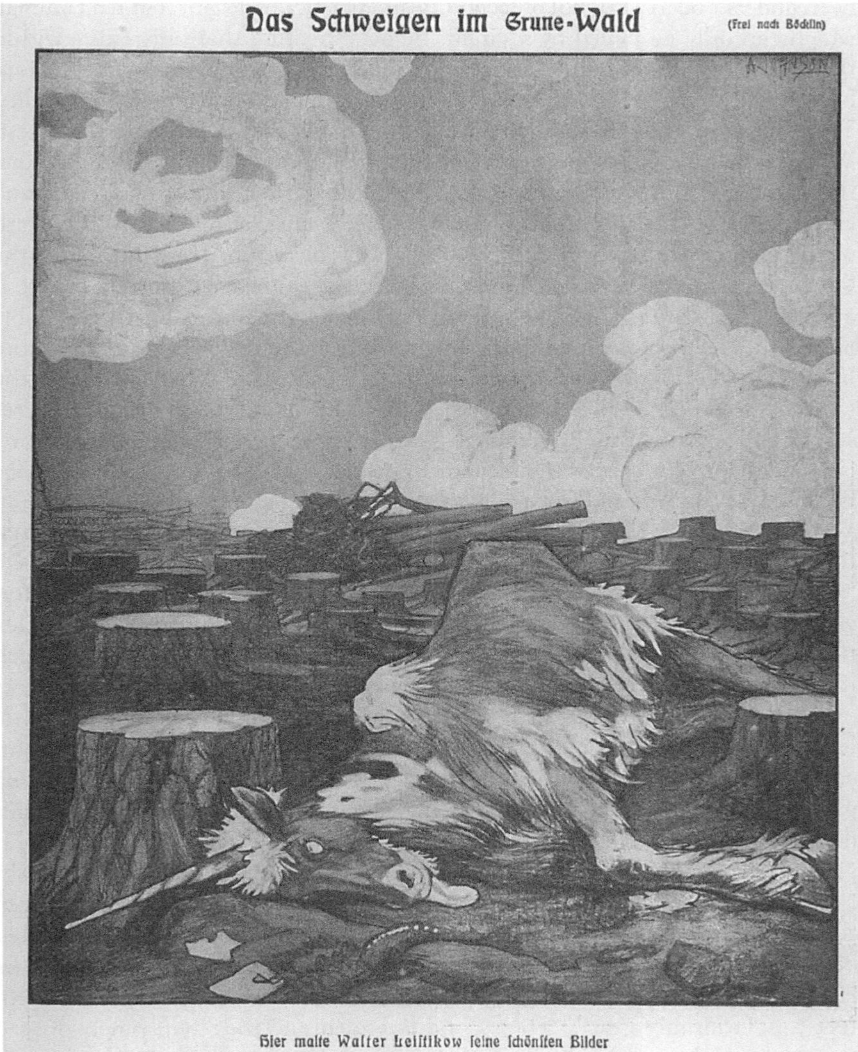

Figure 13.2 Johnson, A. (1908). "Das Schweigen im Grune-Wald (frei nach Böcklin)". *Beiblatt zum Kladderadatsch*, 61(35), 1 (30 August 1908) http://digi.ub.uni-heidelberg.de/diglit/kla1908/0617 [Accessed 1 August 2017].

Johnson's drawing is a parody of the painting *Das Schweigen des Waldes/The Silence of the Forest* created by the famous symbolist Arnold Böcklin in 1885. His painting shows a unicorn strolling through a healthy forest and carrying a crowned woman on his back. In sharp contrast, Johnson's satirical picture shows a dead unicorn lying among the stumps of cut trees. The noble fabulous creature has been understood as a symbol of fecundity and purity, an elegant, but wild animal, which can only be tamed by a virgin. Being a creature that only exists within the realm of myth and art, it ultimately stands for art itself, or the marvellous. Significantly, it is only to be found in the forest, which protects it from reality. Thus, Johnson's picture implies that art inevitably dies along with the forest. While in the title of Böcklin's painting the 'silent forest' refers to a sacral sphere, the silence in Johnson's satirical picture indicates the death of nature, art and culture. Besides the loss of the Grunewald, the picture also bemoans the death of the painter Walter Leistikow, who was well-known for his paintings of the Grunewald and had died just a month earlier. Assuming that Johnson's satire is a friendly obituary for Leistikow, it suggests that the painter's death inevitably means the death of Grunewald painting. On a more complex level the satire implies that Leistikow's Grunewald can never be surpassed artistically, since the forest as he saw it has been destroyed; only in his immortal paintings is it preserved forever. Interestingly, although this was probably not intended by the artist, the unicorn has special significance in the context of environmental damage because its horn was said to be able to clean contaminated wells and waters (cf. Wehrhahn-Stauch 1958: 1534). If the animal dies, environmental contamination cannot be reversed.

At first sight, most caricatures that address the destruction of the German forest and the negative consequences on culture seem semantically less complex than the example discussed previously, but at a second glance they leave much room for (mis-)interpretation and often make it difficult to discern their object of mockery. This, however, is the prerequisite for reaching their double goal of entertaining the readers and influencing their political opinion. Often they cynically comment on a critical issue without offering solutions, like the following page-filling drawing (Fig. 13.3), whose heading ironically proclaims: "Nieder mit dem deutschen Wald!"/ "Down with the German forest!" The picture from 1921, presumably created by Hans Maria Lindloff, shows a man in fancy, but old-fashioned clothes sitting on a lonely oak amidst a field of tree stumps. The first line of the text below the picture says: "Der letzte deutsche Romantiker auf dem letzten deutschen Waldbaum (singend): 'Schirm Dich Gott, du deutscher Wald!'"/ "The last German Romantic on the last German forest tree (singing): 'May God protect you, beautiful forest!'" The second line identifies the stout, timely dressed man beneath the tree holding an axe as a wood speculator. He impatiently yells at the Romantic above: "Sie da, runter! Wird's nu bald!"/ "You there, come down! Hurry up!". He unmistakably tells us that the Romantic era has definitely come to an end, especially now after the defeat in the First World War. Romantics are being chased away, the reign of the Realists and the sell-out of German culture have begun.

We could simply interpret this piece as one of many attacks against speculators, who sacrifice nature for money. Alternatively, we can read it as a critique of an exaggerated emotional attachment to nature and thus laugh about the Romantic who believes in the possibility of a harmonious relationship between humans and

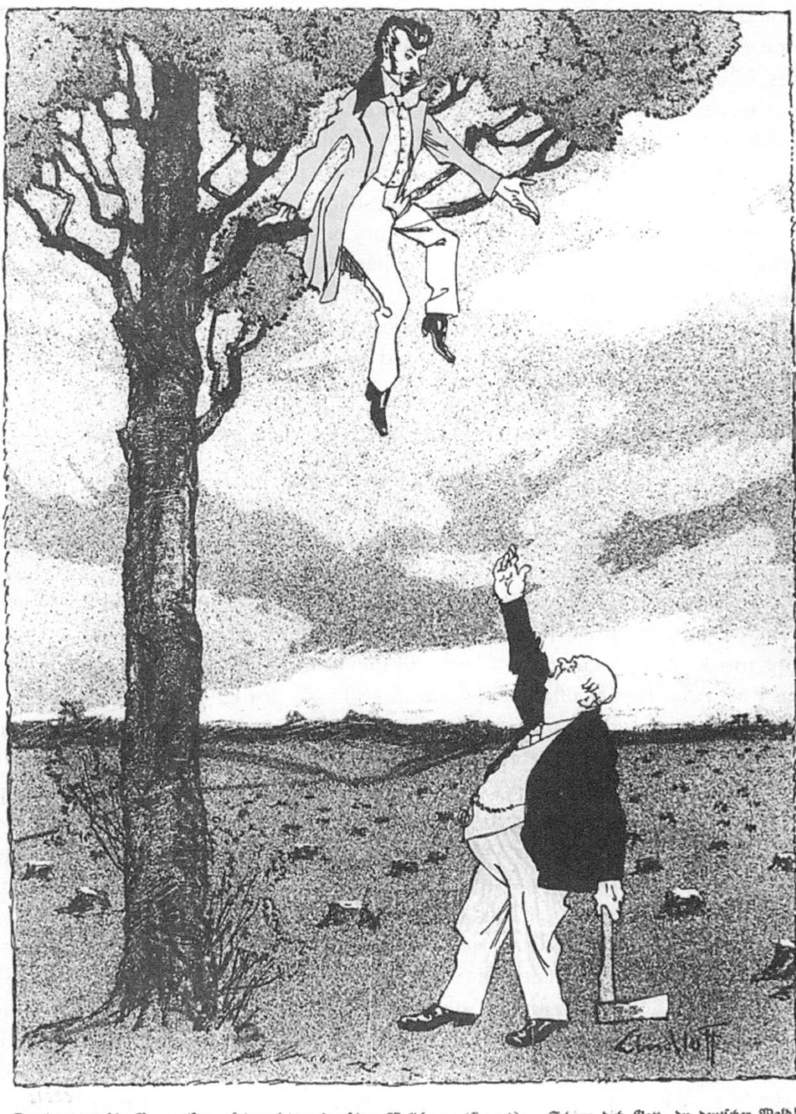

Figure 13.3 Anon. [signature unidentifiable] (1921). "Nieder mit dem deutschen Wald!" *Kladderadatsch*, 74(42), 2 (16 October 1921)

http://digi.ub.uni-heidelberg.de/diglit/kla1921/0585 [Accessed 1 August 2017].

nature. In this sense, it mocks the whole Romantic discourse represented by the other satires discussed earlier. Or, the ridicule could be directed against the revival of such an emotional attachment in a climate of growing nationalism. It should be noted that many caricatures from this time comment on the First World War reparations demanded from Germany by the Allied Powers in the Treaty of Versailles. Chancellor Joseph Wirth's policy of complying with the Allies' demands was criticized by conservatives and industrialists. This caricature appeared one week after his Minister for Reconstruction, Walter Rathenau (who resigned only a few weeks later), had reached an agreement with Loucheur, the French Minister of Reconstruction. The Treaty of Wiesbaden, signed on October 6 and 8 in 1921, allowed Germany to replace monetary payments with material goods such as wood.[22] Its many critics bemoaned the handing-over of the German forest, the symbol of German identity, to the French (cf. Escherich 1924).[23] In any case, the field of stumps and the fight over the forest is intended to mirror the deplorable state of the German nation, once pictured as a mighty oak forest.

In search of cultural identity – caricatures of the German oak

There is more than enough proof that before the mid-nineteenth century the association of the German oak and the German people had become a commonplace. This development is best documented by satire, whether in the form of poetry or verbo-visual caricature. Heinrich Heine, the last poet of Romanticism and the one who overcame it, is well-known for his satirical attacks on German society and morality, having been a critical observer of rising nationalism. The poet of Jewish origins mocks in particular the German's metaphoric self-image as a faithful oak. In his poem *Zur Beruhigung/For Reassurance* (1844), for example, he criticizes the people's lack of a revolutionary spirit by ironically praising its loyalty to its rulers.

As delineated before, the roots of the symbolic German forest with its legendary oaks extend back to pre-modern times, but new emphasis has been put on it since the early nineteenth century, and these symbols became ever more popular up until the Second World War. All journals bear witness to this fact, but satirical magazines especially document that the oak "soon became proverbial as a patriotic icon of bravery, power, and strength", which was held up against domestic political disruptions as well as used as a means of self-assurance in international conflicts (Zechner 2011: 20; cf. Hürlimann 1987). After 1871, many pictures show, without any need for explanations, an oak being trimmed, split and uprooted. In most cases the oak simply stands for the state or nation being shaken by all kinds of political issues.[24] In some cases, however, it explicitly symbolizes a transnational German culture, as in the caricature *Die deutsch-österreichische Eiche* (Fig. 13.4) that appeared on January 27, 1898, in the popular Austrian satirical magazine *Kikeriki*.

Here, the icon of German identity was shared by the Austrians, reminding us of the "Greater German Solution" ("großdeutsche Lösung") once considered, but then abandoned in 1848. In the centre of the picture we see a giant oak, which is being attacked from two sides, but does not succumb to the pressure. The

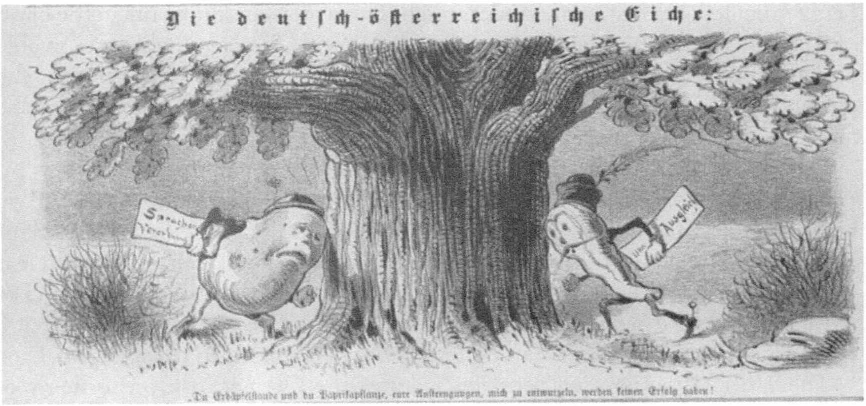

Figure 13.4 Anon. (1898). "Die deutsch-österreichische Eiche". *Kikeriki. Humoristisches Volksblatt*, 38(8), 2 (27 January 1898)

http://anno.onb.ac.at/cgi-content/anno?aid=kik&datum=18980127&seite=2 [Accessed 1 August 2017]

political background of this caricature is the nationalities' struggle in the multinational state Austria-Hungary ("Nationalitätenkampf"). The subtext reads: "Du Erdäpfelstaude und du Paprikapflanze, Eure Anstrengungen, mich zu entwurzeln, werden keinen Erfolg haben!"/ "You potato plant and you pepper plant, you won't succeed in your struggles to uproot me." These words are ascribed to the German-Austrian oak addressing the Slavs (represented by the potato) and the Hungarians (represented by the pepper). The 'pepper' invokes the Austro-Hungarian Compromise of 1867 (the so-called "österreichisch-ungarischen Ausgleich"), which established the dual-monarchy, that is, it re-established Hungarian as one of two official languages. All other ethnic groups of the monarchy envied the Magyars for their partial autonomy. Especially the Czech national movement demanded equalization in view of the Magyars and the Germans, but the requested compromise (the so-called "österreichisch-tschechischer" or "deutsch-tschechischer Ausgleich") soon collapsed. In the last third of the nineteenth century, the German-speaking minority in the Czech lands felt threatened by the Czech culture's fight for autonomy. Bowing to the growing pressure, in April 1897 the Austrian minister-president Count Kasimir Felix Badeni issued an ordinance called the "Badenische Sprachverordnung", which made Czech (besides German) the second official language in Bohemia and Moravia, thus requiring bilingualism in those who wanted to become civil servants. This ordinance – which the 'potato' holds in his hands against the 'oak' – incited protests and boycotts in the parliaments of Prague and Vienna instigated by the German Nationalist Party, and it led to Badeni's resignation at the end of November 1897, two months before the caricature appeared in *Kikeriki*. Since the political turmoil continued, Badeni's successor had to restrict this language ordinance, and in 1899 it was repealed.

In this caricature, it is not the oak and thus the Germans' self-image that is being ridiculed; instead, some readers might laugh at their opponents in the clash of cultures, while others laugh about the whole topic, the seemingly unresolvable cultural conflicts. While this caricature still emphasizes the steadfastness of the oak and thereby the dominance of German-Austrian culture, it foreshadows the fatal development of this conflict and the eventual fall or split of the oak.

Unsurprisingly, during the First World War, German journals show an increased need for self-affirmation and self-encouragement. On the one hand, we find poems praising the invincible oak combined with realistic drawings of giant oak trees, without any humorous note.[25] On the other hand, (self-)doubt and fear concerning the Germans' military strength is expressed via images of bending or mutilated trees. At worst, the unbendable oak is replaced by other tree species, indicating that the Germans can no longer be associated with oaks.[26]

The satirical journal *Simplicissimus* features the German oak on so many of its covers, that they would deserve a separate examination in view of this magazine's own ideological history. During the Wilhelmine Era, the journal was well-known for its satirical attacks on the authoritarian state as a whole, particularly its military and its police. After the declaration of war, however, the journal gave up its focus on internal enemies in favour of external ones and displayed a new patriotism. In the run-up before the First World War, it articulates the domestic political crisis by images of a hollowed oak trunk infested with parasites. Different parties are alternately given the role of the animal pests which gnaw on the oak (Fig. 13.5): the German Centre Party or Catholic Centre Party (Deutsche Zentrumspartei), the Social Democratic Party of Germany (Sozialdemokratische Partei Deutschlands) or the National Liberal Party (Nationalliberale Partei) were in turn attacked satirically for policies considered culturally unsustainable.[27] Of course, all criticism is articulated metaphorically.

During the war, of course, Germany's war opponents figured as the oak's enemies, who, in the first years, did not succeed in bringing down the oak. Interestingly, even in caricatures in which Germany's opponents are depicted as iconic national stereotypes in human form, the Germans are still rendered as a tree that is being attacked with axes, while the peaceful plant itself does not carry a weapon: as in a caricature from 1915, in which the members of the Triple-Entente try in vain to fell the oak (Fig. 13.6). After the lost war, the defeated nation is depicted as a defoliated oak.[28] One year later, a cover (Fig. 13.7) anticipates the ratification of the controversial Treaty of Versailles by showing the felling of the German oak and commenting sarcastically: "Der Friede ist perfekt – der Krieg kann weitergehn"/ "The Peace is perfect – the war can continue". Subsequently, caricatures from the early 1920s mirror the Germans' discontent with their leading politicians' compliance with the Treaty's reparations – a compliance ("Erfüllungspolitik") which was interpreted by many as a surrender of the nation's identity, and a threat to its cultural sustainability.[29] In the following years various crises were met with pictures showing a saw being taken to the oak. In summer 1922, with the onset of hyperinflation, a *Simplicissimus* cover depicted a defoliated oak being felled in order to be processed into paper urgently needed for banknotes.[30]

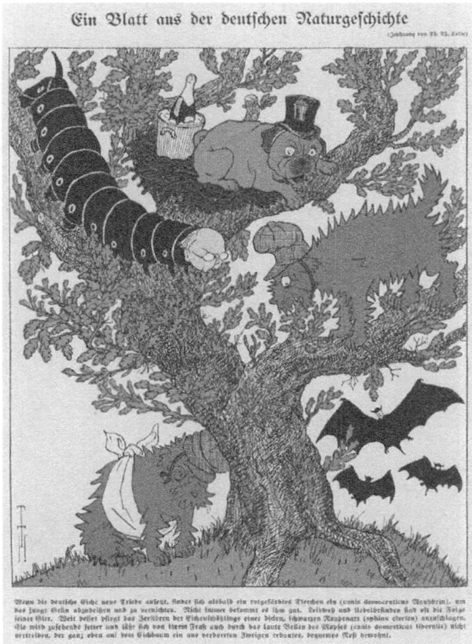

Figure 13.5 Heine, T. (1907). "Ein Blatt aus der deutschen Naturgeschichte". *Simplicissimus*, 11(49), 1 (4 March 1907)

www.simplicissimus.info/uploads/tx_lombkswjournaldb/1/11/11_49_789.jpg [Accessed 1 August 2017].

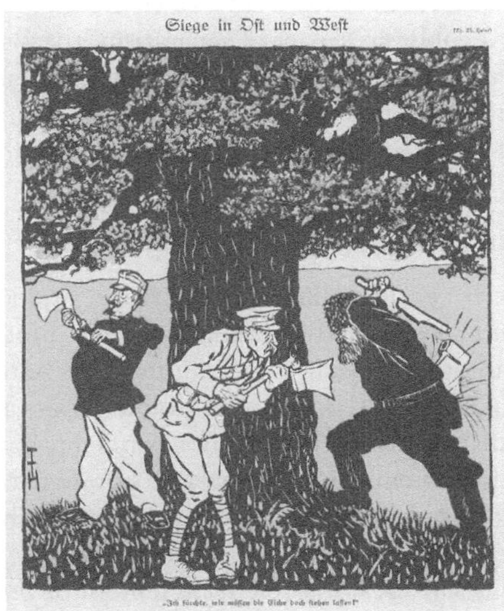

Figure 13.6 Heine, T. (1915). "Siege in Ost und West". *Simplicissimus*, 20(7), 1 (18 May 1915)

www.simplicissimus.info/uploads/tx_lombkswjournaldb/1/20/20_07_073.jpg [Accessed 1 August 2017].

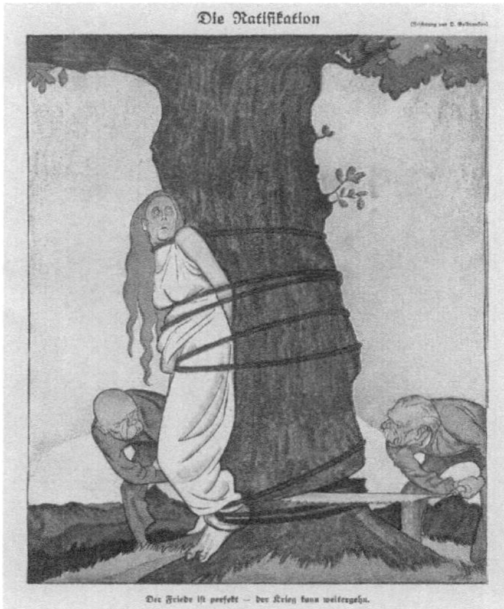

Figure 13.7 Gulbransson, O. (1919). "Die Ratifikation". *Simplicissimus*, 24(31), 1 (28 October 1919).

www.simplicissimus.info/uploads/tx_lombkswjournaldb/1/24/24_31_409.jpg [Accessed 1 August 2017].

And in 1923, when Bavaria declared a state of emergency and turned its back on the young Weimar Republic in an act of separatism, the Bavarians virtually saw off the branch on which they are sitting.[31] Almost ten years later, in late 1932, the oak finally lies on the ground cut into many pieces, the picture visualizing the political fragmentation in the Reichstag.[32]

Racial discourse and its satirical reflection

In the early twentieth century climate of increasing nationalism impregnated by racist ideas, cultural sustainability primarily meant protecting the 'purity' of the German people in order to secure their future. This notion, as one can imagine, provoked various satirical reactions.[33] However, these are not always easy to evaluate. Some poems and caricatures do not condemn nationalism and racism altogether; some of them seem to approve of a moderate nationalism, but ridicule a nationalism based on racist ideology; others solely mock the projection of racial ideas onto the forest.

In view of this essay's interest in the entanglement of verbo-visual discourses on cultural and environmental sustainability, it is worth taking a closer look at the racial connotations attributed to the German forest within anti-Semitic discourse. One very early example that demonstrates the instrumentalization of the German forest for anti-Semitic propaganda appeared in 1896 in the Austrian magazine *Kikeriki*. The journal was founded in 1861 and published in Vienna until 1933. Its founder, Ottokar Franz Ebersberg, was known as a democrat who promoted religious tolerance and stood up for the under-privileged. In its first three decades, the magazine represented liberal and pro-Semitic views, but between the early and mid-1890s, under the growing influence of the Christian Social Party (Christlichsoziale Partei Österreichs) and its founder Karl Lueger, the magazine began to promote anti-Semitic ideas. Thus, until the mid-1890s we find caricatures that mock and critique anti-Semitic discourse, while in the second half of the decade this discourse is affirmed and incited.[34]

As revealed by the names written below the drawing, this caricature, titled *Im Hochgebirge/ In the high mountains* (Fig. 13.8) features Kikeriki, the rooster mascot of the magazine, addressing Karl Lueger, who was at that time vice mayor of Vienna and known for his anti-Semitic politics.[35] As we learn from the subtext of the caricature, the rooster, who represents the naïve lower-class citizen, is astonished to see the vice mayor acting as woodcutter. Lueger is depicted with axe and saw, ready to fell the trees marked by the inscriptions "Jüdische Presse" (Jewish press), "Corruption", and "Großkapital" (high finance). The part of the subtext attributed to Lueger tells the reader: Only after those growing 'trees' are eliminated, will the "German forest" and the "Kircherl" (the small church in the background representing Christianity, or, more precisely, Catholicism) receive enough light and air to grow and thrive again ("Freili – i muß da a so a paar gar z'üppige Stämm' aushauen, damit unser deutscher Wald und das Kircherl drinn' wieder Licht und Luft kriegen!"). Drawing on the stereotypical enemy image of the wealthy and corrupt Jewish elite in control of mass media and thus a powerful cultural force, the caricature articulates the fear of being outnumbered, or, in accordance with the forest metaphor, of being overgrown by foreign peoples, especially Jewry.

As in most caricatures, it is up to the reader to recognize the object of mockery, here, to recognize whether Lueger's anti-Semitism is being affirmed or ridiculed. While we would naturally expect satire to mock Lueger's position, in this case the radical change of the journal's political line and other aggressively anti-Semitic jokes in the same issue argue against a critique of anti-Semitism. It seems to be a rather affirmative pictorial comment on Lueger's propaganda, yet there is a confusing little detail in the centre of the picture. How shall we interpret the fact that the rooster, when speaking to Lueger, drops his cigar, and thus risks setting fire to the whole forest and burning down not only the 'foreign' but the 'domestic' trees as well? In view of this, the caricature could be a warning that this anti-Semitic discourse can, again metaphorically speaking, lead to an uncontrollable 'forest fire'. It remains a matter of speculation how contemporary readers might have interpreted this contribution. In this essay, it serves as another example that locates culture – religious culture to be more precise – within the sphere of

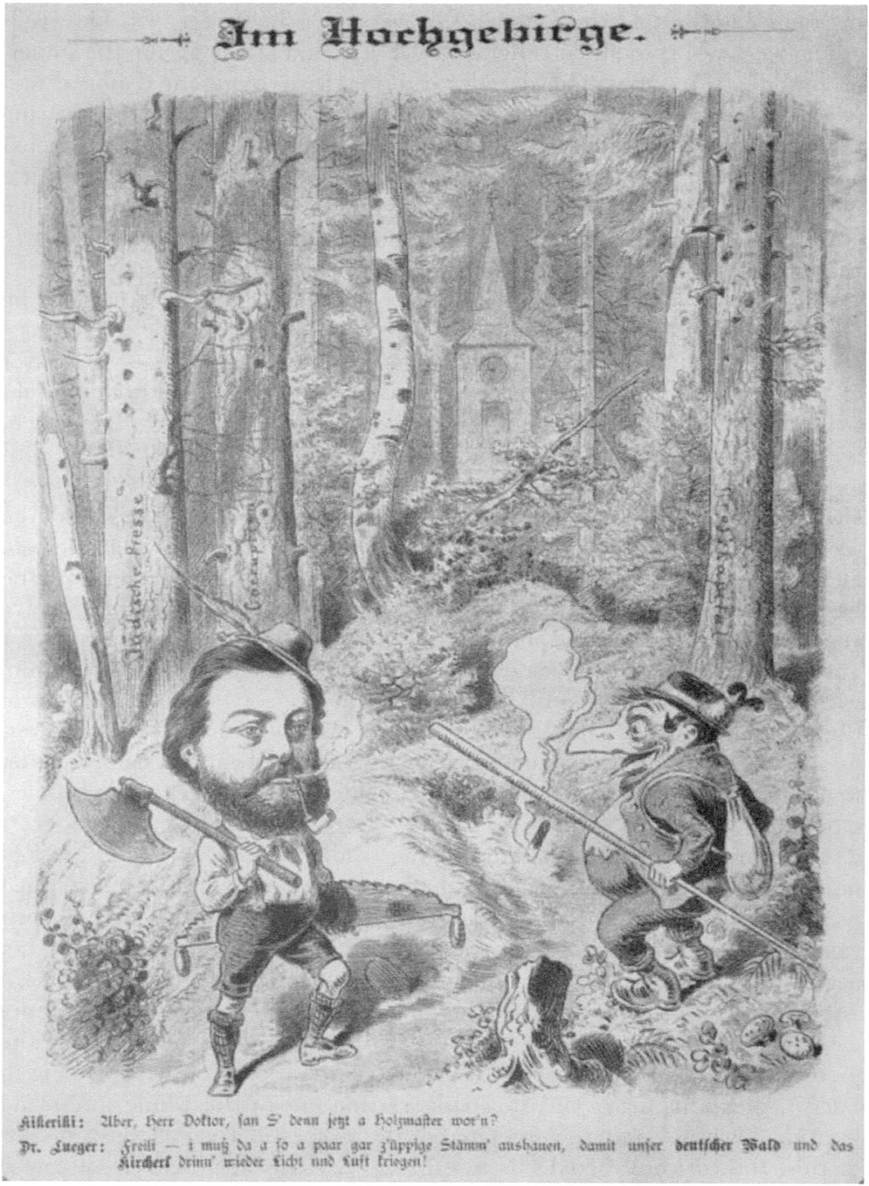

Figure 13.8 Anon. (1896) "Im Hochgebirge". *Kikeriki*, 36(74), 4 (13 September 1896) http://anno.onb.ac.at/cgi-content/anno?aid=kik&datum=18960913&seite=4& [Accessed 1 August 2017]

nature and associates natural heritage (the forest) with cultural heritage (the church). This seems to be underlined by the title *Im Hochgebirge*: The high mountains were considered the last region where one could still find wilderness in the sense of a natural nature – a genuine source of cultural energy that could guarantee cultural sustainability (cf. Riehl 1854: 64–65).

Another caricature that plays with the potential for a racial semantization of the German forest undoubtedly ridicules anti-Semitic discourse: The contribution *Deutscher Wald im Dritten Reich/German Forest in the Third Reich* (Fig. 13.9) appeared more than three decades later, in 1932, shortly before the Nazis' seizure of power, in *Der Wahre Jacob*. The satirical journal, which had been published since 1879, was politically close to the Social Democratic Party; due to its sharp attacks on Nazi politics, it was prohibited in 1933.

This caricature depicts a Nazi and a forester inspecting the forest, standing in front of a beech tree. The Nazi asks the forester to "draw some sap" from the tree in order to "have it tested for 'Semikokken'",[36] the latter being a fictitious term hinting at 'Jewish bacteria'. At the time, this word creation was not entirely new. It had been coined a few years earlier in an anti-Semitic discourse that associated the Eastern Jews with infectious diseases.[37] The caricature mocks the Nazis' pseudo-scientific racism. In the same issue of the journal, another caricature turns the disease metaphor around and uses it against the Nazis themselves: In a drawing of the German Michel (Deutscher Michel), his back is dotted by swastikas indicating that he suffers from a bad rash. The caricature's critique of the Nazis' idiotic extension of their ideology to nature itself was not far from reality: In Nazi documents such as Hermann Göring's famous speech "Ewiger Wald – ewiges Volk"/"Eternal Forest – Eternal People", the military leader describes the forest as the "foundation of German culture" ("Grundlage deutscher Kultur", Göring 1940: 250).

Interestingly, in this caricature the oak is for once replaced by the most common deciduous tree in Germany: the beech, more precisely, the copper beech (or European beech), called "Rotbuche" or "Blutbuche" in German. Using the botanically equally correct term "Blutbuche", the subtext reinforces the pseudo-medical context of the blood test and at the same time hints at the Nazis' blood and soil ideology ("Blut und Boden"). Less known, but also interesting, are other connotations of the beech, which is an ambivalent symbol (cf. Zumsteg 2008). It stands for strength, but also for cultural creativity and for calamity: While praised as *genius loci* in many German poems (e.g. by Eduard Mörike), it also figures as a gallows tree in famous texts (most prominently Annette von Droste-Hülshoff's *Die Judenbuche*). Most importantly, this tree is a basic symbol for text culture due to the etymological link between "Buch" (book) and "Buche" (beech), resting on the fact that the Germanic peoples had used beech rods for writing. Choosing the beech (along with the oak) as a symbol for a people or nation means characterizing it as both closely connected to nature and as highly civilized. In view of that period's brute nationalism, caricatures can only ironize this self-image.

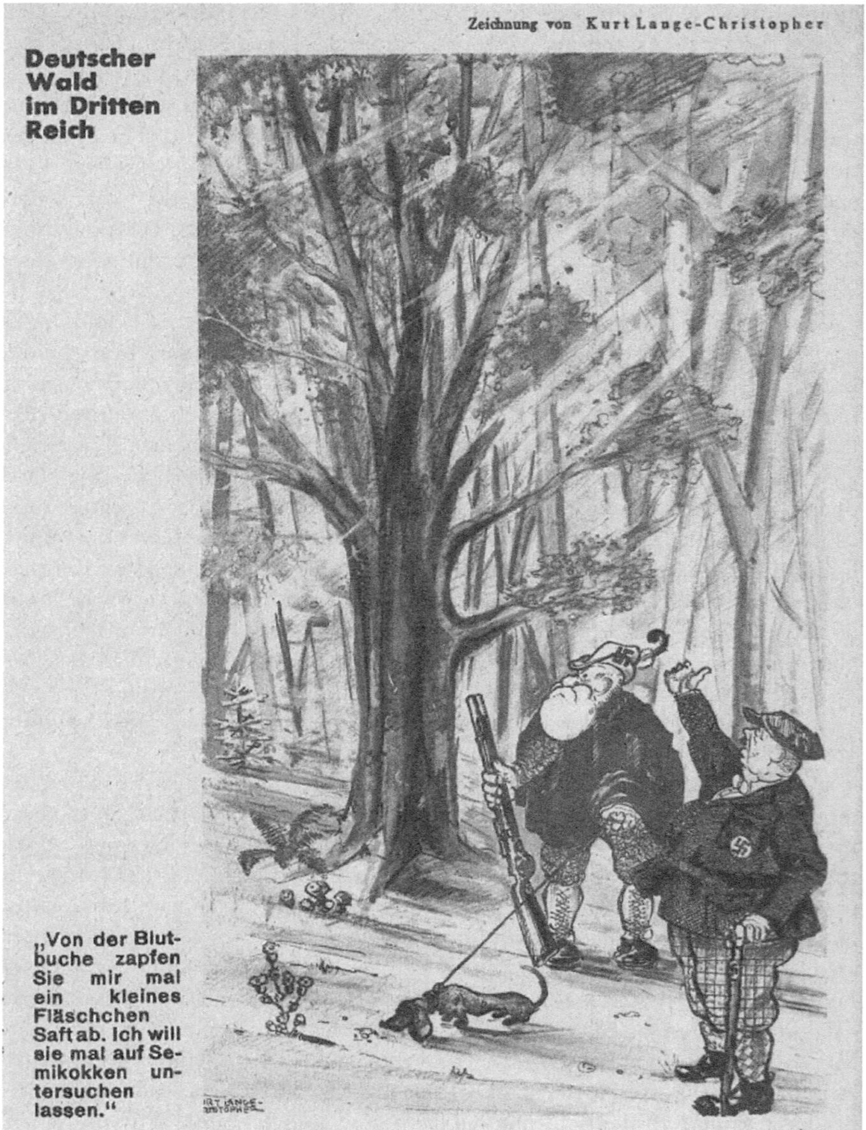

Figure 13.9 Lange-Christopher, K. (1932). "Deutscher Wald im Dritten Reich". *Der Wahre Jacob*, 53(12), 14 (4 June 1932)

http://digi.ub.uni-heidelberg.de/diglit/wj1932/0194 [Accessed 1 August 2017].

Outlook and conclusion

If we accept one of the central premises of Finke's and Zapf's cultural ecology and assume that diversity is a precondition for cultural sustainability, the nationalist and racist approaches doubtlessly pointed in the wrong direction. After the Second World War, the German forest and the German oak as national icons almost completely disappeared from political and cultural discourse, but a special attachment to the forest was still articulated in popular culture such as the post-war Heimatfilm (e.g. *Grün ist die Heide*, 1951; *Der Förster vom Silberwald*, 1954). However, since the key words, German forest and German oak, had no political function for a while, all satire that focused on current issues spared these seemingly discarded icons. It has to be added that satire as a medium did not thrive in the decades following the war. Most satirical magazines had to close down during the Second World War; it was the end of a flourishing German culture of verbal and visual satire. In the second half of the twentieth century, satirical magazines did not play the same role as before. The most prominent German magazine, *Titanic*, was founded as late as 1979 – just in time to witness the public panic over the "Waldsterben" (forest dieback), which brought the German forest back into the centre of public attention.[38] Here is not the place to discuss the countless caricatures and the satirical treatment of this topic, but it should be mentioned that the amount of media attention and concern for this environmental problem can partly be explained by the semantization of the forest in German cultural history. This is in line with the observation that the Germans brought more emotion to the debate on forest damage than their European neighbours, who were more concerned about the general effects of air pollution, not least on humans (cf. Metzger et al. 2009: 43). During this period, German media reports fell back on old ideas, as they tended to interpret the forest dieback as a threat to cultural sustainability.[39] The emerging environmental movement and the Green Party (founded in 1980) greatly profited from the Germans' emotional attachment to their forest. This has been noticed and commented on in caricatures that mock the instrumentalization of the German forest for political goals, especially by the Green Party. In the twenty-first century, Klaus Stuttmann depicts the 'German forest' as a wind farm that, with its many wind turbines standing in dense rows and covering a great part of the land, resembles a forest (Figures 13.10 and 13.11).

The visual analogy hints at the ambiguity between the mythification of the German forest and the pleas to protect it on the one side and its use as a natural resource on the other side. These caricatures suggest that the Germans now worship wind energy instead of the forest; both idealizations are considered stupid, the adoration of the wind turbines perhaps even more so. Above all, the cartoons criticize the presumed ideological and emotional elements in energy politics. The cartoon published on the occasion of the Green Party's twenty-fifth anniversary (Fig. 13.10: "25 Jahre Die Grünen. Schützt den deutschen Wald!"/ "Save the German forest!") implies that in the twenty-first century the Germans' wind farms have become as famous as the German forest had been for two thousand years. But it also criticizes the Germans for concentrating on another object of fancy,

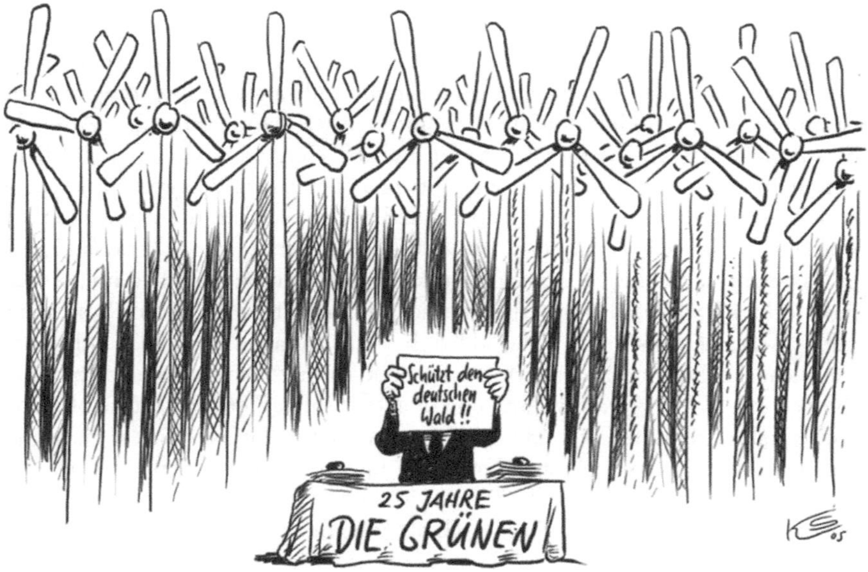

Figure 13.10 Stuttmann, K. (2005). “25 Jahre Die Grünen. Schützt den deutschen Wald!!” *Der Tagesspiegel* (13 January 2005)

www.stuttmann-karikaturen.de/karikatur/660 [Accessed 1 August 2017].

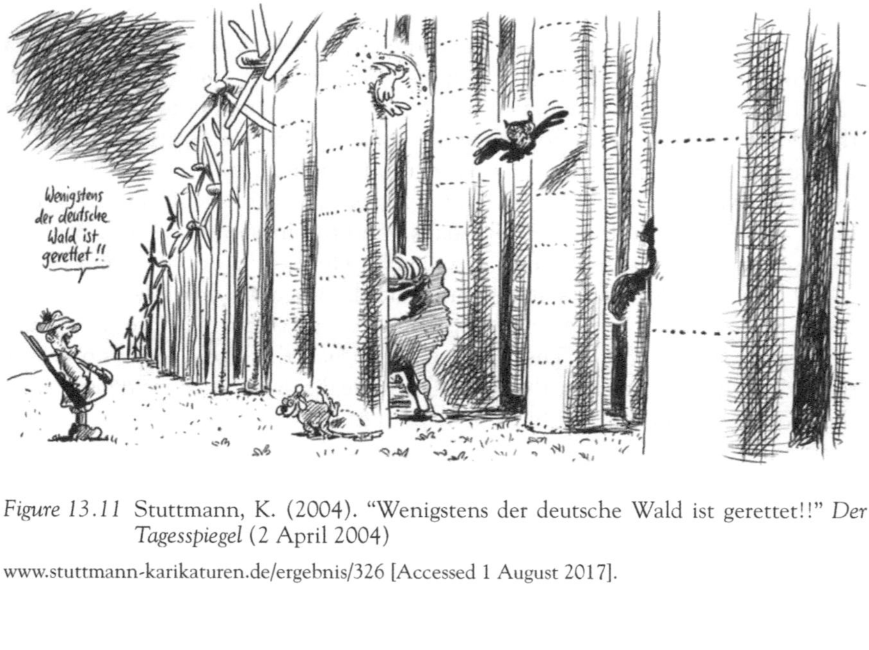

Figure 13.11 Stuttmann, K. (2004). “Wenigstens der deutsche Wald ist gerettet!!” *Der Tagesspiegel* (2 April 2004)

www.stuttmann-karikaturen.de/ergebnis/326 [Accessed 1 August 2017].

thinking that they have already saved their forest by environmental protection policies, while they destroy landscapes by building wind farms, which could be considered as disastrous as forest clearings. The other caricature's message (Fig. 13.11: "Wenigstens der deutsche Wald ist gerettet!"/ "At least the German forest has been saved!") is equally ironic, since it calls attention to the fact that wind turbines, even if they slightly resemble trees from afar, are no adequate substitute and neither fulfil the animals' nor the humans' needs. These new 'German forests' may provide power, but no cultural energy.

In conclusion, it could be noted that in recent satire on environmental transformations the whole concept of a 'German forest' is decidedly exposed to ridicule. Satires that use the term improperly for non-natural anthropogenic phenomena not only point to a change in values, but they also suggest that any cultural mythification of nature is illusionary, and its discursive instrumentalization is reactionary. This does not mean, however, that discourses on environmental sustainability and on cultural sustainability can be disentangled. To the contrary, 'quality media' often quietly presuppose a common understanding that cultural sustainability always depends on environmental sustainability, while 'sensational media' and politics tend to the mythification and instrumentalization mentioned above. Today, the protection of landscapes is still propagated by emphasizing their cultural meaning, but the nationalist view has gradually been replaced by a global perspective – just as sustainability altogether has become a transcultural issue.

Notes

1 All translations from German sources into English are mine unless otherwise noted.
2 See Riehl's critical comments on the forest clear cuttings of the nineteenth century and on the substitution of deciduous forests with conifers as well as his plea for special protection of the oaks (Riehl 1854: 35–37).
3 For more detail, see Zapf (2016b).
4 These remarks can be found at greater length in Zemanek (2017).
5 These remarks can be found at greater length in Zemanek (2017).
6 In our context it suffices to refer to the debate, regardless of the thesis that the degree of the alleged wood shortage did not correspond to reality. Cf. Radkau (2011: 40–42).
7 Zechner (2016) considers it as an 'imagined landscape' and traces its facets from the Romans to the Nazis. My essay owes much background knowledge and many references to this study.
8 See Gaius Plinius Secundus. *Naturalis historiae*, book XVI, ch. 2; Julius Ceasar. *De bello gallico*, book VI, ch. 21, 23, 25–28; and Publius Cornelius Tacitus. Germania, ch. 2–5.
9 Publius Cornelius Tacitus, *Annales*, book II, ch. 88.
10 See Klopstock's odes *Thuiskon*, *Der Hügel und der Hain*, *Unsere Fürsten* and *Mein Wäldchen* (Klopstock 2010: 248, 300–304, 278–279, 365). For more details see Zechner (2016: 21–23).
11 For references to the Grimms' writings, I am indebted to Zechner (2016: 83–104).
12 On the language purism movement, see Gardt (2000).

13 They express this idea in many of their writings; most prominently: Grimm, J. and Grimm, W. (1812): VI. On the Grimms' mythification of the forest, see Ono (2007); Zechner (2013). For an ecocritical interpretation, see Stobbe (2017: 298–304).
14 For forest imaginations in German Romantic art, see Maringer (2009); for the development of the German forest, see Bernhard (2012).
15 For an exceptional focus on environmental history in Eichendorff scholarship, see Frühwald (1994).
16 In Mendelssohn's works the songs are titled "Der Jäger Abschied", also known as "Jägers Abschied" or "Der deutsche Wald" ("Wer hat dich, du schöner Wald") op. 50,2 (1840) and "Abschied vom Wald" ("O Täler weit, o Höhen") op. 59,3 (1843).
17 For a discussion of the picture (Detjens, F. and Wagner, E. [1895]), see Zemanek (2017, 132–134). – All artworks mentioned that appeared in *Fliegende Blätter*, in *Kladderadatsch* and in *Der wahre Jacob* are in the digital library of Heidelberg University available at: http://digi.ub.uni-heidelberg.de.
18 For an explanatory comment on this law, see Amtspresse Preußens (1882).
19 For example: "Das Forst- und Feldpolizeigesetz" (Anon. 1880).
20 There was an outcry against Podbielski articulated in the liberal leftist press inflamed by Berlin workers and citizens who wanted to secure the forest as recreation area. See the caricature "Pod der Waldverwüster" (Stutz 1904). – For a detailed account of the fight over the Grunewald, see Wilson (2012a: esp. 86–131; 2012b).
21 On the building of the Grunewald racecourse, see "Die neue Rennbahn im Grunewald bei Berlin" (Anon. 1907). – Podbielski also played a key role in the construction of the racecourse. The protests against the sell-out of the forest culminated once more in January 1909 at the "Zweiter Berliner Waldschutztag"/"Second Berlin Forest Protection Day". In 1915, finally, the Prussian state sold the Grunewald to Berlin; the city assured its inhabitants that the newly bought 10000 hectares of forest would be preserved and exclusively used as recreation area.
22 Still searching for a solution concerning the conflict over reparations, Walther Rathenau, who as a successful Jewish business man also had to face anti-Semitic attacks, was assassinated on June 24, 1922.
23 This view was also articulated in poems, e.g. in another one that appeared in *Kladderadatsch* in 1923 titled "Abschied vom Walde" (Anon. 1923), which again begins with the parodistic question "Wer hat dich, du schöner Wald,/ Abgeholzt. . . ?" and answers "Frankreich hat Dich, schöner Wald!" ("It is France who has got you now, beautiful forest!").
24 Again, many caricatures could be mentioned here. A typical early example from *Kladderadatsch*: "Wie man das Hindernis wegschafft für eine große liberale Partei" (Anon. 1880).
25 See "Die deutsche Eiche" (Anon. 1917).
26 See the text-image-combination titled "Einst und Jetzt" (Anon. 1911) from *Kikeriki*. All artworks mentioned that appeared in *Kikeriki* are available at: http://anno.onb.ac.at/cgi-content/anno?aid=kik [accessed 1 Aug 2017].
27 For caricatures depicting these parties as animal pests, see also "Die deutsche Eiche" (Schulz 1904). – All artworks mentioned that appeared in *Simplicissimus* are available at: www.simplicissimus.info [Accessed 1 Aug 2017].
28 See the caricature "Deutscher Herbst" (Gulbransson 1918) that appeared one week after the Armistice of 11 November 1918.
29 See, for example, "Erzberger und die deutsche Eiche" (Arnold 1920).
30 See "Dividenden über alles" (Arnold 1922).
31 See the caricature "Bayern und Reich" (Arnold 1923), which appeared two weeks before the Munich Putsch.
32 See "Einstimmigkeit" (Gulbransson 1932).

33 See the poem "Treueschwur", which appeared in *Simplicissimus* in 1907 next to a caricature drawn by Eduard Thöny (Thoma and Thöny 1907). The poem was written by Ludwig Thoma, who was at that time chief editor of the Simplicissimus and known for his biting left-liberal critique, which is why he covered his identity with the pseudonym Peter Schlemihl. His poem mocks a pledge of allegiance to Germania; the statue is part of the Niederwalddenkmal close to Rüdesheim on the shores of the Rhine River.

34 On *Kikeriki*'s political outlook, see Holzinger (2015). On caricatures of Jews, see Schäfer (2005).

35 When this caricature appeared in September 1896, Lueger had just been elected mayor of Vienna, but since the Emperor Franz Joseph I resented Lueger's blatant anti-Semitism and prevented his appointment several times, Lueger had to serve as vice mayor until Franz Joseph finally conceded to a plea by Pope Leo XIII and accepted him as mayor. In 1893, Lueger, a zealous Catholic, had founded the Christlichsoziale Partei, who won their voters from the petite bourgeoisie with anti-capitalistic and anti-Semitic propaganda. During his mayoralty Lueger was very popular thanks to infrastructural reforms that improved everyday life in the metropolis – not the least important of which being the transformation into a green city – but also due to the emergence of a personality cult that would later be imitated by Hitler's entourage.

36 The German subtext reads: "Von der Blutbuche zapfen Sie mir mal ein kleines Fläschchen Saft ab. Ich will sie mal auf Semikokken untersuchen lassen."

37 In 1926, the magazine *Vorwärts* – the official voice of the Social Democratic Party since 1876 – included a caricature in which a doctor examines the members of the Reichstag and tests their blood for "Semikokken". Cf. Nemitz (2000: 67).

38 On this public debate, see Metzger (2015); Bemmann and von Detten (2013).

39 'Old ideas' refers to the attribution of a cultural value to the forest, not to nationalist or racist ideas. Cf. Metzger et al. (2009: 45).

References

Amtspresse Preußens (1882). "Neueste Mittheilungen. No. 27. Verantwortlicher Herausgeber Dr. H. Klee. Berlin, den 31. Juli 1882". *Staatsbibliothek zu Berlin* (31 July 1882/13 May 2013). http://zefys.staatsbibliothek-berlin.de/amtspresse/ansicht/issue/11614109/28 [Accessed 1 August 2017].

Anon. [signature unidentifiable] (1880). "Das Forst- und Feldpolizeigesetz". *Kladderadatsch*, 33(10), 96. http://digi.ub.uni-heidelberg.de/diglit/kla1880/0096 [Accessed 1 August 2017].

Anon. (1880). "Wie man das Hindernis wegschafft für eine große liberale Partei". *Beiblatt zum Kladderadatsch*, 33(35), 1 (1 August 1880). http://digi.ub.uni-heidelberg.de/diglit/kla1880/0375 [Accessed 1 August 2017].

Anon. (1907). "Die neue Rennbahn im Grunewald bei Berlin". *Sport und Salon. Illustrierte Zeitschrift für die vornehme Welt*, 10(49), 15 (7 December 1907). http://anno.onb.ac.at/cgi-content/anno?aid=sus&datum=19071207 [Accessed 1 August 2017].

Anon. [signature unidentifiable] (1911). "Einst und Jetzt". *Kikeriki*, 51(89), 2 (5 November 1911). http://anno.onb.ac.at/cgi-content/anno?aid=kik&datum=19111105&seite=2 [Accessed 1 August 2017].

Anon. (1917). "Die deutsche Eiche". *Beilage zum Wahren Jacob*, 34(817), 2 (9 November 1917). http://digi.ub.uni-heidelberg.de/diglit/wj1917/0218 [Accessed 1 August 2017].

Anon. (1923). "Abschied vom Walde". *Kladderadatsch*, 76(2) (14 January 1923), 6. http://digi.ub.uni-heidelberg.de/diglit/kla1923/0022 [Accessed 1 August 2017].

Arnold, K. (1920). "Erzberger und die deutsche Eiche". *Simplicissimus*, 24(50), 1 (10 March 1920). www.simplicissimus.info/uploads/tx_lombkswjournaldb/1/24/24_50_713.jpg [Accessed 1 August 2017].

Arnold, K. (1922). "Dividenden über alles". *Simplicissimus*, 27(29), 1 (18 October 1922). www.simplicissimus.info/uploads/tx_lombkswjournaldb/1/27/27_29_409.jpg [Accessed 1 August 2017].

Arnold, K. (1923). "Bayern und Reich". *Simplicissimus*, 28(30), 1 (22 October 1923). www.simplicissimus.info/uploads/tx_lombkswjournaldb/1/28/28_30_369.jpg [Accessed 1 August 2017].

Bemmann, M. and von Detten, R. (eds.) (2013). *Das Waldsterben: Rückblick auf einen Ausnahmezustand*. München: Oekom Verlag.

Bernhard, A. (2012). "Der deutsche Wald in Malerei und Graphik". In U. Breymayer and B. Ulrich (eds.). *Unter Bäumen: Die Deutschen und der Wald*. Berlin: Sandstein, 129–160.

Detjens, F. and Wagner, E. (1895). "Abschied vom Walde". *Fliegende Blätter*, 103(2626), 194–195. http://digi.ub.uni-heidelberg.de/diglit/fb103/0194 [Accessed 1 August 2017].

Escherich, G. (1924). *Der deutsche Wald und die feindlichen Mächte*. Hamburg: Deutscher Wald e.V.

Finke, P. (2003). "Kulturökologie". In A. Nünning and V. Nünning (eds). *Einführung in die Kulturwissenschaften: Theoretische Grundlagen – Ansätze – Perspektiven*. Stuttgart – Weimar: Metzler, 248–279.

Frühwald, W. (1994). "'Schlesische Toleranz' und 'preußische Reform': Sozialgeschichtliche Grundlagen einer Jugendbiographie Joseph von Eichendorffs". In G. Hahn and E. Weber (eds.). *Zwischen den Wissenschaften: Beiträge zu einer deutschen Literaturgeschichte*. Regensburg: Pustet, 10–24.

Gardt, A. (2000). "Sprachnationalismus zwischen 1850 und 1945". In A. Gardt (ed.). *Nation und Sprache: Die Diskussion ihres Verhältnisses in Geschichte und Gegenwart*. Berlin – New York: De Gruyter, 247–272.

Göring, H. (1940). "Ewiger Wald – ewiges Volk. Rede auf der Tagung des deutschen Forstvereins". In E. Gritzbach (ed.). *Hermann Göring: Reden und Aufsätze* [1936]. München: Eher, 245–255.

Grimm, J. (1884). "Vorträge auf den Germanistenversammlungen/Über die wechselseitigen Beziehungen und die Verbindung der drei in der Versammlung vertretenen Wissenschaften". In J. Grimm (ed.). *Kleinere Schriften* [1846]. Vol. 7, Part 4. Berlin: Dümmler, 556–563.

Grimm, J. and Grimm, W. (1812). "Vorwort". In *Kinder- und Hausmärchen gesammelt durch die Brüder Grimm*. Vol. 1. Berlin: Realschulbuchhandlung.

Grimm, W. (1881). "Bericht über das Deutsche Wörterbuch". In W. Grimm and G. Hinrichs (eds.). *Kleinere Schriften*. Vol. 1. Berlin: Dümmler, 508–520.

Gulbransson, O. (1918). "Deutscher Herbst". *Simplicissimus*, 23(34), 1 (19 November 1918). www.simplicissimus.info/uploads/tx_lombkswjournaldb/1/23/23_34_413.jpg [Accessed 1 August 2017].

Gulbransson, O. (1932). "Einstimmigkeit". *Simplicissimus*, 37(31), 1 (30 October 1932). www.simplicissimus.info/uploads/tx_lombkswjournaldb/1/37/37_31_361.jpg [Accessed 1 August 2017].

Holzinger, S. (2015). *Die Darstellung von Juden und Jüdinnen im Humoristischen Volksblatt 'Kikeriki'*. MA Thesis. University of Graz. http://unipub.uni-graz.at/obvugrhs/content/titleinfo/485590 [Accessed 1 August 2017].

Hürlimann, A. (1987). "Die Eiche, heiliger Baum deutscher Nation". In B. Weyergraf (ed.). *Waldungen: Die Deutschen und ihr Wald*. Berlin: Nicolai, 62–68.

J. S. [initials unknown] (1909). "Der Grunewald bei Berlin: Ein Zukunftsbild nach bekannter Melodie". *Illustrierte Unterhaltungs-Beilage des Wahren Jacob*, 594 (27 April 1909), 6211. http://digi.ub.uni-heidelberg.de/diglit/wj1909/0143 [Accessed 1 August 2017].

Klopstock, F. G. (1975). *Die deutsche Gelehrtenrepublik*. Vol. 1 (Werke und Briefe. Historisch-kritische Ausgabe, Abt. Werke: VII.1). R.-M. Hurlebusch (ed.). Berlin – New York: De Gruyter.

Klopstock, F. G. (2009). *Hermanns Schlacht*. In M. E. Amtstätter (ed.). *Klopstock: Hermann-Dramen*. Vol. 1 (Werke und Briefe. Historisch-kritische Ausgabe, Abt. Werke: VI.1). Berlin – New York: De Gruyter, 1–154.

Klopstock, F. G. (2010). *Klopstock: Oden*. Vol. 1 (Werke und Briefe: Historisch-kritische Ausgabe, Abt. Werke: I.1). H. Gronemeyer and K. Hurlebusch (eds.). Berlin – New York: De Gruyter.

Maringer, E. (2009). "Natur als Spiegelbild der Seele: Der Wald in der deutschen Malerei der Romantik (1790–1840)". In Landschaftsverband Westfalen-Lippe (ed.). *Mythos Wald*. Bönen: Kettler, 27–34.

Metzger, B. (2015). *"Erst stirbt der Wald, dann Du!" Das Waldsterben als westdeutsches Politikum 1978–1986*. Frankfurt – New York: Campus.

Metzger, B., Bemmann, M. and Schäfer, R. (2009). "'Erst stirbt der Wald, dann stirbt der Mensch.' Was hatte das Waldsterben mit dem deutschen Waldmythos zu tun?". In Landschaftsverband Westfalen-Lippe (ed.). *Mythos Wald*. Bönen: Kettler, 43–53.

Nemitz, K. (2000). *Die Schatten der Vergangenheit: Beiträge zur Lage der intellektuellen deutschen Juden in den 20er und 30er Jahren*. Oldenburg: Bis.

Ono, H. (2007). "Waldsymbolik bei den Brüdern Grimm". *Fabula*, 48(1–2), 73–84.

Radkau, J. (2011). *Die Ära der Ökologie. Eine Weltgeschichte*. Munich: C. H. Beck.

Riehl, W. (1854). *Die Naturgeschichte des Volkes als Grundlage einer deutschen Social-Politik*. Vol. 1: *Land und Leute*. Stuttgart – Tübingen: Cotta.

Schäfer, J. (2005). *Vermessen – gezeichnet – verlacht. Judenbilder in populären Zeitschriften 1918–1933*. Frankfurt: Campus.

Schultze-Naumburg, P. (1916/1917). *Die Gestaltung der Landschaft durch den Menschen*. Vols. 1–3 (Kulturarbeiten Vols. 7–9). 3rd Ed. Munich: Callwey.

Schulz, W. (1904). "Die deutsche Eiche". *Simplicissimus*, 9(28), 271. www.simplicissimus.info/uploads/tx_lombkswjournaldb/1/09/09_28_271.jpg [Accessed 1 August 2017].

Steinsiek, P.-M. (2012). "Nachhaltigkeit. Zur Karriere eines Begriffs". In U. Breymayer and B. Ulrich (eds.). *Unter Bäumen: Die Deutschen und der Wald*. Berlin: Sandstein, 91–97. www.gesetze-im-internet.de/bwaldg/ [Accessed 1 August 2017].

Stobbe, U. (2017). "'Green' Readings: Grimms' Fairy Tales and Christa Wolf's 'Störfall'". In G. Dürbeck, U. Stobbe, H. Zapf and E. Zemanek (eds.). *Ecological Thought in German Literature and Culture*. Lanham, MD: Lexington Books, 297–314.

Stutz, L. (1904). "Pod der Waldverwüster". *Kladderadatsch*, 57(41), 164 (9 October 1904). http://digi.ub.uni-heidelberg.de/diglit/kla1904/0570/image [Accessed 1 August 2017].

Thoma, L. and Thöny, E. (1907). "Treueschwur". *Simplicissimus*, 12(1), 11. www.simplicissimus.info/uploads/tx_lombkswjournaldb/1/12/12_01_011.jpg [Accessed 1 August 2017].

von Carlowitz, H. (1713). *Sylvicultura oeconomica, oder haußwirthliche Nachricht und Naturmäßige Anweisung zur wilden Baum-Zucht*. Leipzig: Braun. http://digital.slub-dresden.de/werkansicht/dlf/85039/127/ [Accessed 1 August 2017].

Wehrhahn-Stauch, L. (1958). "Einhorn". In E. Gall and L. H. Heydenreich (eds.). *Reallexikon zur deutschen Kunstgeschichte*. Vol. 4. Stuttgart: Druckenmüller, 1503–1540.

Wilson, J. (2012a). *The German Forest. Nature, Identity, and the Contestation of a National Symbol 1871–1914*. Toronto: University Press.

Wilson, J. (2012b). "Waldverwüster und Waldverliebte: Der Konflikt um den Berliner Grundwald (1860–1920)". In U. Breymayer and B. Ulrich (eds.). *Unter Bäumen: Die Deutschen und der Wald*. Berlin: Sandstein, 216–222.

Zapf, H. (2016a). "Cultural Ecology, the Environmental Humanities, and the Transdisciplinary Knowledge of Literature". In S. Oppermann and S. Iovino (eds.). *Environmental Humanities: Voices from the Anthropocene*. New York: Rowman and Littlefield, 61–79.

Zapf, H. (2016b). *Literature as Cultural Ecology: Sustainable Texts*. London: Bloomsbury.

Zechner, J. (2011). "Politicized Timber: The 'German Forest' and the Nature of the Nation 1800–1945". *Brock Review*, 11(2), 19–32.

Zechner, J. (2013). "From Poetry to Politics: The Romantic Roots of the 'German Forest'". In W. Beinart, K. Middleton and S. Pooley (eds.). *Wild Things: Nature and the Social Imagination*. Cambridge: White Horse, 197–199.

Zechner, J. (2016). *Der deutsche Wald: Eine Ideengeschichte zwischen Poesie und Ideologie 1800–1945*. Darmstadt: Philipp von Zabern.

Zemanek, E. (2017). "Mocking the Anthropocene: Caricatures of Man-made Landscapes in German Satirical Magazines from the Fin de Siècle". In S. Wilke and J. Johnstone (eds.). *Readings in the Anthropocene: The Environmental Humanities, German Studies, and Beyond*. New York – London: Bloomsbury Academic, 124–147.

Zumsteg, S. (2008). "Buche". In G. Butzer and J. Jacob (eds.). *Metzler Lexikon literarischer Symbole*. Stuttgart: Metzler, 59–60.

14 The cultural sustainability of Victorian waste

Ursula Kluwick

Introduction

Cultural sustainability has the ring of a new buzzword. It is probably most readily associated with the replacement of the three-pillar model of sustainable development with a four-pillar model that includes cultural as an addition to ecological, economic, and social aspects to indicate that the role of culture is increasingly being taken into account as a key component of sustainability. However, and as my phrasing already suggests, the idea that cultural sustainability is in some way new is entirely a matter of perspective. Because if we disregard the fact that the *attention paid* to the role of culture in sustainability is new, we are left with the impression that there is in fact really nothing particularly novel about the involvement of culture in questions of sustainability. Indeed, sustainability has everything to do with culture, be it culture as a way of life (after Raymond Williams) or culture as specific artistic and intellectual practices. My contribution to this volume highlights this through a historical perspective on the concept of cultural sustainability and through a focus on the conversation between the two meanings of culture just indicated. I zoom in on a specific controversy in Victorian Britain that circulated around what I argue is a question of sustainability: waste management, in particular discussions about the utilisation of human excrement as manure and the relative merits of a dry conservancy versus a water-borne system of sewage removal. I look at the interdiscursive engagement of Victorian literature with this issue, arguing that it can be read in terms of a form of cultural sustainability. Literary representations of water in the nineteenth century helped naturalise the connection between water and sewage, thereby to some extent also supporting the actual re-organisation of waste disposal. In this sense, literary production contributed to a climate in which a form of waste disposal was adopted that, as far as the use of resources is concerned, was arguably less sustainable than the traditional dry conservancy method that had hitherto been practiced. In this sense, water-borne sewage disposal was culturally sustainable because in the cultural imaginary it figured as a more endurable method enabling a healthier future. One of the main objectives of my essay, then, is to draw attention to the mechanisms by which a culture which valued sustainability (the recycling of manure) nevertheless made possible a shift to an environmentally less sustainable form of sewage disposal as an imaginatively sustainable change.

Waste management in Victorian Britain

I begin with the controversy surrounding the introduction of a widespread system of water-borne sewage removal in London in the mid-nineteenth century. This is well-trodden ground[1] and I only sketch the discussion here with respect to questions of sustainability. In the briefest of summaries, then, until the early to mid-nineteenth century, human excrement in Britain was mainly collected in cesspits, chambers in the ground into which domestic waste was conducted by means of soil pipes. Cesspools were periodically emptied by nightsoilmen, who removed their solid contents during the night. Nightsoil could be lucratively sold as manure, and it was hence treated not as a waste product, but as something valuable – valuable, that is, as long as it was in the right place and used for the correct purpose.[2]

Outside its designated place, nightsoil was increasingly being viewed as dangerous. Cesspools were often leaky, with their leakage contaminating groundwater as well as direct sources of drinking water, such as wells. Numerous sanitary reports and drawings[3] document the anxieties raised by the possibility of such contamination. In addition, a possibly even greater source of concern was the fact that the sheer moisture created by such leakages was, according to Victorian medical theories, dangerous in itself as the potential breeding ground for disease. As a result, moisture, particularly moisture created by festering organic matter such as excrement, was regarded as a serious health hazard. To prevent the accumulation of moist matter in the vicinity of habitations and within cities accordingly developed into one of the main aims of Victorian health officers, and the search for alternative systems of waste disposal into one of the major preoccupations of the sanitary reform movement.

Two different desiderata, therefore, drove the mid-century controversy over the reformation of the system of waste disposal. On the one hand, we have proponents of a shift to a water-borne system as the most effective and quickest way; these were motivated by health concerns relating to the ongoing presence of sewage near or inside human dwellings. On the other hand, and bearing witness to the ecological and economic attractiveness of excrement, we find proponents of the dry conservancy method, who deplored the potential loss of ingredients of nightsoil valuable for agricultural purposes, and regarded the water-borne system of sewage removal as inherently wasteful.

One outspoken partisan of the dry conservancy method was Baron Justus von Liebig, who in his criticism of the loss of manure through aquatic sewage removal frequently used Great Britain as an example. In letter 11 from his *Letters on Modern Agriculture* (1859), for instance, Liebig denounced the waste of human excrement as manure in Europe and the import of guano – dried bird droppings which were imported from South America – as uneconomic and thriftless:

> In the flesh and the produce of the field we have for centuries supplied to the large towns the constituent elements of guano, and have never brought this guano back again; . . . In the year 1855–1856 above 10 million cwt. of guano

> were imported . . . yet all this mass of manure is not worth mentioning when considered in relation to the arable surface of Great Britain, and is but as a drop when compared to the sea of human excrements carried by the rivers to the ocean.
>
> (1859: 179)[4]

What lies behind Liebig's denunciation of the British system of sewage disposal as wasteful is a circular concept of agriculture which regards the nutrients contained in nightsoil as crucial components of agricultural production. In Liebig's vision of a self-sustaining cycle of nutrition, nutrients were transferred from the soil to the crop or meat (of the cattle feeding on grass), thence to the human bodies feeding on the produce of the fields, and then restored to the fields in the form of human excrement used as manure. The use of rivers as open sewers, and eventually, the institutionalisation of the water-borne system of sewage disposal constituted a breach of this cycle by allowing nutrients to flow away, hence disrupting their circulation within a network of people and land. Significantly, for Liebig, this unsustainability was also a cultural problem, as becomes clear in his praise of the Chinese practice of restoring manure to the fields "in exchange for every sack of corn or hundredweight of rape, turnips, and potatoes sold in the market" (Brock 1997: 257). As Brock points out, Liebig developed an argument of cultural supremacy from this comparison, based on the fact "that the Chinese civilization, unlike Roman civilization, had existed for thousands of years" (1997: 257).

What Liebig's vision of self-sustainability also implies is that the recycling of nutrients is a natural process, any deviation from which is potentially harmful. This line of reasoning was reinforced by an ethical argument voiced by other critics of the water-borne system of sewage disposal, such as Charles Kingsley Jun., whose novel *Yeast: A Problem* (1851) argued that the flushing away of sewage was both unnatural and morally wasteful. In Christopher Hamlin's words, *Yeast* regards the water-borne system as "blatant violations of the Creator's laws of the natural economy of matter", with its careless approach to organic resources resulting in "moral, physical, and mental waste" (Hamlin 1985: 402).

Financially and economically as well, the water-borne method of sewage removal was regarded as reprehensibly thriftless by proponents of dry conservancy. As Luckin explains, "[e]conomic orthodoxy dictated that there must be a profitable interaction between the urban and the rural sectors" (1986: 14), and while the dynamics of exchange in the use of human excrement as manure observed this rule, the interaction between the rural and the urban was disturbed by the dilution of human excrement in water. In this respect, it is important to stress that the water-borne system of sewage disposal as originally envisaged was also cyclical. Just like the adherents of dry conservancy, Edwin Chadwick, author of arguably the most seminal sanitary document, the *Report on the Sanitary Condition of the Labouring Population of Great Britain* (1842), and ideological mastermind behind the lobbying for a water-borne system, imagined a cyclical exchange of nutrients between food, human waste, and soil, even while promoting an aquatic system of sewage removal. In Finer's summary, for Chadwick, water enabled the immediate restoration of the nutrients contained in human excrement to the countryside:

> The whole town was comprehended in one arterial system. . . . From the river, in a constant running stream, [water] flowed into the taps and water-closets of each house. From these it flowed away, together with the house refuse into the sewers . . . these streams of water rushed the sewage smoothly and rapidly to the fringe of the towns; and from there it flowed off to the neighbouring fields to manure the crops, while the river from which it was originally drawn, remained pure and sweet and unpolluted.
>
> (Finer 1952: 224)

The underlying conception of sewage disposal here is not categorically different from the one espoused by Liebig. Once again, a cyclical exchange is implied through which nutrients pass back and forth between bodies and soil. Originally, then, the controversy between adherents to the dry conservancy and supporters of the water-borne method primarily concerned the fact that human excrement was rendered less valuable through its dilution in water, something that for the proponents of the latter method was compensated by the swifter and more hygienic manner of sewage disposal by means of pipes and water. The main thrust of their visions, though, was identical, and this points to the strong cultural resonance of the concept of sustainability.[5]

However, in London in the 1850s, a far less self-sustainable version of the water-borne system of sewage removal eventually carried the day, a system by which it was hoped the city would get rid of its waste, once and for all. John Bazalgette's construction of a sewerage system for the metropolis relied on huge outfalls east of London, where raw sewage was ejected into the River Thames and eventually the sea. *The Times* deplored this as a destruction of financial resources, arguing in November 1859 that "now that ratepayers had paid handsomely for Bazalgette's cloaca maxima, which took sewage out to sea, ratepayers were throwing their money away" (qtd. in Brock 1997: 258) because the loss of the potential revenue of sewage had to be added to the costs of construction. Nevertheless, Bazalgette's drainage of London became a model for other cities, thus initiating a shift towards an arguably less sustainable method of sewage disposal supported by cultural views of water which were extremely powerful, and which continue to influence us today.

Literary water: Charles Dickens's *Our Mutual Friend*

For the rest of this essay, I want to use Charles Dickens as a case study for the discussion of the cultural sustainability of Victorian waste. Dickens was an active supporter of sanitary reform and frequently used his public appearances to endorse water-borne sewage removal. At the same time, however, the literary water imagery he developed in his interdiscursive engagement with sanitary reform is not categorically different from that of other Victorian writers. It can thus be used to gauge the dynamics at work in the replacement of the dry conservancy by the water-borne method as culturally sustainable. I want to pursue this point with the help of Dickens's last finished novel, *Our Mutual Friend*, published in 1865.

With the exception of his novel on education, *Hard Times*, Dickens never wrote thesis novels, and *Our Mutual Friend* is no exception. Nevertheless, the novel engages with sanitary reform in a variety of ways. It starts with the discovery of a corpse in the river, and this beginning has two effects. On the one hand, the Thames is at once connected with crime and with the abject, and this automatic link helps foster the naturalisation of the association between water – particularly the water of this river, which played such a central role in Bazalgette's sewerage scheme – and organic waste matter. The Thames here emerges as the natural receptacle of what has been discarded and from what the city is supposed to be cleansed. At the same time, the manner in which the corpse is discovered also immediately inscribes the Thames into an alternative economy. The corpse is brought in by Jesse – "Gaffer" – Hexam, a dredgerman who searches the river for salvageable goods, making a living from what others have thrown away, as well as from the rewards offered for corpses and from the money found on them. The body retrieved from the river in the opening scene is called Gaffer's "luck" by a fellow dredgerman (Dickens 1997: 15; bk. 1, ch. 1), and this implies that the river is also a valuable, if sinister, source of wealth.

The River Thames is one of two symbols which signal Dickens's interdiscursive reintegration of sanitary discourse and which form the two imaginative centres of his novel; the second one is the dust heap, which partly alludes to the dry conservancy method. In order to contextualise my discussion of these symbols, a brief summary of the plot of the novel is useful at this point. *Our Mutual Friend* contains various interrelated parallel and sub plots, but the main storyline revolves around the apparent death of a young heir upon his return to England, John Harmon, whose supposed corpse is the body discovered in the first chapter. Harmon's recently deceased father, whose property he has come to claim, made his wealth from dust heaps, that is, from the collected rubbish of the city. In addition to all sorts of debris from streets and households, one significant constituent of such dust heaps is likely to have been animal and human excrement,[6] a fact that links Harmon's dust heaps to the dry conservancy method. The latter's economic value is implied in the fabulous wealth these dust heaps have generated: they have turned Harmon into an immensely rich man.

Part of the plot is concerned with the mystery of what really happened to John Harmon. He turns out not to have died after all, and the body mistaken for his own at the start of the novel is revealed to belong to his doppelgänger, a sailor who intended to kill him for his money. The dust heaps, therefore, are not only related to the production of wealth, but they also become the cause of crime. Harmon survives, but assumes a series of false identities in order to be able to test Bella Wilfer, a young woman wholly unknown to him but intended by his father for his bride, without marriage to whom Harmon will not be able to enter into his inheritance. While Harmon is believed dead, his father's old servants succeed to his wealth, and young Harmon, in disguise, becomes their private secretary in order to protect them against fraudsters.

Through Jesse Hexam's discovery of Harmon's presumed corpse, the Hexam family also becomes involved in the plot, and this strand of the story centres on the River Thames as a negatively connoted social space, part of the topography of urban crime. The banks of the river are populated with characters regarded as the debris of society: dredgermen, dishonest paupers, villains, thieves, and even murderers. The riverside characters are represented in terms of a moral pollution explicitly connected with the river as a sewer: they are the "accumulated scum of humanity . . . like so much moral sewage . . . pausing until its own weight forced it over the bank and sunk it in the river" (Dickens 1997: 30; bk. 1, ch. 3). The river, in this sense, constitutes a semi-criminal place in which environmental and moral pollution are equated, and from which morally sound characters need to distance themselves to preserve their integrity. The novel emphasises this in its focus on Lizzie, Hexam's daughter, and her relationship with Eugene Wrayburn, a barrister who wants to seduce her and to whom she is eventually married. Through Lizzie and her struggles to escape the vicinity of the river, the novel delves deeply into the life of and on the Thames, and it is here that Dickens creates some of his most compelling water imagery.

Harmon's dust heaps and the polluted Thames are obviously connected in *Our Mutual Friend*, not only because they are the two single most important symbols of the novel, but also because they stand for two methods of cleansing the city of its filth. Economically and socially, the two are presented as occupying opposite ends of a broad spectrum. As indicated earlier, the dust heaps are lucrative objects, part of a thriving economy,[7] while the Thames is associated with those living on the margins of society, scavengers who carve out a meagre existence from their collection of rubbish. When Mr and Mrs Boffin, Harmon's servants, inherit the money Harmon made from his business as a dust contractor together with the dust heaps themselves, they are immediately accepted as members of genteel society. Their new status as nouveaux riches is expressed in their new appellation: Boffin becomes "The Golden Dustman" (Dickens 1997: 137; bk. 1, ch. 11), suggesting his social rise and good fortune. However, this moniker also implies that Boffin is now wholly associated with his new position and wealth, and thus reduced to the role of a lucky parvenu. His kind disposition is ignored by the people who suddenly fawn over him, and his apparent naivety exploited by sycophants. In this manner, the dust heaps are shown to be a mixed blessing. Most significantly, they carry the negative legacy of the unhappy Harmon family and their history of repression, sterility, and death. As the novel reveals through flashbacks, Harmon Senior turned out both his children, causing his daughter's death in poverty as well as his son's emigration. The servants who inherit Harmon's wealth recall their former master's coldness and Harmon Junior remembers himself as an emotionally starved child. His very inheritance is burdened by his father's despotism through the legal condition that he can only succeed to the family's large fortune if he marries after his father's choosing. Following Harmon

Senior's death, the fortune he earned from dust becomes the centre of a web of betrayal, greed, and murder: It is in order to inherit fraudulently that Harmon Junior's doppelgänger tries to drug and murder him only to perish in the attempt himself. Likewise, old Harmon's eventual heir Boffin also becomes the target of swindlers. Thus the wealth generated by the dust heaps, far from becoming part of a productive form of circulation, appears to corrupt everybody who comes into contact with it. It is only cleansed and can only become the basis of a healthy future for young Harmon and his wife once it has been disowned and lost and freely returned by the honest and kind-hearted Boffins. As a symbol of avarice, exploitation, isolation, and sterility, then, the dust heaps are far from offering a vision of sustainability. Instead, they stand for a manner of accumulating wealth that is divorced from social principles. Sitting in Harmon's garden and attracting nothing but fraudsters and swindlers, they appear as an emotional and social dead end.

Compared with the sterility of the dust heaps, the Thames emerges as extraordinarily fertile, an image of connection that reaches across and links the various characters and plots of the novel and that points to the fact that Dickens himself found water an inexhaustible source of poetic creativity. Though also associated with death, not only at the beginning of the novel, but at various points throughout, the river at the same time functions as a symbol of life. The atmosphere of death, for instance, is already challenged in the very beginning of the text when Gaffer reproaches his daughter Lizzie with her dislike of the river:

> How can you be so thankless to your best friend, Lizzie? The very fire that warmed you when you were a babby [sic!], was picked out of the river alongside the coal barges. The very basket that you slept in, the tide washed ashore. The very rockers that I put upon it to make a cradle of it, I cut out of a piece of wood that drifted from some ship or another.
>
> (Dickens 1997: 15; bk. 1, ch. 1)

As Gaffer insists, the Hexams' existence is intimately intertwined with the life of the river. They are dependent on the goods it brings them, but they also help cleanse it by gathering and removing the debris it carries. Their lives are built on the recycling of waste – and this includes human waste, such as bodies – and it is through the river that they are joined to the rest of society, not only through what this society discards but also through the encounters which the river makes possible between the delicately interconnected strands of Dickens's plot. This vision of riparian life as part of a larger cycle of existence in its turn subtly endorses the concept of circulation associated with Chadwick's conception of water-borne sewage disposal. In *Our Mutual Friend*, the river indeed deposits its waste – members of the "waste population" (Allen 2008: 60), such as Lizzie – in the countryside, only to retrieve it and return it to the city once it has been translated into an invigorating source of sustenance (Lizzie's influence, for instance, purifies and restores the morally weak Eugene). As such, the function of the river in the novel fosters a vision of water-borne waste removal as productive and sustainable.[8]

Finally, and perhaps most relevantly, it is precisely the dirty and sewage-polluted Thames that fires Dickens's imagination, becoming the ground of a powerful aesthetics of water. Throughout his oeuvre, Dickens returns to the topographically, morally, and aesthetically liminal space of the Thames to explore the imaginative possibilities of the interaction between water, riverside, and riverside population. In doing so, he draws closely on literary history, which attests to a vibrant tradition of connecting (mostly dirty) water with negative morality. But while this link is often approached by other writers in a stereotypical fashion, as mere platitudinous repetition, Dickens's water imagery is immensely rich, breathing new life into a stale trope. Thus images as diverse as the description of Gaffer's clothes as seemingly "made out of the mud that begrimed his boat", which characterises him as part of the riverside filth (Dickens 1997: 13; bk. 1, ch. 1), various shadowy female figures flitting along the banks of the river in different Dickens novels (such as *Oliver Twist* [1838], *David Copperfield* [1850], or *Bleak House* [1853]) as signals of the link between female transgression and water, a female servant's barefooted walk through wet grass which indicates her criminal inclinations (Dickens 1996: 299–300; ch. 18), or the discovery of Rogue Riderhood and Bradley Headstone's drowned bodies in the "smooth pit" of a Thames sluice, "lying under the ooze and scum behind one of the rotting gates" in *Our Mutual Friend* (Dickens 1997: 781; bk. 4, ch. 15), all strengthen the aesthetic function of water as a literary indicator of moral corruption and, at the same time, an agent capable of cleansing society of its filth. As an expression of what is regarded as valuable, what as worthless, and what as capable of new inscription and re-evaluation, Dickens's water is thus intimately connected with questions of un/sustainability.

Conclusion: Dickens and cultural sustainability

The specific poetic use of water described here, highly relevant also today, certainly did not originate with the Victorians, but equally certainly it was reinforced in literary texts written during the age of sanitary reform. The very fact that the symbolic link between water and questionable morality is still so active in contemporary literature (which is just as prone to turn bodies of water into settings for crime, or to create sinister atmospheres from aquatic imagery) points to the power of a vision that was popularised and naturalised by writers such as Dickens. In their interdiscursive engagement with sanitary reform these writers brought together literary history and its great water lore with their specific socio-cultural context. One effect of this was a narrativisation of sanitary reform – in Dickens's case particularly through the symbolic convergence of images of rebirth and recycling, through the pervasive influence of Thames imagery on his more general water imagery, and through the function of the river as a link between various characters and plots. In addition, the naturalisation of the connection between water, dirt, pollution, and purification accomplished by Dickens and other writers indirectly also formed an endorsement of the vision offered by the water-borne system of sewage disposal. Victorian writers' interdiscursive engagement with sanitary reform – and Dickens's, as I have argued, is one of the most compelling approaches to this – can thus serve as a case study of the intricate intertwining of culture as a way of life and culture as artistic creative practice that I see as key to questions of cultural sustainability. The acceptance of the imaginative link fostered between water, corruption and pollution, and cleansing and purification in Victorian writing thus also supported the shift in waste disposal methods by making it culturally sustainable. In this way, the force of a compelling cultural imaginary helped render acceptable measures that were probably less sustainable environmentally than the method they displaced.

Notes

1 See, for instance, Michelle Allen (2008), Pamela K. Gilbert (2004, 2007), Stephen Halliday (2009), and Anthony S. Wohl (1983).
2 In fact, how lucrative the removal of nightsoil really was is not quite clear. Edwin Chadwick, for instance, challenged its profitability. What is clear, however, is that "the belief in the economic value of this waste was widespread" (Handy 1995: 121).
3 One eloquent example of the latter is T. Pridgin Teale's illustrated guide to sanitary problems (Teale 1878).
4 Published in 1859, this letter actually predates the opening of John Bazalgette's new sewers for London in 1865, though it was published after Parliament had passed the Act enabling their implementation. Liebig here also deplores the drain of nutrients from the country to the city through demographic shifts. However, even before Bazalgette's great reform of the sewerage of London, rivers played an important role for the disposal of sewage, and the dry conservancy method was on its way out. Edwin Chadwick's sanitary report of 1842 (see below) is a key document for the promotion of the water-borne system of sewage removal. In the decade following this report, various government acts encouraged the disposal of waste in waterways.
5 As Catherine Gallagher points out, the economic model that sanitarian reformers followed formed a counterpoint to Malthusian economics in conceptualising waste as a basis of growth. Thus sanitary reform relied on "a model of self-sustaining growth based on the continual recycling of the population's own remains (the more people, the more waste; the more waste, the more food; the more food, the more people, etc.)" (2006: 104). Food and excrement, in this sense, were literally two sides of the same coin.
6 On the composition of dust heaps, see Allen (2008: 86–87), Cotsell (2009: 30–33), Handy (1995), and House (1960); also Dickens (1997: 805n20).
7 It appears that by the time the novel was published, however, the value of dust heaps had fallen. See Cotsell (2009: 30).
8 The extensive use Dickens makes of the ancient symbolic equation between immersion and (spiritual) rebirth also supports this function of the river. At the same time, most of the criminal characters of the novel are killed by drowning. This suggests that they do not offer society anything worth recycling and are hence discarded in an effort to cleanse society.

References

Allen, M. (2008). *Cleansing the City: Sanitary Geographies in Victorian London*. Athens: Ohio University Press.

Brock, W. H. (1997). *Justus von Liebig: The Chemical Gamekeeper*. Cambridge: Cambridge University Press.

Chadwick, E. (1842). *Report to Her Majesty's Secretary of State for the Home Department form the Poor Law Commissioners, On an Inquiry into the Sanitary Condition of the Labouring Population of Great Britain; With Appendices. Presented to Both Houses of Parliament, By Command of Her Majesty, July 1842*. London: W. Clowes and Sons.

Cotsell, M. (2009). *The Companion to Our Mutual Friend*. [1986]. Abingdon: Routledge.

Dickens, C. (1996). *Bleak House*. [1853]. Ed. and Introd. N. Bradbury. London: Penguin.

Dickens, C. (1997). *Our Mutual Friend*. [1865]. Ed. and Introd. A. Poole. London: Penguin, 1997.

Finer, S. E. (1952). *The Life and Times of Sir Edwin Chadwick*. London: Methuen.

Gallagher, C. (2006). *The Body Economic: Life, Death, and Sensation in Political Economy and the Victorian Novel*. Princeton: Princeton University Press.

Gilbert, P. K. (2004). *Mapping the Victorian Social Body: SUNY Studies in the Long Nineteenth Century*. Albany: State University of New York Press.

Gilbert, P. K. (2007). *The Citizen's Body: Desire, Health, and the Social in Victorian England*. Columbus: Ohio State University Press.

Halliday, S. (2009). *The Great Stink of London: Sir Joseph Bazalgette and the Cleansing of the Victorian Metropolis*. Stroud: The History Press.

Hamlin, C. (1985). "Providence and Putrefaction: Victorian Sanitarians and the Natural Theology of Health and Disease". *Victorian Studies* 28(3), 381–411.

Handy, E. (1995). "Dust Piles and Damp Pavements: Excrement, Repression, and the Victorian City in Photography and Literature". In C. T. Christ, and J. O. Jordan (eds.). *Victorian Literature and the Victorian Literary Imagination*. Berkeley: University of California Press, 111–133.

House, H. (1960). *The Dickens World*. [1941]. London: Oxford University Press.

Kingsley, Ch. Jun. (1851). *Yeast: A Problem*. London: John W. Parker.

Liebig, J., Baron von (1859). *Letters on Modern Agriculture*. Ed. J. Blyth. New York: John Wiley.

Luckin, B. (1986). *Pollution and Control: A Social History of the Thames in the Nineteenth Century*. Bristol: Adam Hilger.

Teale, T. P. (1878). *Dangers to Health: A Pictorial Guide to Domestic Sanitary Defects*. London: J. & A. Churchill; Leeds: Charles Goodall.

Wohl, A. S. (1983). *Endangered Lives: Public Health in Victorian Britain*. London: J.M. Dent & Sons.

15 The moral dilemma of unsustainability

Eco-dystopian fiction as cultural critique

Alexa Weik von Mossner

Introduction

Something remarkable took place in the literary world shortly after Donald J. Trump had been sworn in as the 45th President of the United States. Within hours of Trump adviser Kellyanne Conway's cheerful declaration that press secretary Sean Spicer had expressed "alternative facts" in relation to the size of Trump's inaugural crowd, George Orwell's novel *1984* began its climb to the top of the Amazon bestseller list. Journalists promptly began to analyze the reasons for this sudden surge of popular interest in an almost seventy-year-old dystopian classic. Writing in the *New York Times*, Michiko Kakutani went as far as declaring *1984* a "must-read" in 2017 because the novel's "world of endless war, where fear and hate are drummed up against foreigners, and movies show boatloads of refugees dying at sea" mirrors so closely the political vision of the Trump administration. Kakutani suggests that Orwell's dystopian gaze into a totalitarian future in which 2 + 2 equals 5 "has found a nervous readership in today's 'post-truth' era" in which "the acceptance of bad arithmetic simply becomes a testament to the power of rulers to define reality and the terms of debate" (2017: n.p.). Other commentators have either agreed or singled out other dystopian texts – from Octavia Butler's *Parable of the Sower* (1993) to Margaret Atwood's *The Handmaid's Tale* (1985) and her more recent *MaddAddam* Trilogy (2003, 2009, 2013) – as must-reads, suggesting that if we do not watch out their dark visions of the future may swiftly become our present reality.

In this essay, I want to take the recent resurge of interest in dystopian fiction as a starting point for an investigation into the relationship between speculative storytelling and the cultural discourse on sustainability. While Orwell's novel is more interested in the cultural and psychological fallout of totalitarian regimes (particularly of the Stalinist kind) than in the ecological havoc that such regimes might wreck, Atwood and Butler belong among those writers who have foregrounded in their works the close connection between the personal, the cultural, and the ecological. That connection is not always easily made in scholarly debates on issues of sustainability. In their 2013 study *Ecological Sustainability*, Robert Northrop and Anne Connor posit that "human ecological sustainability" is affected by a wide range of factors, among them human population, freshwater supplies, energy supplies, food supplies, natural resources, the economy, emerging diseases, and pollution (1). Human cultural production is not even on the list, let alone anything so specific as speculative fiction. And yet, it is my contention that such works make important cultural interventions that must be accounted for in any discussion of ecological sustainability. As Sunny Moraine has observed, the alternative worlds we encounter

> As Sunny Moraine has observed, the alternative worlds we encounter "in speculative fiction don't exist apart and separate from the world we live in" (2017: n.p.). Instead, such fictional worlds make claims on a reality they are fundamentally connected to. "As a genre," writes Moraine, "speculative fiction allows us to remake our own present. It allows us to imagine a future for ourselves. It allows us to make a way out of whatever unbearable moment we seem to be stuck in"
> (2017: n.p.)

This makes speculative fiction an important, even vital contribution to the cultural discourse on sustainability.

My focus will be on a subsection of speculative fiction that in previous publications I have called the critical eco-dystopia.[1] Following the lead of Lyman Tower Sargent, the science fiction scholar Tom Moylan defines the critical dystopia as

> a textual mutation that self-reflexively takes on the present system and offers not only astute critiques of the order of things but also explorations of the oppositional spaces and possibilities from which the next round of political activism can derive imaginative sustenance and inspiration.
> (2000: xv)

The critical *eco*-dystopia is a text that fits that definition but is also marked by an interest in ecological sustainability. Set in an uncertain future, it imagines the harsh fates of humans who live in severely degraded environments and are forced to realize that their very existence is predicated on a set of ecological conditions that have been thrown out of balance. Its protagonists often learn the hard way that the nonhuman world possesses multiple forms of agency eluding human control as they try to cope with the unintended consequences of the lifestyles of previous generations. Its authors often admit that their text is meant as warning, which makes it an interesting genre for an investigation of the ways in which cultural texts influence their readers' perceptions of sustainability. I will argue that an ecocritical approach that draws on research in cognitive narratology and neuroscience is particularly helpful for such an investigation, because it can give insight into the ways in which eco-dystopian fiction allows readers to experience not only mentally, but also on the physical level *what it is like* to live in an unsustainable future environment. How do the social, cultural, and ecological dimensions of an environment emerge before our inner eye as we read through a dystopian text? And how do such texts engage us on the affective and cognitive level? These are some of the questions I aim to address in this essay. Butler's *Parable of the Sower* will serve as my prime example for a speculative text that invites us to a visceral imaginary experience of an unsustainable future world and asks us whether we would want to live in such a world.[2]

Experiencing imaginary environments: how literary texts involve us in a world

Like any literary text, speculative fiction sheds light on the human condition and on the social, cultural, and ecological conditions of its survival. Hubert Zapf defines literature as a cultural form in which the living interrelationship between nature and culture "is explored in specifically productive ways, providing a site of critical reflection of modern civilization as well as a source of creative cultural self-renewal" (2016: 3). As Zapf's contribution to this volume demonstrates, understanding literature as a vital part of a larger cultural ecology is productive not only for ecocritical analysis, but also for an interdisciplinary discussion about the role and function of literary texts in the quest toward more sustainable societies. Many ecocritics will readily assert that literature plays an important role in how we see, feel, and act in the extra-textual world. But not only ecocritics or even literary scholars have made such assertions. In *Cultivating Humanity* (1996), the philosopher Martha Nussbaum states that "narrative imagination is an essential preparation for moral interaction" (1997: 90) and goes on to assert that literature can play a vital role in creating more empathetic and morally responsible citizens. Social science scholars David Lewis, Dennis Rogers, and Michael Woolcock have acknowledged that "[n]ot only are certain works of fiction 'better' than academic or policy research in representing central issues relating to development, but they also frequently reach a wider audience and are therefore more influential" (2008: 198). What unites these two positions is the conviction that literary texts, particularly fiction, offer something that other forms of human communication are missing and that this uniqueness makes them an important contribution to the cultural discourse on issues of concern. "A novel which exposes the reader to a sense of injustice or to a dilemma", write sociologists Shai Dromi and Eva Illouz, "and which imbues these dilemmas and injustices with emotional value is not only a work of fiction but what we may call a critique" (2010: 352). It is the combination of ethical, experiential, and affective appeals, then, that makes the effects of fiction unique, and eco-dystopian texts are *critiques* in this sense: they present readers with a dilemma and then imbue that dilemma with emotional value.

But how do we analyze the narrative structures and strategies involved in both the presentation of a dilemma and the subsequent saturation with emotional value? And how do we find out what effect they may have on readers? On the theoretical side, there is a long history of approaches that focus on the relationship between narrative structure and its reception, among them reception aesthetics, reader response theory, psychoanalytical and other psychological approaches, as well as some strands of postclassical narratology, including cognitive narratology. On the empirical side, we can consult a range of studies in linguistics, social psychology, cognitive psychology, and, in more recent years, neuroscience.

A particularly exciting development has been the emergence of *cognitive cultural studies*, a relatively young field of research that Lisa Zunshine has defined as a combination of "literary and cultural analysis with insights from neuroscience, discursive psychology, cognitive evolutionary psychology and anthropology, cognitive linguistics, and philosophy of mind" (2010: 1). Cognitive cultural studies, it becomes clear from this definition, is a highly interdisciplinary affair that draws on cognitive science and related fields for a better understanding of cultural texts, one that combines an astute interest in the narratological properties of a text with an attention to psychological and contextual factors. In my book, *Affective Ecologies* (2017), I have argued that this interdisciplinary approach is particularly useful for ecocritical investigations (2). Not only can it give us a better understanding of how cultural texts create virtual environments in our minds, but it also helps us analyze how they invite us to care about both those environments and the characters that populate them.

I want to single out one concept from cognitive cultural studies that I consider particularly fruitful for the analysis of eco-dystopian fiction because it helps us understand how we experience a speculative future world in ways that are vivid and emotionally engaging. The concept of *liberated embodied simulation* has been developed by the literary scholar Hannah Wojciehowski and the neuroscientist Vittorio Gallese, one of the leading figures in mirror neuron research. Gallese has argued that "[t]he sharp distinction, classically drawn between the first- and third-person experience of acting and experiencing emotions and sensations appears to be much more blurred at the level of the neural mechanisms mapping it" (2011 442). This blurring of the distinction between one's own body and that of another is what he calls embodied simulation. Complex processes of "neuronal reuse" allow us to map the motions of other agents onto the respective areas of our brains, although our own body is not currently engaged in the same motion and might not even be capable of it (for example, when we watch a professional ballet dancer). The same is true for the *perceptions* and *emotions* of other agents, regardless whether we perceive actual agents in our immediate environment or fictitious agents in a film or a novel. However, there is nevertheless a difference. In a co-authored article, Wojciehowski and Gallese argue that when we engage with fictional characters

> our embodied simulation becomes *liberated*, that is, it is freed from the burden of modelling our actual presence in daily life . . . Through an immersive state in which our attention is focused on the narrated virtual world, we can fully deploy our simulative resources, letting our defensive guard against daily reality slip for a while.
>
> (2011: n.p., emphasis mine)

Important in our context here is that embodied simulation not only allows us to understand the actions and emotions of an experiencing agent in a narrative text, but also plays a crucial role in our understanding of the larger storyworld that surrounds that agent.

Following the lead of cognitive narratologists such as Richard Gerrig and David Herman, ecocritic Erin James defines the storyworld as "a mental model of context and environment within which a narrative's characters function" (2015: x).[3] In *Affective Ecologies*, I use the example of Cormac McCarthy's post-apocalyptic novel *The Road* (2006) for an illustration of the processes of embodied simulation that feed that model of context and environment, and I also make clear that it is not only a *mental* model but one that resonates in the reader's *body* as well (Weik von Mossner 2017: 1). When we read that McCarthy's nameless protagonist "pulled down the cotton mask from his face and wiped his nose on the back of his wrist" (2006: 4), we literally map those movements onto the motor cortices of our brains as the mental processing of action verbs activates the respective neurons.[4] When we read that the protagonist gazes through his binoculars at "segments of road down there among the dead trees" and watches "the ashen daylight congeal over the land" (4) we are cued to simulate his visual perception as well as his affective state. Depending on the narrative situation (for example in the case of omniscient narration), a story might also give us information about its storyworld independently of the character's perception of it, but that doesn't change the fact that experiencing agents placed within the storyworld tend to strongly influence our understanding and emotional relationship to it.

The question that poses itself is whether dystopian fiction that contains experiencing agents – as virtually every dystopia does – can offer something to the experiential systems of our brains that makes it valuable for a better understanding of sustainability. If it is the combination of a dilemma and our emotional engagement in a character's confrontation with that dilemma that makes a novel a cultural critique, as Dromi and Illouz have claimed (2010: 352), then what role does embodied simulation play in that engagement? Ecocritic Eric Otto reminds us that environmental speculative fiction shares with environmental nonfiction "a rhetoric of estrangement and extrapolation that compels readers toward critical reflection on seemingly invisible everyday attitudes and habits" (2012: 7). The elucidating effect of speculative fiction depends on a carefully calibrated balance between cognition and estrangement, argues Otto with reference to the influential work of science fiction scholar Darko Suvin. The features of the speculative world must be familiar enough to be understandable for the reader, and yet strange enough to point their attention toward something that they may have overlooked or taken for granted (Otto 2012: 7). It is within this force field between the familiar and the unfamiliar that Moylan locates the potential for "an enlightening triangulation between an individual reader's limited perspective, the estranged re-vision of the alternative world on the pages of a given text, and the actually existing society" (2000: xvii). Having been confronted with a recognizable and yet estranged version of one's actually existing society in a dystopian novel, one might behold that actually existing society not only with new eyes, but with new feelings as well.

With all that in mind, I want to turn to Butler's *Parable of the Sower*, a novel that presents us with a world that isn't quite as devastated as the one we find in *The Road*, yet is scary nevertheless. Butler is subtler than McCarthy in her portrayal of ecological unsustainability and at the same time she is much more direct in her imagination of the relationship between human culture and ecological carrying capacity. As Chelsea Frazier points out, Butler's dystopia "transports us to a not-so-distant future in which the world has slowly but steadily descended into social, environmental, and economic chaos" (2016: 40). Just like *The Road*, *Parable of the Sower* follows a desperate quest of vulnerable humans on broken roads. And just like *The Road*, it asks existential questions about how to survive and act morally in an anarchic world that is dominated by scarcity and violence, a world almost devoid of oil, clean water, food, and any other number of resources that we now take for granted. But this is where the parallels end. *Parable of the Sower* does not give us the "clean slate" scenario of a global apocalypse of unknown origin that we find in McCarthy; instead, it is set in the climate-changed near-future world of 2025 and in a Southern California that is dark but decipherable. At the outset of the story, California still is a vaguely coherent socioeconomic and political entity, but collapse is imminent due to unsustainable environmental, economic, and social practices. It is a vile world that Butler has created for us, one that is even more difficult to bear because she aligns us with a young woman who suffers from *hyperempathy* – an incurable condition that was induced by the drug abuse of her mother and that forces her to share the pain of others. Lauren Olamina is a 16-year-old African American girl who narrates her story in the first person through her journal entries. It is through Lauren's perceptual apparatus, her cultural knowledge, and her hyperempathetic mind that all elements of this future world are channelled. My analysis will pay attention to the ways in which the text links readers' visceral imaginative experience of that world to Lauren's evolving positions on questions of ethics, environmental justice, and sustainability.

Experiencing unsustainability: simulation and speculation in *Parable of the Sower*

From the outset, the central theme of Lauren Olamina's journal entries is change. "All that you touch/You Change", read the first lines of the novel, "All that you Change/Changes you. The only lasting truth/Is Change. God Is Change" (1993: 3). In addition to a pattern of repetition, it is the idiosyncratic capitalization in these verses that emphasizes the importance of Change in Lauren's world view. It is a world view that will eventually crystalize into Earthseed, a new religion that is meant to promote and enable a sustainable, just, and peaceful life on and with the Earth. Throughout the novel, Laura mixes the verses that are the cornerstones of her evolving religion with the recounting of her past experiences and future expectations. Her style of narration is itself subject to change and adapts to whatever situation she finds herself in. At times it is pensive, reflexive, even ruminant. In other moments, she is a breathless and overwhelmed narrator who jots things down to keep a record of the events that change her life in often disruptive and shocking ways. Lauren's narration is unreliable not only because of her limited knowledge of the world, but also because she is often highly emotional and at an age where she is still trying to figure things out. The fact that we read her journal entries makes her account very personal, even intimate. This quality of the novel's narrative situation in turn determines our experience of its dystopian world.

I have suggested that characters are of central importance for our understanding of a storyworld, and this is particularly true in the case of first-person narration. By definition, a first-person narrator is an experiencing agent within the storyworld and therefore subjected to the physical conditions of that world and to the limitations of his or her own body. If it is the only narrator and the main protagonist of the story, as it is the case in Butler's novel, this inevitably means that readers' imaginative worldbuilding entirely depends on this one experiencing agent. We only sense what Lauren senses and chooses to share with us, and while we are free to judge her subjective impressions, we have no way of going beyond them in our experience of the storyworld. External and potentially conflicting views on that world can only be related through written documents or reported dialogues of other characters, and even the content of those utterances is filtered through Lauren's consciousness. The first incident she entrusts to her journal is her recurring dream about trying to fly without instruction. "Not a very subtle image", she acknowledges, "but a persistent one" (1993: 4). And indeed, the metaphor is pertinent since so much in Lauren's life depends on her being able to master difficult, even seemingly impossible tasks without much preparation or help.

At the beginning, Lauren's circumstances are difficult but stable, and they are defined by the presence of walls. Together with her father, her half-siblings, and her stepmother Cory, she lives in the walled-in community of Robledo just outside the anarchic urban space of Los Angeles. "None of us goes to school anymore", she writes, "Adults get nervous about kids going outside" where "things are so dangerous and crazy" (1993: 7). Lauren's father is a Baptist minister who "once had a church just a few blocks outside our wall" (1993: 8), but who now has resigned himself to holding his services for his congregation in the front rooms of the family home. At this point in the story, Lauren's life is marked by a sense of enclosure and omnipresent risk. According to her father, Los Angeles "is a carcass covered with too many maggots", but Lauren is painfully aware that "not all the maggots are in L.A. They're here, too" (1993: 9). Addicted to a drug that makes them love fire and living in a constant state of poverty, hunger, and despair, these "maggots" shave their heads and paint their faces, and then mug, rape, and murder people right outside of the walls that protect her.

That is why the members of her community are heavily armed. "By now there are at least two guns in every household", notes Lauren, and even her father "carries a nine millimeter automatic pistol whenever he leaves the neighborhood" and hides an illegal "silenced nine millimeter submachine gun" at home (1993: 38–39). The minister also teaches his children to shoot, in accordance with this conviction that "[a]rmed people do get killed – most often in crossfires or by snipers – but unarmed people get killed a lot more often" (1993: 38). Lauren has internalized this brutal logic, but she is also aware that the people her father calls maggots are the direct outcome of steadily worsening environmental conditions. She warns her friend Joanne that "[p]eople have changed the climate of the world" (1993: 57), leading to devastating tornadoes, blizzards, and other seemingly "natural" catastrophes throughout the Midwest, while "[i]n New York and New Jersey, a measles epidemic is killing people" (1993: 54). When Joanne gives back that they "can't do anything" (1993: 57) and that not even her own father believes the scientists, Lauren loses patience. She tells her friend that, whether they see it coming or not, their community will be "hit and hit and hit, then the big hit will come" (1993: 55) and so they better prepare themselves for the worst. By the end of the chapter, Joanne is frightened enough to take home the books Lauren has given her, books that "might help [her] to survive out there" (1993: 58) once the sturdy walls that protect their community are breached.

Throughout the first half of the novel, Lauren's first-person narration thus cues readers to imagine the storyworld as a strict dichotomy between a mortally dangerous outside and a relatively safe but unsustainable inside. "Robledo's too big, too poor, too black and too Hispanic to be of interest to anyone", she asserts. "What it does have is street poor, body dumps, and a memory of once being well-off – of shade trees, big houses, hills, and canyons" (1993: 120). The walled-in community offers more than just a memory of these attractive environmental assets, however. In the very next sentence, Lauren acknowledges that "[m]ost of those things are still here" (1993: 120), and she also repeatedly mentions the community's many crop-bearing gardens, which provide the basis for their survival. Its inhabitants might not belong among the super-rich, but compared to the misery that surrounds it, Robledo's green and fertile space nevertheless constitutes what the science fiction writer Kim Stanley Robinson has called "a pocket utopia" (1988: 60). And the problem with all pocket utopias, Robinson's protagonist Tom Barnard recognizes in his novel *Pacific Edge* (1988), is that they cannot be sustained. "We stand on little islands of luxury", Barnard writes down in the prosperous and seemingly safe city of Zurich in Switzerland, "while the rest – great oceans of abject misery, bitter wars, endless hunger" (1988: 60). But such islands of luxury cannot exist in isolation, no matter how high they build their walls, and no matter how many weapons they carry. At some point those walls will come crumbling down and their weapons will be overpowered by the sheer onslaught of desperation. Both Robinson and Butler therefore problematize the idea of the isolated and enclosed pocket utopia in their novels and have their protagonists embrace more equitable and ecologically sustainable visions of human conviviality.[5] But whereas Robinson dares to imagine a utopian world in which humanity has found political solutions to a breakdown in the past, *Parable of the Sower* focuses on the time before and immediately after the breakdown, and on the morally fraught question of what we should do when our bulwarks against the "oceans of misery" give way.

Whenever Lauren leaves the enclosed space of her community, her subsequent journal entries allow readers to viscerally experience, with and through her, what it *feels like* to be exposed to that ocean of misery: "Most of the street poor – squatters, winos, junkies, homeless people in general – are dangerous", she tells us. "They are desperate or crazy or both. . . . Worse for me, they often have things wrong with them. They cut off each other's ears, arms, legs. . . . They carry untreated diseases and festering wounds" (1993: 10–11). These words cue a sense of uneasiness in readers and not only because Lauren calls the situation dangerous. Gallese has shown that processes of embodied simulation are triggered by "words referring to face, arm, or leg actions, or listening to [or reading] sentences expressing actions performed with the mouth, the hand, and the foot" (2011: 443). Reading about people cutting off each other's limbs, about diseases and "festering wounds" leads us to vividly imagine those actions and injuries and automatically triggers feelings of horror and disgust, feelings that parallel those felt by the protagonist. Lauren, however, has an additional problem: "As I rode, I tried not to look around at them, but I couldn't help seeing – collecting – some of their general misery" (Butler 1993: 11). Because of her hyperempathy syndrome, Lauren does to an extreme degree what everyone would do in her situation: she not only witnesses other people's suffering but empathically *feels* it in her own body. "I can take a lot of pain without falling apart", she asserts in her journal entry, "I've had to learn how to do that" (1993: 11). She is used to dealing with her condition, but she nevertheless knows it is a dangerous liability that will likely leave her incapacitated when she is forced to absorb too much of the pain of others.

What exactly this means for Lauren's survival in the outside environment becomes clear in the moment when her worst fears come true and the walls around her community can no longer protect her and her family:

> Last night, when I escaped from the neighborhood, it was burning, the trees, the people: Burning.
>
> Smoke awoke me, and I shouted down the hall to Cory and the boys. I grabbed my clothes and emergency pack and followed Cory as she herded the boys out. The [alarm] bell never rang. Our watchers must have been killed before they could ring it.
>
> Everything was chaos. People running, screaming, shooting. The gate had been destroyed. Our attackers had driven an ancient truck through it. . . . They must have been pyro addicts – bald people with painted heads, faces, and hands. Red faces; blue faces; green faces; screaming mouths; avid, crazy eyes, glittering in the firelight.
>
> They shot us and shot us and shot us. I saw [our neighbor] Natalie Moss, running, screaming, then pitching backward, her face half gone, her body still impelled forward. She fell flat on her back, and did not move again.
>
> (Butler 1993: 154)

In this fateful moment of violent intrusion, the reader is cued to share Lauren's perceptive experience of it. Rather than telling us how anxious and horrified she felt, thereby describing her emotions, Lauren goes into "showing mode" and gives us an impressionistic sketch of the situation that cues us to vividly imagine the scene and react emotionally to it. Her description is highly economical and effective, relying on the repetition of action words ("They shot us and shot us and shot us"), visual impressions ("red faces; blue faces; green faces") and the combination of the two ("I saw Natalie Moss, running, screaming, then pitching backward, her face half gone, her body still impelled forward"). It is worth noting that this is remembered perception and action rather than currently occurring perception and action. Lauren uses past tense rather than present tense, and she makes clear that she is writing down these words after the fact, in a quieter moment that allows her to take a breath and attend to her journal.[6] And yet, she chooses to relate the events in short, breathless sentences, thereby stimulating the reader to adapt to her pace and to imagine the events almost like a sequence from an action film as they simulate in their minds Lauren's perception and movements.

In the same breathless style of narration, Lauren also shares her own emotional and physical reactions to what she sees:

> I fell with her, caught up in her death. I lay there, dazed, struggling to move, to get up. Cory and the boys, running ahead of me, never noticed. They ran on.
>
> I got up, felt for my pack, found it, and ran. I tried not to see what was happening around me. Hearing the gunfire and the screams didn't stop me. A dead body – Edwin Dunn – didn't stop me. I bent, snatched up his gun, and kept running.
>
> (Butler 1993: 154)

Lauren here relates an experience that the reader cannot fully share because they do not share her hyperempathy condition. However, as the cognitive narratologist Patrick Hogan has argued, we need not be *like* a fictional character to be able to empathize with them. By empathizing with the character's *situation*, we can approximate the character's feelings and concerns, even if that character is different from ourselves. Such *situational empathy* can lead us to feel with members of an out-group with whom we do not share a great number of group-defining features (Hogan 2003: 140–46). This explains why we can not only empathize with characters from a different gender, ethnical group, or sexual orientation, but also with characters in speculative forms of fiction that may have superpowers or other abilities that we would not generally expect in our fellow humans. Once we have accepted Lauren's hyperempathy condition as a fact of the narrative world, we will also be able to empathize with her in this as well as in ensuing situations when she finds herself physically incapacitated for the sole reason that she sees the pain of a suffering other and feels that pain with the same intensity. The moment – brief as it is in Lauren's record – gives us a first inkling of what it means for her to see other people hurting, but it also informs us that she is nevertheless able to function enough to save her own life. In addition, it becomes clear here that dead people – even dead people she knows and cares for – do not affect her in the way that injured or dying people do because they no longer feel any pain or emotions she could share. In a darkly ironic twist, Lauren will soon find out that it is best for her to kill attackers rather than just shooting to injure, since the latter would negatively affect her own ability to survive.

As Kimberley Ruffin has pointed out, Butler uses "Christian and other religious discourse as a means to transform the negative human behaviours that have resulted in endangered human and nonhuman lives" (2010: 95). Lauren's attitude and behaviour, however, is at odds with the Christian moral imperative "Thou shalt not kill", and it also sheds an interesting light on the relationship between empathy, morality, and prosocial behaviour. Scholars in the fields of philosophy, psychology, political theory, and religious studies have offered different takes on that relationship. Psychologist Martin Hoffman, for instance, has argued that empathy – and related processes of embodied simulation – is of central importance to moral reasoning and prosocial behaviour (2011: 231). Philosopher Jesse Prinz has taken the opposite position, arguing that empathy "is so vulnerable to bias and selectivity that it fails to provide a broad umbrella of moral concern" (2011: 227). Butler's novel sensitizes us to another important dimension in the relationship between empathy, morality, and prosocial behaviour, namely the role of egoism, an extreme form of bias. Even at the outset of the narrative, Lauren's community demonstrates the bias that Prinz has in mind: the urge to protect the interests of one's in-group against the interests of an out-group. Once Lauren has been forced to leave that community, however, her journal entries suggest that killing others is a necessity for her own survival. Christian ideals of pacifism and selfless altruism have become unsustainable in a world full of "people who've made such a hell of life here on Earth" (Butler 1993: 21). Lauren's creation of Earthseed is an attempt at providing both hope and better guidance in those circumstances. It states that "All successful life is Adaptable,/Opportunistic,/Tenacious,/Interconnected, and/Fecund" (1993: 124–25) thereby stressing ecological principles.[7] Only lifeforms that are willing to connect with others, to be creative, and to adapt to changing circumstances will be ultimately sustainable in a deeply dystopian world. These are the things that Lauren tells everyone who is willing to listen, and she quickly starts collecting a group of followers that share or take on her beliefs.

At first, these followers are other people from her community who accompany her on her way "[u]p toward Canada" where Lauren hopes to "get to a place where the water doesn't cost more than food, and where work brings salary" (Butler 1993: 170–71). But it does not take long until the first strangers join the group to escape the hell of Southern California. Lauren is not interested in their ethnicity or personal background, her only conditions for joining are that they make themselves useful, stay loyal to the group, and accept the moral principles of Earthseed. Their long walk as "part of a broad river of people walking west [and later north] on the freeway" (1993: 176) is slow, difficult, and involves many sacrifices. Mixed race relationships of all kinds are highly stigmatized outside of Robledo and so the ethnically diverse group quickly attracts the attention of all kinds of "predators" (1993: 177) who want their water, their belongings, or simply their lives. Again and again, Lauren is engulfed in intolerable pain when group members or even their enemies get hurt. More than once, she comes close to dying. More than once, she decides to kill rather than get killed or see someone in her group die. In order to make the Earthseed community thrive, her ethics has become at once inclusive and deeply parochial.

Without question, the vision of the future expressed through *Parable of the Sower* is very dark. Jerry Phillips has gone so far as to ascertain that in Butler's future America "patterns of race and class dominance have hardened to the point that they have genocidal implications" (2002: 305). And yet, he also locates utopian impulses in the novel's defense of "human agency as a necessary existential value" (2002: 299). Despite all its darkness, *Parable of the Sower* is marked by what Raffaella Baccolini calls "an opening for utopian elements in . . . dystopian fiction" (2000: 13), and I would argue that Butler's defence of human agency is inextricably linked to notions of community and sustainability. As Butler herself has put it in an interview, Lauren "can see that the people around her [can] not sustain themselves if they d[o] not find ways to work together" (Conversations 1993: 340). So, when the community she has grown up in gets destroyed, she sets out to build another, more sustainable one on a remote piece of land in Oregon that has its own water and is fit for farming. The very first journal entry Lauren shares with us tells us about a dream in which she is attempting to fly without proper instruction or guidance. Human beings are not equipped with wings and yet at the end of the novel Lauren has fulfilled the other dream that has driven her up the coast toward Canada: the ecotopian dream of "doing something purposeful and constructive" (Butler 1993: 275) by building an intentional community that lives off the land. "Without ecotopian dreaming", Eric Otto reminds us, "damaging systems are allowed to flourish at the expense of sustainable ecological and social possibilities" (2012: 73). In *Parable of the Sower*, Butler suggests that ecotopian dreams can – at least temporarily – take off and soar.

Butler had originally planned to write five Parable books. The second novel, *Parable of the Talents* (1998), picks up where the first one ended and shows once again that pocket utopias are unsustainable as Lauren and her people are engaged in a desperate fight to defend their ecotopian community against an increasingly fascistic tendency in society, which is occurring within a swiftly disintegrating environment. Only in its epilogue does the thoroughly dystopian *Parable of the Talents* offer another utopian opening as it shows an aged Lauren Olamina witnessing the launch of the first Earthseed ship carrying interstellar colonists off the planet in accordance with her prophecy that the final "destiny of Earthseed is to take root among the stars" (1998: 77). However, even that utopian moment of hope includes its spoilers. As Gary Canavan points out, "the name of the spaceship gives us pause: against Olamina's wishes the ship has been named the Christopher Columbus, suggesting that perhaps the Earthseeders aren't escaping the nightmare of history at all, but bringing it with them instead" (2014: n.p.). Imagining an interstellar repetition of terrestrial colonial history with all its cruelty, injustice, and adverse effect on the planetary environment seems have been too much even for Butler, who never managed to complete the third novel. What we know from her notes and early drafts, however, is that it was meant to be set in a far-away world that "is gray and dank, and utterly miserable" (Canavan 2014: n.p.) and, in its lifelessness, not even a close match to the once beautiful and bountiful planet that was lost.

Conclusion

Because they paint pessimistic images of our future, the imaginary worlds we experience during the reading of dystopian texts can be unsettling, even disturbing. Critics have argued that this approach is problematic and ineffective since an overabundance of negative emotions can push readers easily to the point where they simply close the book and turn to something else. However, the recent revival of Orwell's *1984* reminds us that we should not so quickly disparage the evocation of negative and uncomfortable emotions. As Nussbaum puts it so pointedly, "for literature to play its function [in society] it must be permitted, and indeed invited, to disturb us" (1997: 98). That *Parable of the Sower* has the capacity to disturb becomes clear when we look at readers' reactions. "I don't know about you", writes Malinda Lo in her review, "but I get extremely involved in the books I read, and being forced to live in such a horrible environment, mentally, is not something I usually like to do. I mean, it gives me nightmares" (2010: n.p.). This highly emotional reaction to the speculative world of a novel is a testimony to the visceral power of liberated embodied simulation, and it demonstrates that the enlightening triangulation that Nussbaum, Moylan, and others are hoping for, can occur because of such affective engagement, even if it is painful. "[P]art of my ambiguous reaction to *Parable of the Sower* is defensive", acknowledges Lo. "I had to read it at arm's length, if you know what I mean. . . . Nonetheless, I'm glad I read the book" (2010: n.p.). This reminds us that pleasure is not the only emotion that keeps readers engaged – especially not when they are picking up a dystopia – and that novelists in fact *mean* to instill negative emotions such as fear and anxiety when they are voicing warnings because they hope that these emotions will compel others to action.

In a 2004 interview with Joshunda Sanders, Butler stated explicitly that she was trying to warn her readers with her novels: "One of the kinds of research I did was to read a lot of stuff about World War II. Not the war itself, but I wanted to know in particular how a country goes fascist. So, I have this country, in *Parable of the Sower*, and especially *Parable of the Talents*, sliding in that direction. And I really was not trying to prophesize that somehow we would do that but" (2004: n.p.). The "but" in Butler's unfinished sentence hints at political tendencies she noticed during the Bush administration. One cannot help but wonder what warnings she would give today in the climate change-denying, post-truth world of the Trump administration, if she were still among us. It is not 2025 yet, and as far-fetched and extreme as her speculative scenario of an anarchic, unjust, and unsustainable world may have seemed at the time of its publication, it becomes hauntingly premonitory as we approach that date.

Notes

1 My previous publications on the critical eco-dystopia include "Hope in Dark Times" (2014) and chapter 5 of *Affective Ecologies* (2017).
2 On *Parable of the Sower*'s classification as a critical dystopia, see also Miller (1998).
3 For a detailed narratological discussion of the storyworld concept, see Herman (2005).
4 For a discussion of readers' physical responses during literary reading, see Kuijpers and Miall (2011).
5 I discuss Robinson's *Pacific Edge* in more detail in chapter 6 of *Affective Ecologies*.
6 Although authors do write first-person narratives in the present tense and often succeed in creating a sense of immediacy through this technique, narratologists have pointed out that it is a mode of narration is logically impossible since no one can write down a story while they are acting and experiencing it (Phelan 2013).
7 Butler has stated that Lauren's philosophical views closely mirror her own. In the Fry interview, Butler states that Lauren's philosophical views come "[f]rom me, really. One nice thing about writing is that it forces you to look at your own beliefs. My character got her Books of the Living through a lot of religious books and philosophical writings and stopping whenever I found myself in agreement or violent disagreement. Figuring out what I believed helped me figure out what she believed" (Fry 1994: n.p.).

References

Baccolini, R. (2000). "Gender and Genre in the Feminist Critical Dystopias of Katherine Burdekin, Margaret Atwood, and Octavia Butler". In M. S. Barr (ed.). *Future Females, The Next Generation: New Voices and Velocities in Feminist Science Fiction Criticism*. Lanham: Rowman & Littlefield, 13–34.

Butler, O. (1993). *Parable of the Sower*. New York: Grand Central Publishing.

Butler, O. (1998). *Parable of the Talents*. New York: Grand Central Publishing.

Canavan, G. (2014). "'There's Nothing New Under the Sun, but There Are New Suns': Recovering Octavia E. Butler's Lost Parables". *Los Angeles Times Review of Books* (9 June 2014). http://lareviewofbooks.org/article/theres-nothing-new-sun-new-suns-recovering-octavia-e-butlers-lost-parables [Accessed 3 April 2017].

Dromi, S. M. and Illouz, E. (2010). "Recovering Morality: Pragmatic Sociology and Literary Studies". *New Literary History*, 41(2), 351–369.

Frazier, C. M. (2016). "Troubling Ecology: Wangechi Mutu, Octavia Butler, and Black Feminist Interventions in Environmentalism". *Critical Ethnic Studies*, 2(1), 40–72.

Fry, J. (1996). "An Interview with Octavia E. Butler". *Poets & Writers* (March/April). www.joanfry.com/congratulations-youve-just-won-295000/ [Accessed 2 April 2017].

Gallese, V. (2011). "Mirror Neurons and Art". In F. Bacci and D. Melcher (eds.). *Art and the Senses*. New York – Oxford: Oxford University Press, 441–449.

Herman, D. (2005). "Storyworld". In D. Herman, M. Jahn and M.-L. Ryan (eds.). *Routledge Encyclopedia of Narrative Theory*. New York: Routledge, 569–570.

Hogan, P. C. (2003). *The Mind and Its Stories: Narrative Universals and Human Emotion*. Cambridge: Cambridge University Press.

James, E. (2015). *The Storyworld Accord: Econarratology and Postcolonial Narratives*. Lincoln: University of Nebraska Press.

Kakutani, M. (2017). "Why '1984' Is a 2017 Must-Read". *New York Times* (26 January 2017). www.nytimes.com/2017/01/26/books/why-1984-is-a-2017-must-read.html?smid=tw-share&_r=0 [Accessed 30 March 2017].

Kuijpers, M. M. and Miall, D. S. (2011). "Bodily Involvement in Literary Reading: An Experimental Study of Readers' Bodily Experiences during Reading". In F. Hakemulder (ed.). *De stralende lezer: Wetenschappelijk onderzoek naar de invloed van het lezen*. Delft: Eburon, 160–182.

Lewis, D., Rodgers, D. and Woolcock, M. (2008). "The Fiction of Development: Literary Representation as a Source of Authoritative Knowledge". *Journal of Development Studies*, 44(2), 198–216.

McCarthy, C. (2006). *The Road*. New York: Vintage.

Miller, J. (1998). "Post-Apocalyptic Hoping: Octavia Butler's Dystopian/Utopian Vision". *Science Fiction Studies*, 25(2), 336–360.

Moraine, S. (2017). "Imagine and Survive: Resistance Through Speculative Fiction". *Tor.com* (27 March 2017). www.tor.com/2017/03/27/imagine-and-survive-resistance-through-speculative-fiction [Accessed 1 April 2017].

Northrop, R. B. and Connor, A. N. (2013). *Ecological Sustainability: Understanding Complex Issues*. Baton Rouge: CRC Press.

Nussbaum, M. (1997). *Cultivating Humanity: A Classical Defense of Reform in Liberal Education*. Cambridge, MA: Harvard University Press.

Otto, E. (2012). *Green Speculations: Science Fiction and Transformative Environmentalism*. Columbus: Ohio State University Press.

Phelan, J. (2013). "Improbabilities, Crossovers, and Impossibilities: A Rhetorical Approach to Breaks in Mimetic Narration". In J. Alber, H. S. Nielsen and B. Richardson (eds.). *A Poetics of Unnatural Narrative*. Columbus: The Ohio State University Press, 165–184.

Phillips, J. (2002). "The Intuition of the Future: Utopia and Catastrophe in Octavia Butler's *Parable of the Sower*". *NOVEL: A Forum on Fiction*, 35(2/3), 299–311.

Prinz, J. J. (2011). "Is Empathy Necessary for Morality?". In A. Coplan and P. Goldie (eds.). *Empathy: Philosophical and Psychological Perspectives*. New York: Oxford University Press, 211–229.

Robinson, K. S. (1988). *Pacific Edge*. New York: Tom Doherty.

Ruffin, K. N. (2010). *Black on Earth: African American Ecoliterary Traditions*. Athens: University of Georgia Press.

Sanders, J. (2004). "Interview with Octavia Butler". *In Motion Magazine* (14 March 2004). www.inmotionmagazine.com/ac04/obutler.html [Accessed 2 April 2017].

Weik von Mossner, A. (2017). *Affective Ecologies: Empathy, Emotion, and Environmental Narrative*. Columbus: Ohio State University Press.

Wojciehowski, H. and Gallese, V. (2011). "How Stories Make Us Feel: Toward an Embodied Narratology". *California Italian Studies*, 2(1). http://escholarship.org/uc/item/3jg726c2 [Accessed 7 January 2017].

Zapf, H. (2016). *Literature as Cultural Ecology: Sustainable Texts*. London: Bloomsbury.

Zunshine, L. (ed.). (2010). "Introduction". In *Introduction to Cognitive Cultural Studies*. Baltimore: Johns Hopkins University Press.

16 A tale of love and fallout

Lauren Redniss' graphic narrative and the ecological imaginary

Torsten Meireis and Gabriele Rippl

Introduction: "Absolutely Dazzling"

"Absolutely dazzling. *Radioactive* is both vibrant history and a work of art. Like radium itself, it glows with energy." (Paul's Corner, 2015)[1] Obviously, Pulitzer Prize-winner Richard Rhodes could not resist acclaiming the graphic narrative *Radioactive – Marie & Pierre Curie. A Tale of Love and Fallout* (2010) by American artist and cartoonist Lauren Redniss with an adjective that describes the state the recipient finds himself/herself in, when reading this amazing book: S/he is left dazzled, stunned and overwhelmed. By characterizing his reading experience of *Radioactive* thusly, Rhodes attests to the impact of such narratives on readers' imaginations and feelings. Describing it is one thing but categorizing this intermedial narrative is harder: It combines text and images, yet it is not exactly a graphic novel – the comic strip format with its panels and speech balloons is missing. It seems more correct to speak of a graphic narrative, as this term, used henceforth, simply refers to storytelling done not exclusively with words.[2]

To explore the relationship between fiction/visual works of art and cultural sustainability, graphic narratives, artist's books and graphic novels offer themselves as excellent cases in point. We argue that they are eminent objects of investigation because they combine often striking visual artworks with intricate plots and stories. Moreover, due to their popularity and accessibility, they have become genres with considerable cultural momentum. Numerous graphic narratives deal with radioactivity and nuclear weapons, thus giving a voice to cultural anxieties as well as to environmental concern. As cultural products that reach a large readership, they warrant discussion in an essay collection dedicated to cultural sustainability. In this context, Lummina Horlings' term 'inner dimension of sustainability' is helpful since it accounts for the fact that "change toward sustainability" is accelerated not only by political systems or new laws, but also by "individual and shared beliefs, values, worldviews and paradigms that influence attitudes and actions" (Horlings 2015: 163). To substantiate our claim that graphic narratives have an impact on the inner dimensions of sustainability and thus serve as examples of its cultural aspects, we will investigate Redniss' innovative graphic narrative. We will first explore the cultural dimensions and patterns of ecological sustainability and then move on to analyse the ways in which

Redniss' images and word-image-configurations feed into the social imaginary and how they function as narrative ethics.

Cultural dimensions and patterns of ecological sustainability

When discussing cultural patterns of ecological sustainability, it is first of all necessary to distinguish a wider, implicit and a narrower, explicit cultural dimension (cf. Meireis, Chapter 5). While a wider concept, prominent in academic debates, takes culture as a comprehensive term which represents the 'symbolic universe' (Cassirer 1944) as product of man's symbolizing faculty, a narrower concept, explicit and prominent in everyday language, identifies 'culture' with a social field of action, i.e. with the performing and visual arts, with music and literature, sculpture and design in various media (Williams 1989). As this symbolizing faculty is applied to different geographic, social, economical and historical scenarios, different symbolic worlds ('cultures') coexist, and – given a personal or medial exchange – interact and intermingle. While cultural products in the context of an 'explicit dimension' are intentionally devised as contributions to a given society's culture, those in the 'implicit dimension' are not even present as such in common understanding, they are just the 'things people normally do.'

Lauren Redniss' graphic narrative *Radioactive* belongs, of course, to literature and hence to the explicit cultural dimension, to culture in the narrower sense. Even though the downsides of science and radioactivity are demonstrated, one does not find an overt conservationist stance in *Radioactive*. Instead, questions of ecological sustainability are raised implicitly, for instance when plant mutation (leading to sterility) is depicted as a consequence of the nuclear meltdown in Harrisburg's Three Mile Island nuclear power plant, and harmful effects on the fauna around the Chernobyl site are shown. Additionally, Redniss neither omits the toll of radiation sicknesses in Curie's family as a by-product of nuclear science, nor does she conceal the effects of nuclear weaponry. However, at no point is conspicuous moralizing to be found. Redniss also depicts the immense contribution of X-Rays to medicine and the potential of radon-treatment. She describes scientists' aspirations for nuclear starships and the colonization of extraterrestrial space, ending on a vision of a lunar crystal city (Redniss 2010: 184–185) made possible by nuclear energy.

Radioactivity and graphic narratives

The ambivalence of radioactivity – and of modern scientific technology in general – is an idea widely mirrored in popular culture at large. It plays an important part in popular music – for instance in songs like Imagine Dragon's "Radioactive" (2012) or Kraftwerk's "Radio-Aktivität/Radio-Activity" (1975). The latter song is performed bilingually and plays on the word 'radioactivity', referring to broadcasts of music as well as radiation. Kraftwerk presents lyrics such as, "radioactivity is in the air for you and me", with a mellow voice and harmonic style. This style, however, contrasts with the content and must have sounded gloomy even before the Harrisburg (1979), Chernobyl (1986) or Fukushima (2011) accidents, especially for listeners in the 1970s in countries where concern over the safety of nuclear energy was running high. While the possible consequences of nuclear experiments, accidents and wars have been explored in popular comic series like Stan Lee's *The Incredible Hulk* (1962) or *Spider-Man* (1962), they are also dealt with in award-winning graphic novels such as Raymond Briggs' *When The Wind Blows* (1982) (cf. Hoppeler and Rippl 2014). The exact impact of graphic and literary images of radioactivity on political decision-making and the formation of social movements is difficult to measure using the qualitative methods scholars usually apply. Yet in the case of the no-nukes movement of the 1970s and 1980s, it is evident that a close relationship between social movement, practical action and cultural products such as literature and pop songs existed. Gudrun Pausewang's young adult novel *Die Wolke* (1987, English translation *Fall-Out*, 1997) serves as an example. The latter imagines nuclear fallout after a meltdown in a power plant. We argue that Redniss' graphic narrative, just as Pausewang's novel, are prime examples of an explicitly cultural dimension of ecological sustainability, i.e. they demonstrate how ecological questions are represented and reflected intentionally in cultural production. The contribution of such works may be understood as comprising three elements: Firstly, they present a set of cultural patterns (semantic, semiotic, iconic) to express propositions, sentiments or even judgements on ecological questions. Secondly, they introduce ecological concepts into the 'social imaginary' (Taylor 2004) by allowing readers to experience emotional immersion in a (fictitious) dangerous environment. Lastly, they thus form a type of narrative ethics, enabling and challenging readers to reflect for themselves on questions of ecologically and culturally sustainable technologies and paradigms. Works like Redniss' *Radioactive* stage other lives imaginatively, thus providing "a forum for the enactment of the dialogical interdependence between self and other" (Zapf 2008a: 173). They always present views on alterity/the Other (Claviez 2008: 447–451), thus allowing the reader to make their own ethical/moral judgement, instead of prescribing a specific one. In its irreducibility of the individual and the particular, this is the common ground they share with religious documents (Schleiermacher 1967: 489–491).

Graphic narratives with their appealing pictures and powerful word-image configurations lend themselves particularly well to the presentation of fictive worlds which, due to their immersive power, drag the reader into the story. Pictures (and depictions of facial expressions and bodily postures in particular) are "fast tracks to narrative empathy" because "they convey emotional states [. . .] [which] call[] upon readers' neural systems for recognition of basic emotions" (Keen 2011, 135 and 137).[3] Through visual codes and colour symbolism they are able to augment and enrich emotions in the recipients and to enhance their participation. They also have the capacity to express trauma and cultural anxieties in ways that black letters on white paper cannot (Hoppeler and Rippl 2014).[4] Colour, the contrast of light and dark, drawing style and layout play an important role in engaging the reader imaginatively. Since graphic narratives such as comic books and graphic novels often participate in counter-discourses and follow cultural-critical impulses, they remind us of past catastrophes and warn us of possible future ones. Like literature in general, graphic narratives have social implications and are a means of political participation: They are "a sensorium for the deficits and imbalances of the larger culture" (Zapf 2006: 49); they fathom latent and displaced individual and collective fears and anxieties and stimulate us to imagine possible alternative worlds and to identify with others. As Henry John Pratt convincingly argues, mass-marketed superhero comic books and graphic novels

> can provide rich aesthetic and ethical experiences. The narratives presented [. . .] allow readers to exercise their imaginative capacities, to contemplate contrasts between good and evil, to think about prejudice and the sociocultural nature of the self (issues nearly all superheroes face), and perhaps, to an even greater extent than the novel, to engage in emphatic exploration of the minds of others.
>
> (Pratt 2009: 103)

This also applies to Redniss' *Radioactive*.

A tale of love and fallout

Lauren Redniss' *Radioactive – Marie & Pierre Curie. A Tale of Love and Fallout* is a stunning biographical graphic narrative-cum history lesson-cum love story which recounts the lives of two brilliant and devoted scientists and their contributions to science and medicine (cf. Hoppeler and Rippl 2014). Redniss explains the concepts of radioactivity, half-life and nuclear fission and explores the implications, both positive and negative, of Marie Curie's ground-breaking discovery of radium. Her images as well as text-picture configurations are rich and highly complex – as, for example, the technopaignium in prose (cf. Redniss 2010: 124–125) proves. Instead of being mere illustrations, her amazing pictures are the backbone of her intermedial storytelling: Images are the hub of *Radioactive* and appear on almost every page or serve as background for the text. By mixing a love story with the scientific history of radioactivity, Redniss documents a 'tale of love and fallout' with archival and published sources, including Eve Curie's biography of her mother, *Madame Curie* (originally published in 1937), and Susan Quinn's 1995 *Marie Curie: A Life* and blends historical photographs, facsimiles of original documents and maps with Redniss' own text and images, i.e. drawings as well as rich and colourful illustrations of elements, people and places, which collectively render the story of Marie and Pierre Curie.

Radioactive is a work of art which combines impressive visual material with equally great prose writing. On a bedside table, the book will again surprise its readers because it 'radiates' in the dark due to the phosphorescent ink that was used for the cover. As readers we relish each picture and its accompanying text which often runs across and over the pages, hence constantly blurring the boundaries between word and image. The aesthetic power of the images perhaps adds to the heightened sense of fear experienced by readers in relation to this particularly unsettling topic. Redniss' graphic narrative intricately interweaves the fascination of scientific discovery and personal loss with subplots dealing with the consequences of Marie Curie's discoveries, i.e. nuclear war and ecological catastrophe. The story is enhanced through the skilful application of those tools available to graphic narratives (colouring, imagery) and those of the classic book (the haptic and optical effects of the cover), all of which combines to depict an ambivalent picture of radioactivity (and love). A sense of promise and fascination is directly juxtaposed by one of threat and horror. Redniss makes use of tropes and iconic images associated with radioactivity – the 'nuclear mushroom', the iridescent blue light of visible radiation. She also refers to characters of popular culture like the comic hero Spiderman, the monster Godzilla or 'Bert the Turtle' a character appearing in US Cold War civil defence films.

Radioactive comprises two parts: part I, consisting of Chapters 1–3, is mainly biographical; the titles of the chapters refer to characteristics of the material the Curies worked with, the discovery of radioactivity as well as the mutual emotional attraction of Marie and Pierre. Part II (Chapters 4–9) illuminates the Curies' revelations around radium, their rise in scientific circles, the effect radiation had on their own health and the consequences of their discovery for the world. Redniss elongates her protagonists' faces in a way reminiscent of Modigliani's portraits (Garner 2010), while also giving them hatched eyelid rims and pale outlines, which seem to hint at the dangers of radioactivity from the outset. Pierre studies crystalline structures and offers to share his laboratory space with Marie. They fall in love, get married and spend their honeymoon in 1895 on a bicycle tour, "riding along the coast of Brittany and into the French countryside" (Redniss 2010: 34). The warm earthen colours of brown, yellow and dark orange used here not only symbolize the feelings of the couple, but they are repeated just a few pages later in a photograph of a nuclear explosion. The author's intricate use of colour patterns, as displayed in the first few pages, makes the reader immediately aware of the ambiguous qualities of radioactivity: the "spontaneous luminosity" (Redniss 2010: 51) of the two compounds of radium (chloride and bromide, Redniss 2010: 54–55), which discharges a faint-blue light, is represented by Redniss' use of blues, and the radiance and heat emitted by radioactive elements are alluded to by the bright orange and red colours in other images. This colour scheme is present throughout the book.

Discoveries and scientific developments at the turn of the twentieth century resonate in Redniss' graphic narrative. The Curies' discovery of two new elements, radium and polonium, heralded this new scientific age. But while new discoveries such as X-rays, radium and radioactivity fascinated people at the turn of the twentieth century, Redniss shows that from the start the scientists were aware of the ambivalence of their discoveries, i.e. medical advantages for surgery and cancer, but also of the potentially dangerous effects and the possibilities of misuse. The tension between life and death is indeed conspicuously present in the rest of the book, for instance in the passage about World War I during which Marie Curie developed X-ray field labs so that the doctors no longer were "performing blind exploratory surgeries on already damaged bodies" (Redniss 2010:

156). The precariousness of human life exposed to radioactivity is showcased in Marie Curie's medical records held at the Curie Archives, on which the following passage is based:

> Years of radiation exposure had ravaged Marie's health. Her fingers were barnacled with fibrous lesions from handling radium. She chronicled her own deterioration as laboratory data in neat columns on graph paper. Body temperature, color and amount of urine discharged, 'crises and pus,' were logged at multiple intervals throughout each day. In the margins, she tracked the pain in her body. [. . .] At dawn on July 4, 1934, Marie Curie passed away. The cause of death was 'aplastic pernicious anemia' due to prolonged radiation exposure. She was sixty-six years old.
>
> (Redniss 2010: 168 and 171)

This is one of several sections which cover the effects of long-term radium exposure, from which Marie and Pierre Curie both suffered. In addition, in Chapter 6 (titled "Half-Life") Redniss presents the story of young women who worked with radium for the US Radium Corporation in New Jersey in the 1920s and experienced "long series of crippling symptoms – decomposing jaws, bleeding gums, severe anaemia, immobilizing weakness" (Redniss 2010: 90).

Initially, Redniss was struck by the fact that today's weaponry, medicine, energy and their implications, all link back to the Curies' romance in nineteenth-century Paris, and her graphic narrative mentions some more recent consequences of the Curies' discovery of radium, such as the cranial radiation treatments that enabled a 14-year-old Rhode Island boy named Daniel Fass to survive his non-Hodgkins lymphoma (a massive tumour next to his heart and lungs, Redniss 2010: 70–71). But again, Redniss also presents the precarious disadvantages of radioactivity, which range from the partial meltdown of two nuclear reactors at Three Mile Island near Harrisburg in 1979 (Redniss 2010: 102–103), that produced mutant plant specimens, to the disaster of Chernobyl in 1986 (Redniss 2010: 114–115). There are several one- or two-page peripheral stories, such as one on Irving S. Lowen (Redniss 2010: 77–79), a theoretical physicist who worked on the Manhattan project and became paranoid about Germany's progress in the nuclear arms war, as is documented in his declassified FBI file (Redniss 2010: 78–79); or the mini-narratives on atomic tests in the Pacific Ocean and Nevada during the Cold War period (Redniss 2010: 138–139 and 140–141). In spite of *Radioactive*'s strikingly beautiful visual art, two aspects of radioactivity, its power to both save and utterly destroy life, remain present.

Redniss' *Radioactive*, the social imaginary and 'Narrative Ethics'

'Social imaginary' is a term coined by Charles Taylor. It denotes the everyday view on the mechanisms of society (rather than a scholarly devised social theory), consisting of narratives, images, etc., shared by many people and enabling social practices (Taylor 2004: 23–24). Literature, images and other forms of cultural production shape our social imaginary. Redniss' narrative can be read as a contribution to the social imaginary as she, by focussing on iconic characters like the Curies, skilfully combines and interweaves the lure and the threat of love and scientific progress. As has been demonstrated, *Radioactive* expressly reflects the role of radioactivity in the social imaginary by collecting echoes of radioactivity in popular iconography and culture. From North American comic book heroes like the Hulk and Spiderman to Japanese movie characters like Godzilla, who embody the ambivalent promise and threat of radioactivity, and from the famous (if faulty) concept of the cockroach as last survivor of a nuclear blast (which puts images of human superiority to the test) to mutant plants and the notorious US propaganda figure of 'Bert the Turtle' who simply ducks and covers to escape the harmful effects of a nuclear blast, the presence of radioactivity in the consciousness of the community is encapsulated in the novel. Of course, this also includes religious imagery – the crystal city on the moon (Redniss 2010: 184–185) strongly evokes – at least in the religiously literary – the luminous city of God built from gems, the New Jerusalem envisioned in Rev. 21, 10–19. Moreover, by connecting personal and scientific life, Redniss balances the popular iconic image of the – socially highly detached – 'mad scientist' in order to demonstrate that it is nothing but a projection: While people who are commonly recognized as scientific genii become more like average human beings, the collective fantasies attached to radioactivity's promises may be understood to be part of the 'mad scientist' in all of us.

'Narrative ethics' investigates the ethical content of such cultural products which – by reflecting on (un)sustainability – are themselves important agents for the negotiation of values (Hofheinz et al. 2009). Culture in the narrow definition certainly impacts upon public perceptions of environmental issues. Many contemporary narratives in literature and films covering nuclear disaster and climate change help us to envisage the precarity of both locally and globally significant scenarios, thus making us understand what it means to be interconnected with far away regions and people (Zapf 2008b; Heise 2006). They invite us to think about values such as respect for nature, ecological and cultural sustainability and have hence the potential to trigger action and engagement in sustainable development projects. Since works that deal with radioactivity or climate change disaster show us an increasingly hostile nature that is oblivious to national boundaries, they help us envisage globalization. What is more, they attract global audiences, and have thus a worldwide significance and impact on the social imaginaries of many.

Since the 1980s philosophers in particular have argued that "literature, especially [. . .] literary narrative, as a field of demonstration and testing ground for responsible and rewarding human behavior" is "superior to the abstract argumentation of the ethical discourse of philosophy" (Grabes 2008: 1). The ethical turn reached literary studies in the 1980s and was then taken up by cultural studies in the mid-1990s. It triggered a discussion of the role of literature and other media for the dissemination of ethical values within a culture and even beyond. Together with the ecological turn, this ethical turn is at the heart of today's research in the humanities, and debates about human values are hence no longer exclusively the prerogative of philosophy. Scholars of narrative fiction have focused on anthropocentric ethical issues such as encounters with otherness, the self, values, responsibility, relatedness, community/conviviality and compassion. The research focus on narrative fiction is usually justified with the argument that it represents "precisely what ethics is about, namely: a reflection on human action and character, conflicting drives, desires, and choices evolving in time, offered for the reader's appreciation or judgement from different perspectives" (Korthals Altes 2008: 142). Philosophers such as Emmanuel Levinas, Jacques Derrida, Paul Ricoeur and Martha Nussbaum understand literature as having an ethical role in that it offers "the experience in radical strangeness of the other, the self, and the world, and in the final undecidability of meaning and values" (Korthals Altes 2008: 143). According to Martha Nussbaum (1995), by immersing readers into imaginary worlds, narrative fiction has the power to teach readers empathy and to broaden their horizons by offering emotional experiences and alternative visions, thus inviting readers to transform their traditional take on the world and to train their ethical positioning (cf. Locatelli 2008). Since literature models the emotions, self-conception and attitudes of its readers, it plays an important role in their moral development: "By engaging us in situations of value-conflicts, narrative exercises our practical moral sense, allowing for vicarious experiential learning" (Korthals Altes 2008: 143).

To summarize: besides contributing to the social imaginary and influencing the cultural language of sustainability, works like *Radioactive* lend themselves to being interpreted as a certain type of 'narrative ethics' characterized by an open approach. Rather than presenting a fixed set of moral values, they challenge the reading public to reflect their own by creating a pre-normative 'ethical model' that recalls experiences or creates them through fiction, bundles imagery (by showing X-ray technology alongside the atomic bomb) without fixing it terminologically and remains open for conflicting interpretations (Mieth 2007: 223–224). The point of this type of narrative ethics – and works like *Radioactive* – lies in allowing for a recurring exploration of evaluations concerning actions and characters. They are "exploration travels into the realm of good and evil" (Ricoeur 2005: 201). We may admire Marie Curie's relentless stamina and power of volition that carries her from a poor family ravaged by disease and an occupied country to the Nobel prize-winning scientist and professor at a time when women were universally considered to be mentally inferior to men, but we may also abhor Pierre Curie's radium experiments on himself, leading to radiation sickness and deep lesions. We may be fascinated by the combination of mutual love and utter devotion to science and yet be deterred by the unintended consequences of that research, which, as we are constantly reminded, are progressions toward the atomic bomb and radioactive pollution. The Curies' discoveries are contextualized within a violent and politically divided world, with the narration of Marie Curie's biography intermingling with these dystopian images of fear and destruction. Her life thus may be understood as a human micro-cosmos depicting mankind's story with radioactivity: "Radioactivity had made the Curies immortal.

Now it was killing them" (Redniss 2010: 74). In her emotionally cogent tale, Redniss raises questions regarding moral principles. However, the reader is neither forced to answer these questions, nor are answers delivered by an omniscient narrator. The challenge to the reader, who might be undergoing some emotional turmoil while following the story of Curie's biography and her biological and intellectual offspring, is not to be underestimated. Readers are invited to contemplate central questions like these: Do we have the right to conduct research without regard to consequences? Should we be enamoured by our work as we are by another person? Is it acceptable to draw our children into the risks we are taking, even if such risk constitutes our way of life, and further, should we follow our dreams regardless of their effects on other people's lives? If a person belongs to an oppressed or persecuted group or is discriminated against, may she employ *any* means to achieve freedom and recognition? Which risks should society take in regard to technologies? How do we judge the close relationship of science and the military? Are the promises of nuclear technology worth its risks? The seemingly inexhaustible list of questions arising from the narrative refers to the interior, subjective dimension of sustainability. Indeed, the connection between sustainability and the role values play in forming individual motivations, beliefs and actions has not yet been thoroughly dealt with. Philosophy, literary and cultural studies, as well as theology and religious belief systems (cf. Ricoeur 2005: 200–206) discuss human values and play an increasingly crucial role in debates on how to implement them. Since narrative ethics focuses on questions of communication, relationships, motivation, principles, preferences and emotions in individuals and communities, it helps to sharpen people's awareness of their own actions and feelings of responsibility for their environment.

Conclusion: *Radioactive* as a contribution to cultural sustainability

As elaborated earlier, 'culture' may be understood in at least two ways: as the universal 'symbolic universe' typical for humankind, always formed in a particular way depending on geographic, social and historical conditions, and as a social realm characterized by the intentional production of symbolic imagery. Since sustainability may be applied to questions outside of ecology, at least three aspects of the term cultural sustainability may be named:

- First of all, and most important in the context of this volume, 'cultural sustainability' may denote the cultural dimension of ecological sustainability which always implies a degree of normativity. On the basis of the Brundtland report's (1987) influential definition linking sustainability and development, cultural sustainability may be understood as concerning the formation of particular symbolic universes in regard to the natural environment and its media and resources. That also includes a certain normative language of talking about the environment and the social imaginary connected to it. As human behaviour, preferences and lifestyle are closely connected to questions of culture in the senses outlined previously (i.e. the language in which ecological questions are phrased, the social imaginary in which they are envisioned and the narrations in which they are framed), the normative analysis and reflection in the cultural field are tantamount to the social consensus underlying any political strategy to ensure ecologically, economically and socially sustainable development.

- Secondly, 'sustainable culture' may refer to the sustainability of explicit cultural activity, i.e. the public esteem of the performing and visual arts, religion, education and so on.
- Thirdly, 'cultural sustainability' may refer to the particular forms of the symbolic universe, and denote material 'cultures' that are sustainable, i.e. not bent towards self-destruction through problematic practices or inflexibility as, for instance, the culture of the Easter Islands or the Viking colonists on Greenland as depicted in Jared Diamond's study *Collapse* (2005).

Redniss' graphic narrative helps to exemplify these three aspects of the term cultural sustainability:

- To begin with, the narrative ingrains the ambivalence of radioactivity in the social imaginary by calling up adverse images, invoking and reflecting social imaginaries and delivering narrative ethics, thus contributing to the cultural dimension of ecological sustainability.
- Since Redniss' *Radioactive* is a national book award finalist and was nominated for the Pulitzer Prize while amassing much public acclaim, it, secondly, demonstrates the possibilities of cultural sustainment: public medial recognition, public recognition through prizes and acclaim.
- Thirdly, as a work discussing the ambivalences of radioactivity, it contributes to debates on ecological sustainability and may also strengthen the long-term chances of the development of a sustainable culture.

Graphic narratives like *Radioactive* can have an impact on the interior dimensions of sustainability. Combining strong popular appeal with the provision of imagery for the social imaginary, *Radioactive*'s ethical stance fosters reflection in readers, rather than suffocating them with moral judgements.

Notes

1 Richard Rhodes is quoted in a review in *Prairie Lights* on 8 August 2015. Cf. www.prairielights.com/pauls-corner/pauls-corner-radioactive-lauren-redniss.
2 Reviewers have called *Radioactive* an illustrated biography (Garner 2010), a *bande dessinée* and a graphic novel (Nye 2011; Harayda 2011).
3 For a discussion of the creation of empathy in novels, see Keen (2010).
4 It is important to point out that there has not yet been sufficient (empirical) research into how exactly the combination of text and pictures in comic books and graphic novels trigger emotions and empathy in recipients, and whether intermedial modes of storytelling are emotionally enhanced ways of narration which produce more intensive

responses in their readers than exclusively verbal storytelling (cf. Keen 2011: 152–153). Due to the fact that pictures are processed in the right hemisphere of the human brain, which also controls emotions, psychologists and narratologists have claimed that pictures trigger more attention and emotions in recipients (the emotional impact increases with the iconicity of the pictures) and are more memorable than stories mediated via black letters on white paper (Nöth 2000: 467–477 and 481). But questions relating to the interrelationship between pictures and the creation of emotions can only be answered by further quantitative and qualitative research which, as Arvid Kappas and Marion G. Müller have recently suggested, requires collaboration between psychologists of emotion, visual communication researchers and visual culture experts (Kappas and Müller 2006: 3–23).

References

Cassirer, E. (1944). *An Essay on Man: An Introduction to a Philosophy of Human Culture*. Yale, New Haven: Yale University Press.

Claviez, T. (2008). *Aesthetics and Ethics: Otherness and Moral Imagination from Aristotle to Levinas and from Uncle Tom's Cabin to House Made of Dawn*. Heidelberg: Winter.

Diamond, J. (2005). *Collapse: How Societies Choose to Fail or Succeed*. New York: Viking.

Garner, D. (2010). "The Curies, Seen Through an Artist's Eyes". *The New York Times* (21 December 2010). www.nytimes.com/2010/12/22/books/22book.html?_r=0 [Accessed 24 February 2018].

Grabes, H. (2008). "Introduction". In A. Erll, H. Grabes and A. Nünning (eds.). *Ethics in Culture: The Dissemination of Values through Literature and Other Media*. Berlin – New York: De Gruyter, 1–15.

Harayda, J. (2011). "Marie and Pierre Curie, Exposed – Lauren Redniss's *Radioactive*". *One-Minute Book Reviews* (26 November 2011). http://oneminutebookreviews.wordpress.com/2011/11/26/marie-and-pierre-curie-exposed-lauren-rednisss-radioactive/ [Accessed 24 February 2018].

Heise, U. K. (2006). "Afterglow: Chernobyl and the Everyday". In C. Gersdorf and S. Mayer (eds.). *Nature in Literary and Cultural Studies: Transatlantic Conversations on Ecocriticism*. Amsterdam: Rodopi, 177–207.

Hofheinz, M., Mathwig, F. and Zeindler, M. (eds.) (2009). *Ethik und Erzählung: Theologische und philosophische Beiträge zur narrativen Ethik*. Zurich: TVZ.

Hoppeler, S. and Rippl, G. (2014). "Narrating Radioactivity: Representations of Nuclear Disasters and Precarious Lives in Anglophone Comic Books and Graphic Novels". In B. Korte (ed.). *Precarious Lives*. Heidelberg: Winter, 55–70.

Horlings, L. G. (2015). "The Inner Dimensions of Sustainability: Personal and Cultural Values". *Current Opinions in Environmental Sustainability*, 14, 163–169.

Imagine Dragons. (2012). "Radioactive". In *Imagine Dragons*. *Night Visions*. KidinaKorner/Interscope.

Kappas, A. and Müller, M. G. (2006). "Bild und Emotion – ein neues Forschungsfeld". *Publizistik*, 51(1), 3–23.

Keen, S. (2010). *Empathy and the Novel*. Oxford: Oxford University Press.

Keen, S. (2011). "Fast Tracks to Narrative Empathy: Anthropomorphism and Dehumanization in Graphic Narratives". *SubStance*, 124(40.1), 135–155.

Korthals Altes, L. (2008). "Ethical Turn". In D. Herman, M. Jahn and M.-L. Ryan (eds.). *Routledge Encyclopedia of Narrative Theory*. London – New York: Routledge, 142–146.

Kraftwerk (1975). "Radio-Aktivität". In Kraftwerk. *Radio-Aktivität*. EMI/Capitol Records.

Locatelli, A. (2008). "Literature's Versions of Its Own Transmission of Values". In A. Erll, H. Grabes and A. Nünning (eds.). *Ethics in Culture: The Dissemination of Values through Literature and Other Media.* Berlin – New York: De Gruyter, 19–33.

Mieth, D. (2007). "Literaturethik als narrative Ethik". In K. Joisten (ed.). *Narrative Ethik: Das Gute und das Böse erzählen.* Berlin: Akademie Verlag, 215–234.

Nöth, W. (ed.) (2000). *Handbuch der Semiotik.* Stuttgart – Weimar: Metzler, 2nd ed.

Nussbaum, M. C. (1995). *Poetic Justice: The Literary Imagination and Public Life.* Boston, MA: Beacon Press.

Nye, M. J. (2011). "Of Passion and Polonium". *American Scientist* (May–June 2011). www.americanscientist.org/bookshelf/pub/of-passion-and-polonium. [Accessed 24 February 2018].

Paul's Corner (2015). "Paul's Corner: *Radioactive* by Lauren Redniss". *Prairie Lights* (8 August 2015). www.prairielights.com/pauls-corner/pauls-corner-radioactive-lauren-redniss [Accessed 24 February 2018].

Pratt, H. J. (2009). "Medium Specificity and the Ethics of Comics". *Storyworlds*, 97–113.

Redniss, L. (2010). *Radioactive – Marie & Pierre Curie. A Tale of Love and Fallout.* New York: Harper Collins.

Ricoeur, P. (2005). *Das Selbst als ein Anderer.* Munich: Wilhelm Fink.

Schleiermacher, F. D. E. (1967). "Über den Begriff des höchsten Gutes: Zweite Abhandlung. Gelesen am 24. Juni 1830". In *Abhandlungen, gelesen in der Königlichen Akademie der Wissenschaften.* In *Schleiermachers Werke, Erster Band*, ed. and introduced by O. Braun and J. Bauer, Leipzig 2nd ed. 1928, Reprint Aalen, 468–493.

Taylor, C. (2004). *Modern Social Imaginaries.* Durham – London: Duke University Press.

Williams, R. (1989). "Culture is Ordinary". In *Resources of Hope: Culture, Democracy, Socialism.* London: Verso, 3–14.

Zapf, H. (2006). "The State of Ecocriticism and the Function of Literature as Cultural Ecology". In C. Gersdorf and S. Mayer (eds.). *Nature in Literary and Cultural Studies: Transatlantic Conversations on Ecocriticism.* Amsterdam: Rodopi, 49–69.

Zapf, H. (2008a). "Narrative, Ethics, and Postmodern Art in Siri Hustvedt's *What I Loved*". In A. Erll, H. Grabes and A. Nünning (eds.). *Ethics in Culture: The Dissemination of Values through Literature and Other Media.* Berlin – New York: De Gruyter, 171–194.

Zapf, H. (2008b). "Literary Ecology and the Ethics of Texts". *NLH*, 39(4), 847–868.

17 Sustainability, eco-ekphrasis and the ethics of literary description

Gabriele Rippl

Introduction: literature and cultural sustainability

The starting point for this chapter is the insight that literature, much like other cultural products, has the potential to contribute to a central, if under-theorized dimension of sustainability, namely the cultural.[1] In current academic and political debates, the visibility of the important contribution to sustainability made by the humanities – and their objects of investigation – is diminished. Important concepts developed in the humanities such as Hubert Zapf's 'literary ecology' have neither been recognized by the agendas of global politics nor by natural scientists and economists working in the field of sustainability and sustainable development (WCED 1987). This is highly problematic since societal change toward more sustainable lifestyles is by no means solely engendered and implemented by political decisions, new laws or scientific discoveries regarding climate change, but linked directly to people's world views, beliefs and cultural and ethical values: the so-called 'inner dimension of sustainability' (Horlings 2015). These world views, beliefs and values, which are shaped by, amongst other things, cultural products such as literature, impact how we act. Literature must be and has been understood as an agent of cultural change (Berensmeyer et al. 2016). In the tradition of Hubert Zapf's work,[2] I claim that cultural products such as literature have an enormous impact on people's world views and values and, thus, potentially also on the sustainability of their lifestyles. They can introduce ecological concepts into the 'social imaginary' (Taylor 2004)[3] by allowing readers to (mentally/emotionally) immerse themselves in a fictitious environment, depicted as dangerous or wracked with unsustainable practices. This challenges readers to reflect on questions of ecologically and culturally sustainable technologies and paradigms. Fiction by definition stages other lives and forms of community imaginatively, thus providing "a forum for the enactment of the dialogical interdependence between self and other" (Zapf 2008c: 173) and for negotiating societal problems and cultural anxieties. Literature and other cultural products have the potential to sharpen people's awareness of their own actions and feelings of responsibility for their environment. What is more, they enable us to experiment and test alternative scenarios in our imagination; specific literary genres such as utopian and dystopian novels stimulate us to imagine possible alternative worlds and to identify with others. Many of these possible worlds are characterized by unsustainable, 'anthropocenic' human-ecosystems and dystopian socio-political settings. By inviting us to contemplate values such as empathy, respect for nature and ecological as well as cultural sustainability, and by inviting us to conceive of the human-nature relationship in a new and non-anthropocentric way (Horn 2017), literary texts can trigger action and engagement in sustainable development projects (cf. Erll et al. 2008).

Sustainability and the ethics of literary description

Literary description is a defining feature in an array of twentieth- and twenty-first-century utopian and dystopian novels in English, novels which investigate alternative, often post-catastrophic worlds characterized by the negation of human rights, societal collapse, unsustainable and precarious environments, ecological disasters resulting from (anthropocenic) climate change and nuclear fallout (cf. Voigts 2015: 1–3). Aldous Huxley's *Ape and Essence* (1948), *Brave New World* (1932) and *Island* (1962), Cormac McCarthy's *The Road* (2006) and Margaret Atwood's *MaddAddam* trilogy (2003, 2009, 2013) are such novels. They are characterized by a tendency to include descriptive passages, creating intense emotional atmospheres. This in turn increases their ability to address issues around anthropocenic unsustainability. This chapter focuses on Atwood's *The Year of the Flood* (2009), the second novel in her trilogy. Replete with at times short and succinct, but more often extended, detailed, graphic and emotionally charged descriptions of nature after ecological catastrophes and disastrous developments in the socio-political sphere, the novel also presents some memorable eco-ekphrases. This focus on textual details of descriptions will help to analyze not only thematic and content-related matters of environmental literature but also its (often neglected) aesthetic and formal features. Description, understood as a conceptual space for presenting the downsides of unsustainable actions, offers a means of presenting sustainable visions and alternative ways of living. While ecologically inclined literature often immerses the reader in an impending apocalypse or a post-apocalypse of an ecological catastrophe, and while precarious life and human suffering after nuclear disaster, challenge representation (Hoppeler and Rippl 2014), writers of eco-dystopias such as Margaret Atwood have succeeded in finding adequate literary forms to express what is hard to imagine and even harder to describe. Critical eco-dystopias such as Atwood's *MaddAddam* trilogy are a "vital contribution to the cultural discourse on sustainability" because "such works make important cultural interventions that must be accounted for in any discussion of ecological sustainability" (Weik von Mossner 2018: 22, Chapter 15).

Margaret Atwood's MaddAddam *trilogy and sustainability*

Margaret Atwood's *MaddAddam* trilogy consists of three novels, *Oryx and Crake* (2003), *The Year of the Flood* (2009) and *MaddAddam* (2013), all of which deal with issues central to our posthuman age. She speaks of her eco-feminist trilogy as "speculative fiction" and not of science fiction, because it deals with "things that really could happen" (Atwood 2011: 6). The novels are based on multiperspectivism and present male and female human as well as posthuman protagonists, the vegetarian, non-aggressive Crakers. The setting of the three novels is an American post-catastrophic, post-pandemic society of the twenty-first century characterized by extreme social inequalities, alleged bio-scientific advances, eugenics and cloning. The story world is characterized by an intensified form of late capitalism and a corrupt totalitarian regime of technological corporations that "rule the post-ecocide world, divided into communities of the wealthy and the 'pleeblands'" (Ferreira 2015: 44) through "their collective security arm, the CorpSeCorps" (Atwood 2014: xiii). Atwood convincingly describes the temperate northern hemisphere in the grip of brutal weather conditions such as a merciless burning sun and daily thunderstorms. At the same time, Atwood negotiates "the overall ethical question what it means to be human" (Mohr 2015: 285) and discusses the role of science, religion, storytelling and the visual arts in a post-pandemic world.

Sustainability – eco-ekphrasis – descriptive ethics

The theoretical backdrop against which I develop my ideas of eco-ekphrasis in twentieth- and twenty-first-century Anglophone literature is a concept I call 'descriptive ethics.' In the late 1980s, literary scholars began to develop the concept of 'narrative ethics' as an important new field of research.[4] Among other things, they investigated literature's role in implementing cultural and ethical values. When literary critics speak of 'narrative ethics' in connection with literary texts, they do not do so from a normative point of view. Rather, they analyze how morally demanding situations are presented and negotiated aesthetically in narrative texts. Presenting alternative worlds and views on alterity, literary texts remain open to virtually endless interpretations, allowing the reader to make their own ethical and moral judgement, instead of prescribing a specific one. Since the ethics imbedded in the narratives focus on questions of communication, relations, motivations, principles, preferences and emotions in individuals and communities, it helps to sharpen people's awareness of their own actions. This includes their feelings of responsibility for their environment.

What has not been addressed properly in the long debate on narrative ethics is literature's 'descriptive ethics' (Rippl 2015a). This is an environmental ethics whose focus lies no longer exclusively on the human and its human other but is interested also in non-human nature. My use of the term descriptive ethics varies from its usage in philosophy, where it may also be referred to as comparative ethics and is an empirical study of people's beliefs about morality. It contrasts with normative ethics, which is the study of ethical theories that prescribe how people ought to act. Due to its principal interest in humans and autonomous subjects, pragmatic and humanist strands of ethics have been criticized as being out of step with the "new awareness of nature" (Zapf 2009: 853). This awareness is part and parcel of the move away from "the universalist, subject-centered, and exclusionary anthropocentric bias of traditional ethics" (Zapf 2009: 854). Emmanuel Levinas, Jacques Derrida and Serenella Iovino (2010) have suggested new ways of thinking and speaking in non-anthropocentric ways, while Thomas Claviez (2006), starting from Levinas's ethics of radical otherness, has demonstrated how literature's aesthetic mode has a special power in representing this new ecologically oriented ethics that includes the non-human world. What interests me is how literature, with the help of descriptions of nature and descriptions of ecological works of art, evokes ecologically threatened environments, thus raising questions akin to what researchers in ethics call an ecocentric ethics. Since readers' emotional and ethical involvement depends very much on descriptive techniques which mediate environmental issues, the way catastrophes such as nuclear fallout and climate change induced tsunamis are described is key. Recently, scholars like Alexa Weik von Mossner (2017) have drawn on research in cognitive narratology, cognitive cultural studies (Zunshine 2010) and neuroscience to explain how "cultural texts influence their readers' perceptions of sustainability", because these approaches "give insight into the ways in which eco-dystopian fiction allows readers to experience not only mentally, but also on the physical level *what it is like* to live in an unsustainable future environment" (Weik von Mossner 2018: 195, Chapter 15). By presenting readers with a dilemma, "cultural texts" engage them affectively and cognitively. The engagement is particularly strong when a first person narrator relates their experiences from his/her subjective position (Weik von Mossner 2018: 200, Chapter 15).[5] Weik von Mossner's analysis of the protagonist's style of narration in Octavia Butler's eco-dystopian novel *Parable of the Sower* (1993) is productive. I, however, will take a different approach and focus on the descriptive features of *The Year of the Flood* and on the way Atwood employs the showing mode opposed to the telling mode. By engaging with close readings of literary descriptions of severely degraded environments, social wastelands and ekphrases of ecological works of art depicting precarious environments, I hope to shift some attention to a non-anthropocentric protagonist in Atwood's novel, namely non-human nature. Atwood's novel relies on rhetorical tools such as vivid and emotionally charged description, in order to immerse the readers in the fictive world it presents, to invite them to imagine what life under a totalitarian regime and in a post-epidemic, environmentally degraded world is like, and to thus make the connection back to the unsustainability of present-day life.

What, then, is description? It is the less recognized stepsister of narration and, as a text type, has not been theorized as extensively. While 'narratology' is an established field of research, 'descriptology' does not exist, even though theoretical debates on the art of description and its potential to create presence and evidence have a long tradition, starting in antiquity (Halsall 1992, Rippl 2005). Rhetoricians of late antiquity studied description using terms such as *evidentia* or *descriptio* in Latin and *ekphrasis* or *hypotyposis* in Greek (Webb 2009). Description was considered a mode of speaking with the capability of bringing absent things, places and persons before the listener's inner eye by aiming at *enargeia*, a technique that produces *evidentia* and *presence* through a lively, precise and detailed verbal depiction. According to Quintilian (1977, vol. 2, Part I, VI: 2, 32, 434), such enargetic descriptions are able to rouse the listener's emotions by making her/him a quasi-eyewitness to the things described (Rippl 2015b). *Enargeia* as a device of detailing and amplifying description enhances *evidentia* and enables writers not only to reach their recipients' minds, i.e. their rational thinking, but also their emotions. While description was extremely important in eighteenth- and nineteenth-century literature – for instance in realist novels – influential modern critics such as Gérard Genette have reduced description to the role of a maid to narration (*ancilla narratienis*, Genette 1976: 6) and consider the descriptive mode nothing but a marginal ornament. Likewise, classical narratology has understood description as secondary and derivative, i.e. as a mere pause in narration, which interrupts the plot. Georg Lukács's 1936 essay "Narrate or Describe" (2005), is still a classic with its warning that novels replete with detailed and extended descriptions turn into static, monotonous still lifes. However, more recently critics such as Harold F. Mosher (1991)[6] and Ruth Ronen (1997) have suggested that description and narration are not to be conceived as an oppositional pair for the very reason that description often triggers narrative impulses and narrative elements may serve description. Even more recently, David Herman has underlined this overlap between narrative and description: he refers to category gradience and other prototypical features shared by description and narrative (2008: 451–452). Thus, when I speak of descriptive ethics, it is not my intention to pit it against narrative ethics. Rather, I want to highlight the – less recognized – importance of description in ecological fiction.

The first example of a description of nature is taken from *The Year of the Flood*: even after the pandemic and its brutal consequences, descriptions often utilize the pastoral mode: "All around her is a sweet scent – the tall clover's in bloom, the Queen Anne's lace, the lavender and marjoram and lemon balm, self-seeded. The field hums with pollinators: bumblebees, shining wasps, iridescent beetles. The sound is lulling" (Atwood 2010: 327). Even though the air is much clearer, "now that man-made pollution has ceased" (Atwood 2010: 371) and the "leaves on the overhanging trees are covered with tiny droplets that shine in the strengthening pink light. Everything looks so fresh, as if newly created" (Atwood 2010: 383). It is Toby, one of the novel's female protagonists, who presents these pastoral descriptions, which are pitted against those of the man-made wasteland, testifying to violence, destruction and death. This is the case when Toby has to leave the safe haven of what used to be a spa where she has lived since the pandemic struck. When she sees shoes lying in the streets, she keeps telling herself to look away, "[n]ot to think about shoes. Not to think about the mouldering handbag she's just glimpsed nearby" (Atwood 2010: 326). Trash clutters the streets, but "[i]t was the small normal things that bothered me the most. Somebody's old diary, with the words melting off the pages. The hats. The shoes – they were worse than the hats, and it was worse if there were two shoes the same. The kids' toys. The strollers minus the babies" (Atwood 2010, 338). Toby and her friends find "bundles of rags and bones. 'Ex-people,' said Croze. They were dried out and picked over, but I didn't like the eyeholes. And the teeth – mouths look a lot worse without lips" (Atwood 2010: 339). This gruesome description of the remains of human beings is topped by other haunting views. When Toby and Ren, another female protagonist, find their friend Oates hanging from a tree, Ren's description reads like an ekphrasis of plate 39 *Grande hazaña! Con muertos!* (*A heroic feat! With dead men!*) of the well-known *The Disasters of War* series by Francisco Goya. These prints depict the atrocities of war[7] and were taken up by Jake and Dinos Chapman in their installation *Great Deeds against the Dead* (1994). For American readers, Atwood's description might also evoke iconic lynching photographs of tortured black bodies dangling from trees:[8]

> We smell the thing before we see it. 'Don't scream,' says Toby.
>
> This is what the crows have been cawing about. 'Oh no,' I whisper.
>
> It's Oates. He's hanging from a tree, twisting slowly. The rope is passed under his arms and knotted at the back. He doesn't have any clothes on except for his socks and shoes. This makes it worse, because he's less like a statue that way. His head is thrown back, too far because his throat has been cut, crows flap around his head, scrabbling for footholds. His blond hair's all matted. There's a gaping wound in his back, like those on the bodies they used to dump in vacant lots after the kidney theft. But these kidneys wouldn't have been stolen for transplants.
>
> (Atwood 2010: 375–376)

It is precisely this pitting of pastoral and bucolic nature against such utterly disturbing images of horror – birds picking on the maimed and tortured body that used to be Oates and the reference to cannibalism – that makes Atwood's descriptions so effective and powerful. While works of art might help the onlooker to distance her-/himself from the gruesome scene depicted, this is not an option for Ren and her friends: the illusion that they deal with a statue is destroyed by the socks and shoes of which Oates's body has not been stripped. What is more, the description activates other images of atrocities and even chains of such images in our cultural memory and has thus the potential to intensify the graphic nature of the scene.

While the quote above functions very much like an ekphrasis of an existing Goya print or lynching photograph, there are also additional ekphrases, what I call 'eco-ekphrases'. These are ekphrases of (fictive or existing) works of art dedicated to ecological topics. Generally speaking, ekphrasis is a special case of literary description, namely, of a work of art. That is to say, it is a second-degree, often highly self-reflexive, mode of representation (cf. Heffernan 1993) which also serves as a space for poetological negotiations. As a textual strategy "for eliciting highlighted visualization" it serves the management of attention and reinforces "comprehension, memory and emotional response" (Brosch 2015: 343). Specifically, however, Atwood's eco-ekphrases discuss ecological issues and take on important cultural-critical, ethical and socio-political functions. In contemporary Anglophone literary texts, ekphrases come in many shapes and are sometimes accompanied by pictures. In Atwood's trilogy, however, all ekphrases are pictureless, a feature they share with the ancient variety. As elaborated earlier, for ancient rhetoricians the term ekphrasis referred to any description that was 'enargetic' (cf. Webb 2009), and enargetic descriptions – that is, detailed, lively and often extended ones – are still an important feature of many contemporary texts. However, the reader often also encounters relatively short, "stenographic" ekphrases which are mere references and non-representational traces, instead of fully fledged descriptions of images.[9] In Atwood's trilogy, ekphrasis' ethical and cultural-critical functions are pivotal. These functions are intrinsically linked to ekphrasis because, for one, ever since John Keats's "Ode to a Grecian Urn", ekphrasis has been closely linked to questions of elegy and death (Kennedy 2012: 5–6) and, second, ekphrasis is a mode of writing which reflects on questions of (re-)representation and showcases epistemological frameworks and cultural hierarchies (Rippl 2015b). Atwood negotiates these issues and employs ekphrasis as a means to halt the narration in order to capture and focus the reader's attention on environmental conditions.

In *The Year of the Flood*, Amanda Payne, a girl from the rough pleeblands, becomes an installation artist. Her friend Ren explains that Amanda does "art involving Creatures or parts of Creatures arranged outdoors on a giant scale" (Atwood 2010: 299) and produces an extended ekphrasis of one of Amanda's art installations with the title 'The Living Word':

> Then I called Amanda. [. . .] Amanda was in the Wisconsin desert, putting together one of the Bioart installations she's been doing now that she's into what she calls the art caper. It was cow bones this time. Wisconsin covered with cow bones, ever since the big drought ten years ago when they'd found it cheaper to butcher the cows there rather than shipping them out – the ones that hadn't died on their own. She had a couple of fuel-cell front-end loaders and two illegal Tex-Mexican refugees she'd hired, and she was dragging the cow bones into a pattern so big it could only be seen from above: huge capital letters, spelling out a word. Later she'd cover it in pancake syrup and wait until the insect life was all over it, and then take videos of it from the air, to put into galleries. She liked to watch things move and grow and disappear.
>
> Amanda always got the money to do her art capers. She was kind of famous in the circles that went in for culture. They weren't big circles, but they were rich circles. This time she had a deal with a top CorpSeCorps guy – he'd get her up in the helicopter, to take the videos. [. . .]
>
> Her Wisconsin thing was part of a series called The Living Word – she said for a joke that it was inspired by the Gardeners because they'd repressed us so much about writing things down. She'd begun with one-letter words – *I* and A and O – and then done two-letter words like *It*, and then three letters, and four, and five. Now she was up to six. They'd been written in all different materials, including fish guts and toxic-spill killed birds and toilets from the building demolition sites filled with used cooking oil and set on fire. Her new work was *kaputt*. When she'd told me that earlier, she'd said she was sending a message.

> "Who to?" I'd said. "The people who go to galleries? The Mr. Rich and Bigs?"
>
> "That's who," she'd said. "And the Mrs. Rich and Bigs. Them too."
>
> "You'll get in trouble, Amanda."
>
> "It's okay," she said. "They won't understand it."
>
> The project was going fine, she said: it had rained, the desert flowers were in bloom, there were a lot of insects, which was good for when she'd pour on the syrup. She already had the *K* done, and she was halfway through the *A*. But the Tex-Mexicans were getting bored.
>
> (Atwood 2010: 56–57)

What used to be the green pastures of Wisconsin, America's dairy producing state, has been turned into a desert strewn with cow bones. Amanda's "Bioart installations" (Atwood 2010: 56) link art and nature through a kind of temporary "cryptographic bio-writing" (Mohr 2015: 295), sending out her environmental message, "*kaputt*" (Atwood 2010: 76), which reflects the state the planet is in and thus functions as a *memento-mori*. The description of Amanda's work of art also refers to illegal refugees, matching with the readers' life-world. Art historians differentiate between land/environmental art and bio art, even if the borderline between the two is not always clear-cut. They tend to reserve the term bio art for laboratory art practices, such as that of Eduardo Kac, where live tissues, bacteria and living organisms are used and often modified (cf. Hildebrandt 2011). Carsten Höller's SOMA installation is another example of bio art. According to the differentiation, Amanda Payne's monumental sculptures do not count as bio art, falling rather into the category of installative land/environmental art (reminding readers perhaps of Robert Smithson's desert land art works).

Another installation from Amanda's outdoor landscape series spells "words out in giant letters, using bioforms to make the words appear and then disappear, just like the words she used to do with ants and syrup when we were kids. Now she said, 'I'm up to the four-letter words.' And I said, 'You mean the dirty ones, like *shit?*' And she laughed and said, 'Worse ones than that.' And I said, 'You mean the c-word and the f-word?' and she said, 'No. Like *love*'" (Atwood 2010: 304). Amanda's four-letter word installation 'Love' is reminiscent of American artist Robert Indiana's sculpture LO/VE (1970, displayed at the Indianapolis Museum of Art), originally done in COR-TEN steel, but since reproduced in a variety of formats, media (painting, polychrome aluminium) and languages. Amanda's 'love' installation is again brought alive via the remains of dead animals – bones, fish guts, toxic-spill killed birds – to be eaten up and disappear. The fact that 'love' has become a four-letter word comments on the '*kaputt*' state of human relations in the social wasteland of the not so 'brave new world'.

Such ekphrastic passages invite the reader to think about the role Atwood ascribes to the arts and to storytelling in the face of climate change, unsustainable lifestyles and aggressive consumerism. One important function of Amanda's monumental land or environmental art installations is to bring nature and culture into close contact; they are no longer separate entities, but are intimately connected, thus overcoming one of the central oppositional pairs of Western philosophy. Amanda's bio art produces an eco-ethical comment on the state the planet earth is in – even if this comment is not of a durable nature. There is also a critical note on the mechanism of the art world, after all Amanda's video-making and photograph-taking of her monumental installations is only made possible by the financial support of a corrupt CorpSeCorps patron who has earned his money unethically.

Conclusion

Atwood's descriptions of unsustainable environments and ekphrases of disturbing ecological works of art provide a space for re-imagining what future human life on a destroyed planet could look like. Her trilogy returns time and again to ethical questions. According to Gerry Canavan, Atwood's "postmodern ecological jeremiads" call for a radical change to "our social relations and anti-ecological lifestyles" (Canavan 2012: 155). At the end of the last novel of the trilogy, *MaddAddam*, Atwood sketches a new, sustainable life style and makes imaginative room for peaceful visions of community, including interspecies contracts. This allows her readers to experience some utopian hope, thus brightening the bleak apocalyptic setting of her trilogy. By bringing cultural and biogenetic perspectives into contact, Atwood demonstrates how important imaginative activities such as the visual arts, story telling, myth-making, speculative narratives and religion are, not only for the survival of humanity, but also for the survival of all other species and indeed the entire biosphere. We can conclude that her novels, with their numerous descriptions of non-human nature and their eco-ekphrases, contribute to sustainability in important ways: for one, they present sets of cultural patterns to express sentiments, propositions or judgements on ecological questions. Second, they introduce ecological concepts into the social imaginary by allowing the audience to emotionally experience alternative ways of thinking and living. Third, they thus form a type of narrative and descriptive ethics, enabling and challenging audiences to reflect on questions of ecologically sustainable technologies and paradigms. The fact that Atwood's novels have received much public attention and acclaim demonstrates, fourth, the possibilities of cultural sustainability: literary prizes and acclaim enhance public (medial) recognition of the problems to which the novels attend. And, five, while putting on stage many ethical conflicts, overall Atwood's trilogy has a didactic impetus: in spite of her polyphonic storytelling, Atwood in her emotionally cogent tale sends out her ecological imperative – 'You must change your lifestyles!'

Notes

1 This text has profited immensely from discussions with colleagues and friends: I would like to thank Toni Hildebrandt, Torsten Meireis, Hubert Zapf, Johanna Hartmann, Christopher Schliephake, Peter Schneemann and Edward Wright for their insightful comments on the first draft of this chapter. My thanks also go to all members of Hubert

Zapf's research colloquium at the University of Augsburg and the attendants of the 2017 Kosmos Summer School "Religion, Sustainability, and Politics" at the Humboldt University Berlin.

2 Hubert Zapf's concept of literary ecology, which he developed in a range of publications over the last 15 years (2002, 2006, 2008a, 2008b, 2016), is an indispensible attempt to demonstrate literature's importance and central position when it comes to questions of sustainability. It highlights the significant ways in which literature and literary studies can contribute to a much-needed transdisciplinary dialogue.

3 'Social imaginary' is a term coined by Charles Taylor. It consists of narratives, images and ideas which are shared by many people and enable social practices (Taylor 2004: 23–24).

4 Cf. Nussbaum (1995); Joisten (2007); Mieth (2007); Grabes (2008); Claviez (2008); Zapf (2008a, 2008c).

5 Characters are "of central importance for our understanding of the story-world, and this is particularly true in the case of first-person narration . . . the readers' imaginative world-building entirely depends on this one experiencing agent" (Weik von Mossner 2018: 200, Chapter 15). The readers' experience of the environment and worsening environmental conditions are filtered through the protagonist's consciousness. Unreliable narration in Octavia Butler, since first-person narrator is highly emotional, very personal, even intimate accounts.

6 Mosher introduced the terms "narratized descriptions" and "descriptized narrations" to demonstrate that pure description and pure narration do not exist (1991: 426).

7 *The Disasters of War* (Spanish: *Los desastres de la guerra*) is a series of 82 prints created between 1810 and 1820 by the Spanish painter and printmaker Francisco Goya (1746–1828). Although Goya did not make known his intention when creating the plates, art historians view them as a visual protest against the violence of the wars taking place during his lifetime.

8 The well-known lynching song "Strange Fruit" has these lyrics: "Southern trees bear strange fruit/Blood on the leaves and blood at the root/Black bodies swinging in the southern breeze/Strange fruit hanging from the poplar trees." Readers of Toni Morrison's novel *Beloved* (1987) will remember Sethe recalling the sights of bodies of lynched black boys hanging from trees.

9 With regard to contemporary literature in English, it makes sense to consider as ekphrasis any explicit or covert verbal/literary reference to images, including descriptive-static ekphrases and those charged with narrative impulses, whether concise and abbreviated (cf. Yacobi 1995).

References

Atwood, M. (2009). *Oryx and Crake*. London: Virago.

Atwood, M. (2010). *The Year of the Flood* [2009]. New York: Anchor Books.

Atwood, M. (2014). *MaddAddam* [2013]. Toronto: Vintage Canada.

Atwood, M. (2011). *In Other Worlds: Science Fiction and the Human Imagination*. Toronto: Nan A. Talese.

Berensmeyer, I., Grabes, H. and Schillings, S. (eds.) (2016). *Literature and Cultural Change. REAL. Yearbook of Research in English and American Literature 32*. Tübingen: Narr Francke Attempo.

Brosch, R. (2015). "Images in Narrative Literature: Cognitive Experience and Iconic Moments". In G. Rippl (ed.). *Handbook of Intermediality*. Berlin – Boston: De Gruyter, 343–360.

Butler, O. (1993). *Parable of the Sower*. New York: Grand Central Publishing.

Canavan, G. (2012). "Hope, But Not for Us: Ecological Science Fiction and the End of the World". *Lit: Literature Interpretation Theory*, 23(2), 138–159.

Chapman, J. and D. (1994). *Great Deeds against the Dead* [Plastic, polyester resin, synthetic fibres, wood and guitar strings]. London: Tate.

Claviez, T. (2006). "Ecology as Moral Stand(s): Environmentalist Ethics, Western Moral Philosophy, and the Problem of the Other". In C. Gersdorf and S. Mayer (eds.). *Nature in Literary and Cultural Studies: Transatlantic Conversations on Ecocriticism*. Amsterdam – New York: Rodopi, 435–454.

Claviez, T. (2008). *Aesthetics and Ethics: Otherness and Moral Imagination from Aristotle to Levinas and from Uncle Tom's Cabin to House Made of Dawn*. Heidelberg: Winter.

Erll, A., Grabes, H. and Nünning, A. (eds.) (2008). *Ethics in Culture: The Dissemination of Values through Literature and Other Media*. Berlin – Boston: De Gruyter.

Ferreira, M. A. (2015). "The Posthumanist and Biopolitical Turn in Post-Postmodernism". *The European Messenger*, 24(2), 42–49.

Genette, G. (1976). "Boundaries of Narrative". *New Literary History*, 8(1), 1–13.

Grabes, H. (2008). "Introduction". In A. Erll, H. Grabes and A. Nünning (eds.). *Ethics in Culture: The Dissemination of Values through Literature and Other Media*. Berlin – New York: De Gruyter, 1–15.

Halsall, A. W. (1992). "Beschreibung". In G. Ueding (ed.). *Historisches Wörterbuch der Rhetorik*. Vol. 1. Tübingen: Niemeyer. Col. 1495–1510.

Heffernan, A. W. (1993). *The Museum of Words: The Poetics of Ekphrasis from Homer to Ashbery*. Chicago: University of Chicago Press.

Herman, D. (2008). "Description, Narrative, and Explanation: Text-Type Categories and the Cognitive Foundations of Discourse Competence". *Poetics Today*, 29(3), 438–472.

Hildebrandt, T. (2011). "World-Making Models". *Rheinsprung 11 – Zeitschrift für Bildkritik*, 2, 154–174.

Hoppeler, S. and Rippl, G. (2014). "Narrating Radioactivity: Representations of Nuclear Disasters and Precarious Lives in Anglophone Comic Books and Graphic Novels". In B. Korte (ed.). *Precarious Lives*. Heidelberg: Winter. 55–70.

Horlings, L. G. (2015). "The Inner Dimensions of Sustainability: Personal and Cultural Values". *Current Opinions in Environmental Sustainability*, 14, 163–169.

Horn, E. (2017). "Jenseits der Kindeskinder: Nachhaltigkeit im Anthropozän". *Merkur*, 71(814), 5–17.

Huxley, A. (1932). *Brave New World*. London: Chatto and Windus.

Huxley, A. (1948). *Ape and Essence*. London: Chatto and Windus.

Huxley, A. (1962). *Island*. New York: Harper and Brothers.

Iovino, S. (2010). "Ecocriticism and a Non-Anthropocentric Humanism: Reflections on Local Natures and Global Responsibilities". In L. Volkmann et al. (eds.). *Local Natures, Global Responsibilities: Ecocritical Perspectives on the New English Literatures*. Amsterdam: Rodopi, 29–53.

Joisten, K. (ed.) (2007). *Narrative Ethik: Das Gute und das Böse erzählen*. Berlin: Akademie Verlag.

Kennedy, D. (2012). *The Ekphrastic Encounter in Contemporary British Poetry and Elsewhere*. Farnham: Ashgate.

Lukács, G. (2005). "Narrate or Describe?" In A. Kahn (ed. and transl.). *Writer and Critic and Other Essays*. Lincoln: iUniverse, 110–148.

McCarthy, C. (2006). *The Road*. New York: Alfred A. Knopf.

Mieth, D. (2007). "Literaturethik als narrative Ethik". In K. Joisten (ed.). *Narrative Ethik: Das Gute und das Böse erzählen*. Berlin: Akademie Verlag, 215–234.

Mohr, D. M. (2015). "Eco-Dystopia and Biotechnology: Margaret Atwood, *Oryx and Crake* (2003), *The Year of the Flood* (2009) and *MaddAddam* (2013)". In E. Voigts and A. Boller (eds.). *Dystopia, Science Fiction, Post-Apocalypse*. Trier: Wissenschaftlicher Verlag, 283–301.

Mosher, H. F. Jr. (1991). "Towards a Poetics of Descriptized Narration". *Poetics Today*, 3, 425–445.

Nussbaum, M. C. (1995). *Poetic Justice: The Literary Imagination and Public Life*. Boston, MA: Beacon Press.

Quintilian, M. F. (1977). *Institutio oratoria* [1921]. Transl. H. E. Butler. 4 vols. Cambridge, MA: Harvard University Press – London: William Heinemann.

Rippl, G. (2005). *Beschreibungs-Kunst: Zur intermedialen Poetik angloamerikanischer Ikon-Texte (1880–2000)*. Munich: Fink.

Rippl, G. (2015a). "Postcolonial Ekphrasis in the Contemporary Anglophone Indian Novel". In G. Rippl (ed.). *Handbook of Intermediality*. Berlin – Boston: De Gruyter, 128–155.

Rippl, G. (2015b). "Description and the Production of Presence: Literary Debates in Eighteenth-Century England and Germany". In B. Neumann (ed.). *Präsenz und Evidenz fremder Dinge im Europa des 18. Jahrhunderts*. Göttingen: Wallstein, 458–474.

Ronen, R. (1997). "Description, Narrative and Representation". *Narrative*, 5(3), 274–286.

Taylor, C. (2004). *Modern Social Imaginaries*. Durham – London: Duke University Press.

Voigts, E. (2015). "Introduction: The Dystopian Imagination – An Overview". In E. Voigts and A. Boller (eds.). *Dystopia, Science Fiction, Post-Apocalypse*. Trier: Wissenschaftlicher Verlag, 1–11.

Webb, R. (2009). *Ekphrasis, Imagination and Persuasion in Ancient Rhetorical Theory and Practice*. Farnham: Ashgate.

Weik von Mossner, A. (2017). *Affective Ecologies: Empathy, Emotion, and Environmental Narrative*. Columbus: Ohio State University Press.

Weik von Mossner, A. (2018). "The Moral Dilemma of Unsustainability: Eco-dystopian Fiction as Cultural Critique". In T. Meireis and G. Rippl (eds.). *Cultural Sustainability: Perspectives from the Humanities and Social Sciences*. London: Routledge, Chapter 15.

World Commission on Environment and Development (WCED) (1987). "Report of the WCED. Our Common Future. Transmitted to the General Assembly as an Annex to document A/42/427 – Development and International Co-operation: Environment". *UN Documents*. www.un-documents.net/wced-ocf.htm [Accessed 15 October 2017].

Yacobi, T. (1995). "Pictorial Models and Narrative Ekphrasis". *Poetics Today*, 16(4), 599–649.

Zapf, H. (2002). *Literatur als kulturelle Ökologie: Zur kulturellen Funktion imaginativer Texte an Beispielen des amerikanischen Romans*. Tübingen: Niemeyer.

Zapf, H. (2006). "The State of Ecocriticism and the Function of Literature as Cultural Ecology". In C. Gersdorf and S. Mayer (eds.). *Nature in Literary and Cultural Studies: Transatlantic Conversations on Ecocriticism*. Amsterdam: Rodopi, 49–69.

Zapf, H. (2008a). "Literary Ecology and the Ethics of Texts". *NLH*, 39(4), 847–868.

Zapf, H. (ed.) (2008b). *Kulturökologie und Literatur: Beiträge zu einem transdisziplinären Paradigma der Literaturwissenschaft*. Heidelberg: Winter.

Zapf, H. (2008c). "Narrative, Ethics, and Postmodern Art in Siri Hustvedt's *What I Loved*". In A. Erll, H. Grabes and A. Nünning (eds.). *Ethics in Culture: The Dissemination of Values through Literature and Other Media*. Berlin – New York: De Gruyter, 171–194.

Zapf, H. (2016). *Literature as Cultural Ecology: Sustainable Texts*. London: Bloomsbury.

Zunshine, L. (2010). *Introduction to Cognitive Cultural Studies*. Baltimore: John Hopkins University Press.

18 Ecogames

Playing to save the planet

Joost Raessens

Introduction

> Today, you are not here as Hamburg citizens and theatre visitors, but as delegates of 196 countries. Every one of you represents a country and its wide range of interests. It depends on you, if we – at the end of this evening – reach an agreement that enables us to limit global temperature rise to two degrees Celsius.

These were the opening words of the *World Climate Conference* (2015, see Figure 18.1), organized by the German theatre company Rimini Protokoll. Theatre visitors were invited to participate in a LARP – a live action role-playing game – and play the roles of delegates of different countries that had to negotiate with each other. The ultimate goal was to have your delegation contribute in a significant way to finding a solution for climate change. These negotiations were alternated with sessions with climate change experts in which participants were given information while they personally experienced the consequences of climate change – such as rising temperatures (Schipper 2015).

As this example shows, contemporary game and play practices – digital as well as non-digital – are increasingly used not only to entertain, but also to make a difference on an individual, community, and/or societal level (Raessens 2014, 2016). These so-called 'social impact games' or 'games for change' deal with themes such as war and violence (*This War of Mine* 2014), poverty (*Poverty Is Not a Game* 2010), and refugee issues (*A Breathtaking Journey* 2015). Increasingly, games encourage support, sympathy, and action for a variety of ecological issues. Ecogames not only seek to contribute to ecological thought but also to turn players into ecological citizens. Global ecological citizenship is defined as "including the right to a non-polluted environment and the responsibility both to refrain from harming the environment and to participate in its preservation and rehabilitation" (MacGregor 2014: 114).

In this chapter, I argue that ecogames seem to be able to overcome the critique of researchers like Per Espen Stoknes (2015) that conventional climate communication does not always work optimally, or, worse still, can even be counter-productive. The most important reason for this failure is that the cultural dimensions of the climate change debate are not taken sufficiently into consideration, a

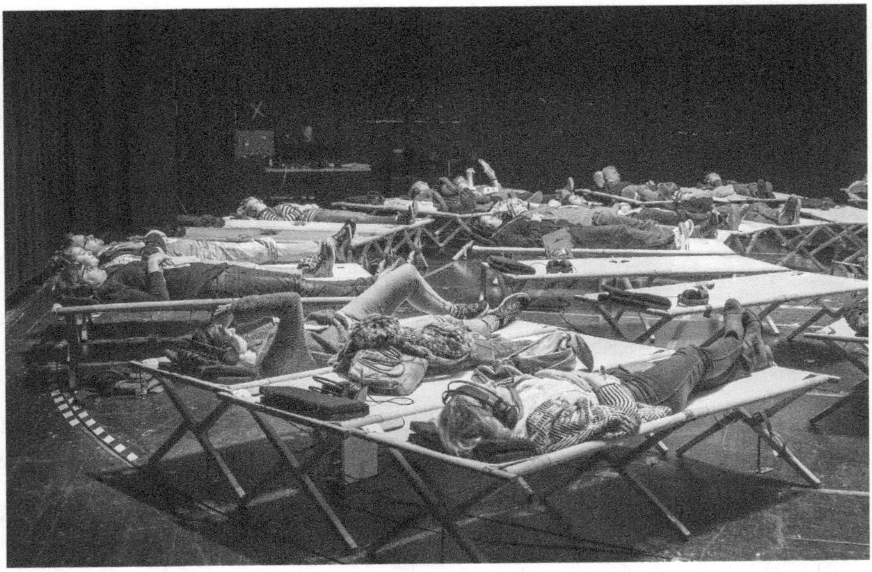

Figure 18.1 World Climate Conference

Source: Benno Tobler

problem ecogames try to counter. I will first briefly discuss the psychological climate paradox and how to solve it. Then I will discuss three complementary strategies to make climate communication more effective, and three important properties of games for change that enable them to facilitate social change. Next, I will present a conceptual model of social change that takes into consideration the micro, meso, and macro levels of social impact that ecogames are aiming for. The last section contains my conclusions. Throughout the text, I will illustrate my argument with short examples of ecogames (cf. Raessens 2017).

The psychological climate paradox and how to solve it

If we want to answer the question if, and if so how, ecogames can be used to engage citizens in the issue of climate change, we first need to improve our understanding of the reasons why conventional climate communication does not always work optimally, or, worse still, can even be counter-productive (erecting barriers), as well as what we should do to overcome these barriers (solutions). Next, we need to understand what strategies ecogames should use to make communication work (strategies) and the characteristics of ecogames that would make them suitable to overcome these barriers (properties); see Table 18.1.

Three barriers might cause conventional climate communications to lead to a state of denial. The first barrier arises when global warming is framed as being

Table 18.1 Barriers, solutions, properties, and strategies

Barriers (this section)	*Solutions (this section)*	*Strategies and properties (next two sections)*
distant in space and time	near, human, personal, urgent	strategies: moral ideas, global citizenship, opposing cultural world views
doom scenario	opportunities, hope	
not compatible with our values	in line with our values	properties: persuasion (procedural rhetoric, narratives), social use, the frame of play

Source: Joost Raessens

distant in space and time. For the majority of us living in the West, the impact of global warming is still relatively far away, both in time and space. That is why Rob Nixon refers to climate change as 'slow violence': "A violence that occurs gradually and out of sight, a violence of delayed destruction that is dispersed across time and space, an attritional violence that is typically not viewed as violence at all" (2011: 2).[1] Research, for example, shows that only a minority of the Dutch population believe that fossil fuel has a big impact on the climate; and only a very small minority see energy transition as an urgent challenge, especially when compared with issues such as the economy and migration (SCP 2016). The second barrier arises when global warming is framed as a doom scenario, an apocalypse-movie mode without any thinkable practical solutions, which generates a depressing wish to avoid the topic. Energy transition is indeed not 'top of mind' for the Dutch (SCP 2016). The third barrier arises when global warming is framed in such a way that it is not compatible with our values or our sense of identity. Dutch citizens who are interested in energy transition are so, not so much because of climate change as because of the idea of energy independency, the stable delivery of energy, and the economic growth that could result from such a transition (SCP 2016).

To make climate communication productive, we need to turn these three barriers into solutions. First, climate change must be framed as being near, human, personal, and urgent. Second, climate messages must be framed in a positive, hopeful way, providing opportunities for consistent and visible action (Popper 1999; Chomsky 2017). Third, climate change must be framed in such a way that solutions are in line with our values. As I discuss elsewhere, people 'play' their identity: people play their values, they play on the basis of who they are, how they understand themselves, the values they want to live by, and who and what they want to become (Raessens 2015).

Strategies to make climate communication work

On the basis of these three barriers with their corresponding solutions, I propose three complementary strategies to make climate communication more effective.

Climate communication must activate and reinforce progressive and positive moral ideas, it must stimulate a post-liberal and global form of citizenship, but it must also – paradoxically – take into account the rhetorical wars over climate change.

Moral ideas: progressive and optimistic

In order to increase our understanding of how ecogames frame climate change, it is productive to use Lakoff's distinction between conservative and progressive moral systems (cf. Flanagan and Nissenbaum 2014). The two systems represent contrasting ideas about environmentalism, which is defined by Lakoff as follows: "The natural world is being destroyed and it is a moral imperative to preserve and reconstitute as much of it as possible as soon as possible" (2010: 80). Whereas the conservative system includes a number of ideas that work against environmentalism – such as nature being there for human use and exploitation, a let-the-market-decide ideology, and the idea that making a profit and economic growth are goals in themselves (Lakoff 2010: 74–75),– the progressive system includes various ideas that support environmentalism, such as empathy linking us with other beings and other things, responsibility for taking care of yourself and others, and the ethic of excellence calling on us to improve the environment (2010: 76). This is in line with Stoknes' argument: "We ought to . . . protect and compassionately care for ourselves, current and future generations, and the other beings we share the planet with" (2015: 118).

Citizenship: post-liberal and global

Stoknes emphasizes that we must act as social citizens, not individuals (2015: 91). Faced with worldwide problems like global warming, pollution, and energy transition, environmental change can only be envisioned when four conditions are met. First, green citizens, post-liberal politics, and environmental social justice movements need to mutually reinforce each other on a local and global scale. Second, green citizens need to behave pro-environmentally and participate in public debate. Third, post-liberal politics need to regulate markets and industries across borders via more stringent environmental legislation. And fourth, environmental social justice movements need to carry out ecological change on a global scale through group action. Changes in personal attitudes and/or behaviour (think of green lifestyles and ethical consumption) should strengthen socio-political solutions, not replace them (Barendregt and Jaffe 2014; Klein 2014; Dyer-Witheford and De Peuter 2009).

Opposing cultural world views

One important thing we need to realize is that culture shapes the climate change debate (Hoffman 2015). The reason why so many people avoid or deny the scientific information and consensus on climate change is rooted in our culture and psychology. We all use cognitive filters – that reflect our cultural identity – to make sense of scientific reasoning. We are only able to go from a scientific consensus to a social consensus on climate change when we, somehow, bridge the cultural schism between progressives and conservatives, when we present and frame climate change solutions in ways that are consistent with both world views. We can do so by appealing to "ethical first principles" that bridge the opposition of left and right (Hoffman 2015: 64), or by taking into account the value patterns of different communities by changing messages accordingly, for example, by focusing on how climate change solutions are influencing people's immediate vicinity (such as their neighbourhood), are offering new job opportunities (green economic growth), and are making life less expensive (savings on energy and costs) and safer (energy independency, the stable delivery of energy, safe water) (cf. Motivaction 2013).

As we will see in the rest of this chapter, ecogames try to overcome the barriers of conventional climate communication by focusing on environmental problems that are near, human, personal, and urgent, offering hopeful opportunities to solve them, framing climate change – intentionally or unintentionally – in such a way that they recognize people's different world views and beliefs systems. These ecogames make use of certain game-specific properties, three of which I will discuss in the next section.

Properties of ecogames

In this section I will discuss the persuasive power of games, the social aspects of gaming (multiplayer, social game play), and the framing of games as a playful activity.

Persuasion: narratives and procedural rhetoric

To understand how persuasiveness is embedded in ecogames' design, I will focus on the narratives they portray – or allow the player to develop – and the rules, or the formal structure of the game, describing how a game functions.

Narratives

In his study of persuasive games, Ruud Jacobs (2017) refers to the importance of character-based narrative persuasion. Research done on screen-based media in general, such as film, television, or games, shows that attitude changes in users – leading, for example, to civic engagement – are more likely to occur when they identify with role models or protagonists who go through the same stages of change (on screen) as these users are supposed to do. Narratives must also foster creativity: "There must be room for humour, emotion, visualization, point of view, climax, surprise, plot, drama. Above all, make it personal and personified" (Stoknes 2015: 148).

Condensed into approximately six hours of playing time, *Walden, a Game* (2017, see Figure 18.2) challenges players to experience how Henry Thoreau

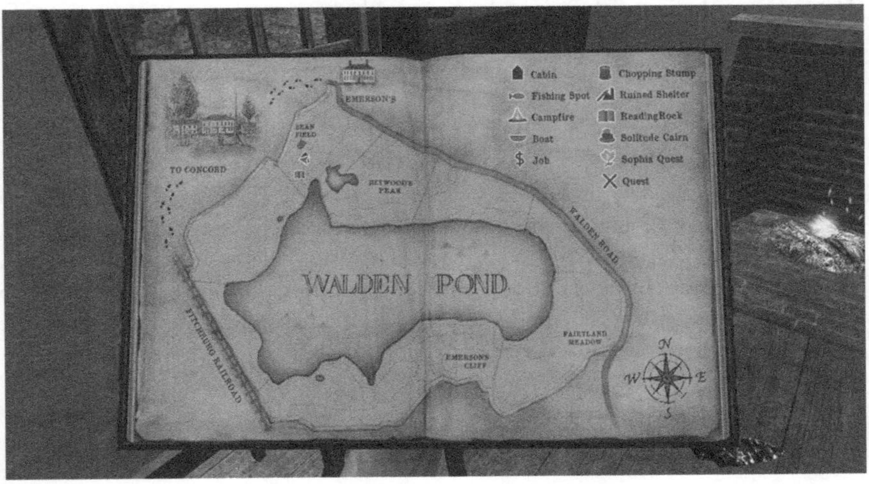

Figure 18.2 Walden, a Game

Source: USC Game Innovation Lab

lived in the woods of Walden Pond in the summer of 1845. By using a first-person point of view, the game embodies Thoreau's own words: "Could a greater miracle take place than for us to look through each other's eyes for an instant?" (2016: 8). The game makes Thoreau's classical work *Walden; or Life in the Woods* (1854) accessible for a generation of gamers and is a tribute to this precursor – born 200 years ago on 12 July 1817 – who is a key source of inspiration for today's ecological movement. Drawing on Thoreau's life story, *Walden, a Game* lets players reflect upon how it is to live in harmony with nature.

Procedural rhetoric

According to Ian Bogost (2007), persuasive games have the unique capability of employing what he calls 'procedural rhetoric' to address serious topics, such as global warming. The goal, argument, and meaning of a game – for example supporting environmentalism – can be found in its formal system, more specifically in the properties of the rules. Players are presumed to surrender to the seduction of a game by playing it as suggested and being guided by the rules. Ecogames influence players through the moral and political values expressed in their rule-based interaction.[2]

DTV Consultants and the Rotterdam-based multimedia company Organiq developed *From5to4*, an online game world for employers and employees. This game, which was used in 2017 in nine different countries, challenges employees to reduce congestion levels and make their commuting more sustainable by leaving their car at home for one day in the week. On that day, employees can choose to use public transport or their bike, but they can also choose to work from home. The smarter their travel and work choices, the more points they can earn. To stimulate employees to participate, a competitive element is built into the rules of the game: teams, departments, and offices compete against each other.

Social use

Stoknes emphasizes the use of the power of social game play. We should not only use one-to-many communication models (such as television documentaries), but also interactive many-to-many communication models (such as meetups, whether online or offline). This is because group behaviour can be a powerful way to influence individual people's behaviour: "Being part of an eco-network is one of the biggest determinants of pro-environmental behavior" (Stoknes 2015: 105). As research shows, this social aspect of media use – known as 'social facilitation,' either online or offline in a physical setting – can enrich reflection on the topic and positively affect persuasion or learning (Neys and Jansz 2010).

> Welcome to headquarters. From here, the superheroes DT, X^2 and GMAN fight Toxico, a company producing pesticides and only aiming for money and power at the expense of plants and animals. To stop Toxico, new superheroes are needed. "Is this maybe something for you? We really need you to help us out."

This appeal appears at the start of *LOI Kidzz ThuisTrainer* ('LOI Kidzz Home-Trainer'), an online course that has been available since 2014 and was developed by the Dutch online school LOI Kiddz. By means of minigames, thousands of 11- and 12-year-old children jointly practice the material for their final examination. This form of ecological education is a form of stealth learning. Children prepare in a playful manner for their final examination while learning how they can become fighters to save the planet by carrying out several environmental missions.

The frame of play

Dutch play theorist Johan Huizinga presents the notion of 'play' as just pretending, as "make-believe" (1955: 25) – and for that reason as inconsequential to real life. We could call this the 'as-if-ness' character of games. Framing a game as 'play' allows an individual to enter specific realms of the imagination; it facilitates "role-play in contexts that would not be feasible, appropriate, or desirable in non-playful action" (Klimmt 2009: 253). This frame of play is characteristic for almost all ecogames, allowing players to try out different roles and scenarios and gaining insight into what consequences – negative and/or positive – they might have in real life.

According to Jane McGonigal (2011), these ecogame worlds are better than the real world because they entail clear, motivating, and often epic missions that can be completed in a practical and satisfactory way. Players feel a great sense of solidarity because they have to collaborate to reach their goals. Making use of this power of games, McGonigal designs so-called 'alternate reality games' (rather than games in which players escape reality). These are games that improve the real lives of people and aim at solving real-world issues, such as in the domain of climate change. One of these games is *World Without Oil*, an experiment that took place in the year 2007. Players had to role-play their real lives in an alternate reality in which they had ended up in a world without oil. The participants and their communities dealt with food and gasoline shortages by growing their own food, to diminish commuter traffic and to generate their own energy. The game strengthened the awareness of our oil dependency and stimulated the individuals and communities to come up with their own solutions for living in a world without oil.

Ecogames and social change

Impact is a shared goal of every social impact game and can occur at three different levels: a micro level (individual cognition and behaviour), a meso level (organizations, groups and communities, such as schools, companies, and neighbourhoods), and a macro level (politics, policy agendas, a shift in public discourse) (Stokes 2016, cf. Burak and Parker 2017). The problem is that there is more and more research on how games can facilitate social change on an individual level (Jacobs 2017), but that meso- and macro-level perspectives are often not taken into consideration (Klimmt 2009). To be able to understand how games can impact action on a micro, meso, *and* macro level, I will present a model for the encouragement of sustainable action.[3] This model makes it possible to understand how these three different levels of social change interact with each other and what role ecogames could play (see Figure 18.3).

The goal of ecogames is to encourage *sustainable action*, be it on a micro-individual or meso-organizational level. Learning, reflective, and persuasive games try to inform players, raise their awareness, and reinforce pro-environmental attitudes and behaviour, at an individual level (*Walden, a Game*) and/or organizational level, where the organizations can be schools (*LOI Kidzz ThuisTrainer*), companies (*From5to4*), or neighbourhoods (*World Without Oil*), as we can see in the following example.

The Dutch water company Vitens approached the game production studio Grendel Games to develop the *Wijk & Water Battle* ('Neighbourhood & Water Battle' 2015), a game where the goal is to save energy and costs by getting customers to avoid peaks in water consumption by spreading their consumption more evenly. The objective of the game was to provide participants with an insight into their water and energy consumption, and ultimately to bring about behaviour change. Children from two primary schools in different neighbourhoods of the Dutch city of Leeuwarden took part in a 'battle' for three months. In the game, whimsical water creatures live in Leeuwarden's water network. Their tiny homes flood regularly because of the inhabitants' intensive and peak water consumption. The schoolchildren participating in the battle could prevent these

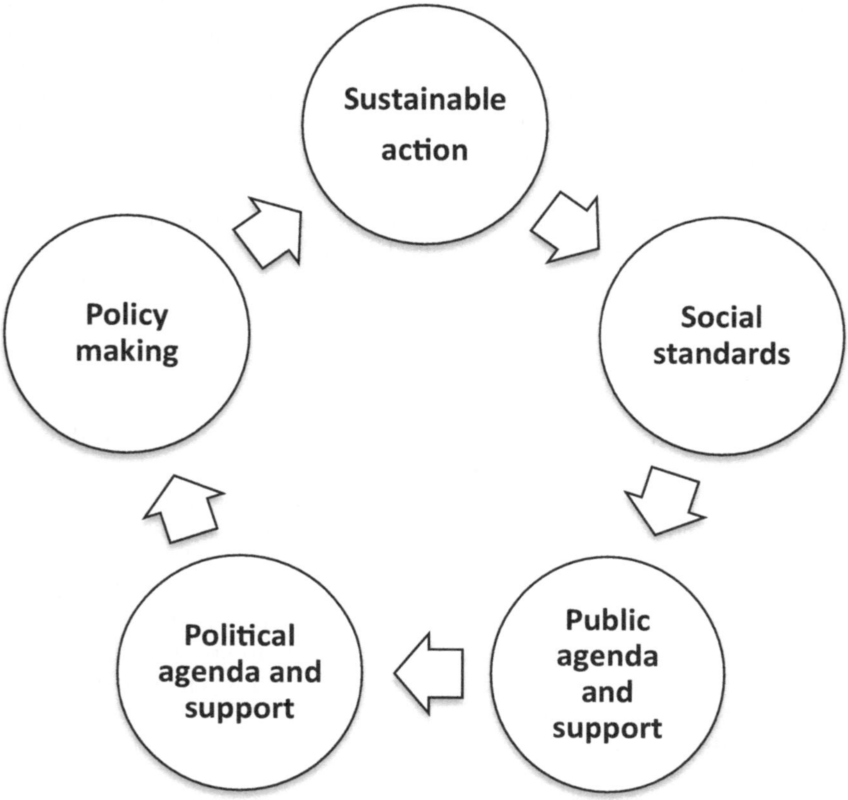

Figure 18.3 Model for the encouragement of sustainable action
Source: Joost Raessens

little creatures from drowning and thereby win the game by measuring their water consumption with a smart metre and an app, and spreading and diminishing their own water usage during the day – and preferably also the water usage of their family members, friends, and neighbours. As they changed their real-world behaviour, the creatures in the game were faced with less flooding.

Ecogames can impact *social standards* as well, and by doing so, influence the *public agenda and support* for political measures (Scheufele and Tewksbury 2007). To be able to do this, ecogames spread climate communication messages in collaboration with mass media and via large-scale and ambitious campaigns by non-governmental environmental organizations such as Greenpeace. *Collapsus-Energy Risk Conspiracy* (2010) was developed in collaboration with Dutch broadcaster VPRO. The game was an online production engaging players – 200,000 in the first few months after the launch of the game – with realistic future scenarios

(2012–2025) about the expected energy crisis and the necessity of energy transition from fossil fuel to alternative energy sources.[4]

In their campaign *Join the Rebellion and Turn VW Away From the Dark Side* (2011, see Figure 18.4), Greenpeace called on their sympathizers to protect the planet in a playful way from Volkswagen (VW), a company resisting legal measures to reduce the emission of greenhouse gases. "Only a rebellion can stop them – a mass of people from all across the globe – demanding that they change. We need you to be part of that rebellion."[5] This Greenpeace campaign started with a remake of a Volkswagen advertisement in which a boy in a Darth Vader costume tried to start a Volkswagen Passat. The battle for sustainability is being represented in this remake as the symbolic battle between the rebellion alliance (representing Good) and Darth Vader and his storm troopers (representing Evil), as if it is being played out in the *Star Wars* series. The Greenpeace activists dressed up as *Star Wars* characters in an epic role-playing game to expose Volkswagen as having been seduced by the Dark Side.

Ecogames also have an impact – directly or indirectly – on the *political agenda and support* and *policy making*, for example in the domains of the economy and technology. On a macro level, they can try to convey and let players experience the incredible complexities and difficulties that UN climate conferences face in assessing progress in climate change issues (*World Climate Conference*). On a meso level, they can engage local residents and politicians and encourage their participation in the circular economy (which is the aim of the Games for Cities

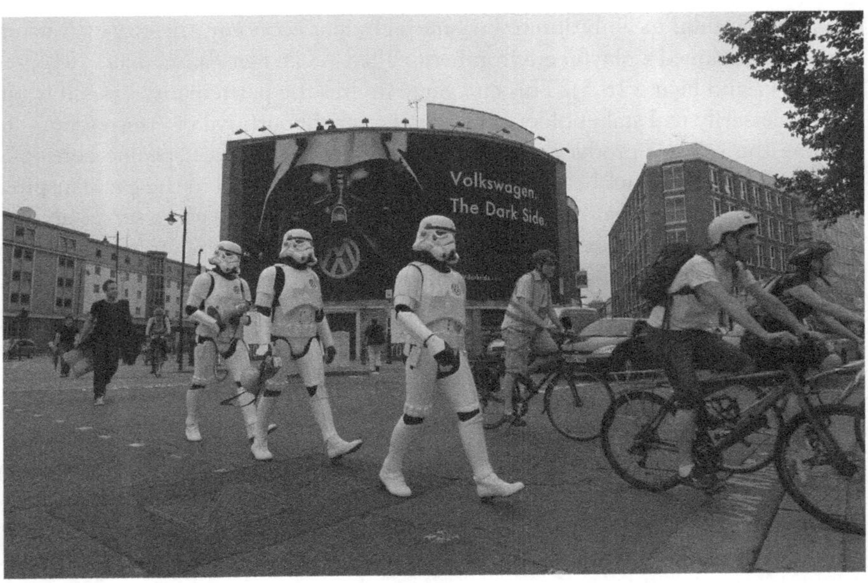

Figure 18.4 Join the Rebellion and Turn VW Away from the Dark Side

Source: Greenpeace

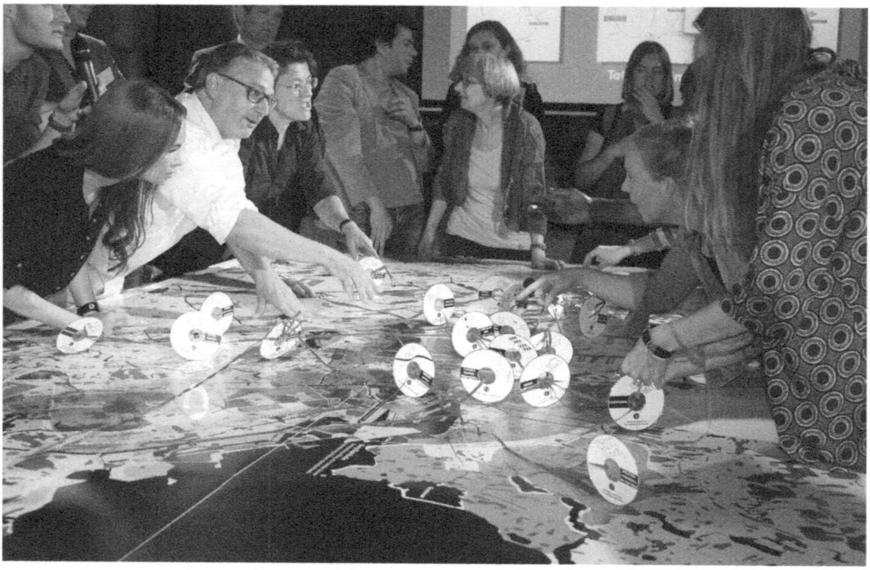

Figure 18.5 A *Circular Amsterdam*
Source: Games for Cities

project). To fulfil its ambition to become a circular economy, the city of Amsterdam co-developed a playful environment called *A Circular Amsterdam* (2016, see Tan 2017 and Figure 18.5). This city game invites the participants – possibly but not necessarily real stakeholders in a specific neighbourhood of Amsterdam – to assume the roles of producers, distributers, and consumers of specific consumer products. Relationships between players that are established in the game inspired real-world collaboration between the players involved. The game serves an educational function for those unfamiliar with the concept of the circular economy while simultaneously enabling real stakeholders to explore the potential of collaborating with other businesses in a space where there are no real or lasting consequences to their decisions and partnerships.

Conclusion

As a sub-category of games for change – or social impact games – ecogames try to make a difference in a contested world order, its underlying normative differences and conflicts. As we have seen in this chapter, ecogames try to have micro, meso, and macro impacts in their own, game-specific ways. The examples I discussed in this chapter engage their players via the specific organization of their content – using the power of narratives and procedural rhetoric – by addressing and organizing collaborations between these players via the power of social game play mechanisms, and by framing their activities as 'playing,' making use of the power of make-believe. Together with other cultural products such as literature (Buell 2005), theatre (Lavery 2016), film (Ingram 2004), and installations (White 2014)[6], ecogames can have a profound impact on people's world views, beliefs, and values. Academic research shows that ecogames, provided that they are well designed and used, can inform players, can incite players to cooperation, and can change or reinforce attitudes and behaviour. The games mentioned in this chapter tell different stories and come up with different solutions, each one of which can contribute to concrete improvements in resolving the complex problem of climate change. Via careful consideration or as a battle on a local or global level, ecogames are being played and used by individual citizens, by social movements, by government agencies, and by private enterprises. Ideally, changes in personal attitudes and behaviour on a micro and meso level should reinforce social-political solutions on a macro level rather than replace them.

Acknowledgments

This research is part of the project *Persuasive gaming. From theory-based design to validation and back*, funded by the Netherlands Organisation for Scientific Research (NWO; 2013–2018; project number 314–99–106). See www.persuasivegaming.nl.

Notes

1 This is of course not the case for those communities that are directly threatened by global warming (Klein 2014; Lewis 2015), "particularly (though not exclusively) across the so-called global South" (Nixon 2011: 4). The most visible impact of global warming in our everyday lives in the West is in extreme weather conditions, like heatwaves, hurricanes, drought, wildfires, and floods.
2 The limits of procedural rhetoric are discussed by Sicart (2011, 2014).
3 This model was developed within the field of tobacco control research (Willemsen 2016: 15) and is analogous to a flywheel. Once the wheel is set in motion, the different parts reinforce each other. The historical analogy of the ongoing shift from a scientific to a social consensus in the debate on cigarette smoking and cancer has been made before (cf. Hoffman 2015: 71–74).
4 The theoretical framework of Stoknes and *Collapsus* as a case study are dealt with more extensively in Raessens (in press).
5 See www.greenpeace.org.
6 The Columbus Earth Center (Kerkrade, the Netherlands) has an installation that gives larger audiences an impression of the Overview Effect. The Overview Effect is said to have transformative power for sustainability thinking and social responsibility, see columbusearththeater.nl.

References

Barendregt, B. and Jaffe, R. (2014). *Green Consumption: The Global Rise of Eco-chic*. London: Bloomsbury.

Bogost, I. (2007). *Persuasive Games: The Expressive Power of Videogames*. Cambridge, MA: The MIT Press.

Buell, L. (2005). *The Future of Environmental Criticism: Environmental Crisis and Literary Imagination*. Malden, MA: Blackwell Publishing.

Burak, A. and Parker, L. (2017). *Power Play: How Video Games Can Save the World*. New York: St. Martin's Press.

Chomsky, N. (2017). *Optimism over Despair: On Capitalism, Empire, and Social Change*. Chicago: Haymarket Books.

Dyer-Witheford, N. and de Peuter, G. (2009). *Games of Empire: Global Capitalism and Video Games*. Minneapolis: University of Minnesota Press.

Flanagan, M., and Nissenbaum, H. (2014). *Values at Play in Digital Games*. Cambridge, MA: The MIT Press.

Hoffman, A. J. (2015). *How Culture Shapes the Climate Change Debate*. Stanford: Stanford University Press.

Huizinga, J. (1955). *Homo Ludens: A Study of the Play-element in Culture*. Boston, MA: Beacon Press.

Ingram, D. (2004). *Green Screen: Environmentalism and Hollywood Cinema*. Exeter: University of Exeter Press.

Jacobs, R. S. (2017). *Playing to Win Over: Validating Persuasive Games*. PhD thesis. Rotterdam: Erasmus University Rotterdam.

Klein, N. (2014). *This Changes Everything: Capitalism vs. the Climate*. New York: Simon & Schuster.

Klimmt, C. (2009). "Serious Games and Social Change: Why They (Should) Work". In U. Ritterfeld, M. Cody and P. Vorderer (eds.). *Serious Games: Mechanisms and Effects*. New York: Routledge, 248–270.

Lakoff, G. (2010). "Why it Matters How We Frame the Environment". *Environmental Communication*, 4 (1), 70–81.

Lavery, C. (2016). "Introduction: Performance and Ecology – What Can Theatre Do?" *Green Letters: Studies in Ecocriticism*, 20 (3), 229–236.

Lewis, A. (dir.) (2015). *This Changes Everything*. Bertha Foundation. Film.

MacGregor, S. (2014). "Ecological Citizenship". In H.-A. van der Heijden (ed.). *Handbook of Political Citizenship and Social Movements*. Cheltenham: Edward Elgar Publishing, 107–132.

McGonigal, J. (2011). *Reality is Broken: Why Games Make Us Better and How They Can Change the World*. New York: The Penguin Press.

Motivaction (2013). *Vijf tinten groen: Input voor effectieve duurzaamheidsstrategieën*. www.motivaction.nl [Accessed 18 February 2018].

Neys, J. and Jansz, J. (2010). "Political Internet Games: Engaging an Audience". *European Journal of Communication*, 25(3), 1–15.

Nixon, R. (2011). *Slow Violence and the Environmentalism of the Poor*. Cambridge, MA: Harvard University Press.

Popper, K. (1999). *All Life is Problem Solving*. New York: Routledge.

Raessens, J. (2014). "The Ludification of Culture". In M. Fuchs, S. Fizek, P. Ruffino and N. Schrape (eds.). *Rethinking Gamification*. Lüneburg: Meson Press, 91–114.

Raessens, J. (2015). "Playful Identity Politics: How Refugee Games Affect the Player's Identity". In V. Frissen, M. de Lange, J. Raessens, J. de Mul and S. Lammes (eds.). *Playful Identities: The Ludification of Digital Media Cultures*. Amsterdam: AUP, 245–260.

Raessens, J. (2016). "Game Studies". *The International Encyclopedia of Communication Theory and Philosophy*, 1–5.

Raessens, J. (2017). "Gamen om de planeet te redden: De kracht van het spel". In E. Meijers (ed.). *De Helling*. Utrecht: Bureau De Helling, 62–65.

Raessens, J. (In press). "*Collapsus*, or How to Make Players Become Ecological Citizens". In R. Glas et al. (eds.). *The Playful Citizen: Civic Engagement in a Mediatized Culture*. Amsterdam: AUP.

Scheufele, D. A. and Tewksbury, D. (2007). "Framing, Agenda Setting, and Priming: The Evolution of Three Media Effects Models". *Journal of Communication*, 57 (1), 9–20.

Schipper, I. (2015). "Den Zuschauer inszenieren – Die Hamburger *Welt-Klimakonferenz*". In A. Fournier, P. Gilardi, A. Härter and C. Maeder (eds.). *Rimini Protokoll*. Bern: Peter Lang, 107–113.

SCP (The Netherlands Institute for Social Research) (2016). *Burgerperspectieven* 4. www.scp.nl/Publicaties/Alle_publicaties/Publicaties_2016/Burgerperspectieven_2016_4 [Accessed 18 February 2018].

Sicart, M. (2011). "Against procedurality". *Game Studies: The International Journal of Computer Game Research*, 11 (3). gamestudies.org/1103/articles/sicart_ap [Accessed 18 February 2018].

Sicart, M. (2014). *Play Matters*. Cambridge, MA: The MIT Press.

Stokes, B. et al. (2016). *Impact with Games: A Fragmented Field*. gameimpact.net/reports/fragmented-field [Accessed 18 February 2018].

Stoknes, P. E. (2015). *What We Think about When We Try not to Think about Global Warming: Toward a New Psychology of Climate Action*. White River Junction: Chelsea Green Publishing.

Tan, E. (2017). *Play the City: Games Informing the Urban Development*. Heijningen: Jap Sam Books.

Thoreau, H. D. (2016). *The Illustrated Walden: Thoreau Bicentennial Edition*. New York: TarcherPerigee.

White, F. (2014). *The Overview Effect: Space Exploration and Human Evolution*. Bethesda: Library of Flight.

Willemsen, M. C. (2016). *Roken in Nederland: De keerzijde van tolerantie*. Inaugural lecture. Maastricht: Maastricht University.

Games

A Breathtaking Journey. 2015. abreathtakingjourney.com.

A Circular Amsterdam. 2016. gamesforcities.com.

Collapsus-Energy Risk Conspiracy. 2010. collapsus.com.

From5to4. 2014. organiq.nl.

Games for Change. gamesforchange.org.

Join the Rebellion and Turn VW Away From the Dark Side. 2011. greenpeace.org.

LOI Kidzz ThuisTrainer ('LOI Kidzz HomeTrainer'). 2014. loikidzz.nl.

Poverty Is Not a Game. 2010. povertyisnotagame.com

This War of Mine. 2014. thiswarofmine.com.

Walden, a Game. 2017. waldengame.com.

Wijk & Water Battle ('Neighborhood & Water Battle'). 2015. grendel-games.com.

World Climate Conference. 2015. rimini-protokoll.de.

World Without Oil. 2007. janemcgonigal.com.

Part IV

Where do we go from here?

19 Conclusion

Torsten Meireis and Gabriele Rippl

Lessons to be learned: the contribution of the humanities and the social sciences

When looking at discourses on the role of culture in the quest for sustainable development, the predominance of an *instrumental understanding* of culture and cultural sustainability can hardly be overlooked. Instrumentality, however, can play out in different ways. Culture may be understood as a fourth pillar of the sustainability discourse inasmuch as it plays a basic role in the upholding of ecological, social and economic sustainability: “If a society's culture disintegrates, so will everything else” (Hawkes 2001: 12). Public cultural policy as well as educational programmes harness narrow understandings of ‘culture’, which refer mainly to “art and creative activities” (Dessein et al. 2015: 29), in order to promote sustainable development (Dessein et al. 2015: 30). A second way of understanding culture instrumentally looks at its mediating function: “In terms of sustainability's three pillars, culture can be the way to balance competing or conflicting demands and work through communication to give human and social meaning to sustainable development” (Dessein et al. 2015: 30). A third way to conceive of culture's instrumentality for sustainable development is to see it as an immaterial basis for a sustainable society, i.e. as a comprehensive bundle of ideas and attitudes furthering sustainable development: “A truly evolutionary culture, or an eco-cultural civilisation, involves practicing a new understanding of the human place in the world, and recognising that humans are an inseparable part of the more-than-human world” (Dessein et al. 2015: 31). In these instrumental takes on culture, “in order to promote the operationalization of culture in sustainability research and policy” (Dessein and Soini 2016, 2) ‘cultural sustainability’ refers ultimately to the possible function of cultural paradigms, practices and concepts for promoting sustainable development or transformation.

However, matters are not that simple. As the contributions in this volume show, culture has a dynamics of its own that is not easily harnessed to presupposed goals. Rather, cultural processes have a way of creating and recreating such goals in ways that cannot be planned by policy makers – especially as such policy makers are drenched in cultural conflicts themselves, as politician and diplomat Michael Gerber's depiction of the fate of the concept of cultural sustainability in the international political sustainable development discourse clarifies: it is exactly the lack of top-down 'operationability' that is one of the main obstacles to the stronger political integration of the concept. Marius Christen, Peter Seele and Lucas Zapf's contribution points to that dynamics in presenting an overview of the change that the sustainability paradigm has undergone concerning the shape of the problems addressed and the solutions offered. Those dynamics are not necessarily favourable to an instrumental approach, but need special attention, as Torsten Meireis contends in his description of culture as an agonal arena, thus pointing to the contingencies of the interpretation and creation of cultural traditions and social imaginaries. As Sacha Kagan demonstrates, cultural production in the performing and visual arts transcends instrumentality for sustainable development through the establishment of new approaches and methodologies, thereby transforming the sustainability discourse as a whole. It is here that a second meaning of cultural sustainability emerges, that takes these dynamics seriously by addressing the *sustainability of culture*. This view differs from the fourth column approach in that culture is not reduced to the arts and creativity sector, is not understood as respect for heritage, nor seen as purely instrumental for sustainable development. Rather, it is understood as a sphere in which the sustainability discourse takes place, allowing for different and even opposing sustainability discourses – as Anton Leist's contribution illustrates by highlighting a politically conservative sustainability concept. The challenge then is to address these different discourses and to promote debates across the boundaries erected by diverging world views, by religion, political opinion and social context. Thus, respect for diversity becomes important not only as an abstract recognition of the difference of cultures, but also as an effort to actively foster interaction between the diverse cultures, as Wolfgang Huber cogently demonstrates.

An initial insight provided by this book, then, would be that any reflection on the function of culture for sustainable development has to take those intrinsic dynamics of culture into account and to explore them in depth. Taking both aspects, the instrumentality of culture for sustainability and the sustainability of cultural dynamics, into account, a perspective emerges in which the dynamics of culture may be assessed in their significance for sustainable development. In that regard, especially the so-called interior dimension of sustainability (Horlings 2015) bears closer scrutiny. Aleida Assmann's approach to cultural sustainability from the perspective of cultural studies' puts emphasis on time regimes, encompassing both the future and the past, by linking cultural sustainability to the concepts of cultural memory and cultural heritage, and thus to questions of values. Additional meanings of culture are subsequently evoked, which open the semantic range of the term considerably. In the same vein, Galit P. Wellner investigates social happiness as a cultural value, thus adding to the ethical debate necessary for sustainable development by overtly addressing shared values for ecosystem assessment which substitute assessment attempts to 'monetize' the impact of human actions.

The sociological distinction of a short-term effective constraining function of culture embodied in restrictive social norms and a long-term enabling function present in rituals and practices, as presented by Ulf Liebe, elucidates the way in which expectations of 'operationability' directed at cultural narratives need to be corrected. Andreas Losch's exposition of the significance of cultural imagery for projects of scientific space exploration points to the range of cultural framing. Gabriele Rippl and Torsten Meireis highlight the fact that cultural sustainability cannot be conceived of as a simple transfer of ideas, but rather as the initiation of reflection. They demonstrate this by depicting the way ethics are at work in popular graphic narratives and how Lauren Redniss' specific narrative, without moralizing, challenges readers to contemplate their own ideas, social imaginaries and stereotypes. In a similar vein, Joost Raessens' depiction of the way ecogames engage the player's imagination shows the initiation of change on a micro-, meso- and macro-level by enabling experiences, rather than by persuasion.

Other cultural productions such as literary texts and visual works of art also play an important role in aesthetically, emotionally and cognitively involving recipients in reflecting upon sustainability issues. Eco-fiction and climate change fiction, for example, open up fictive worlds and serve as important means for recharging our social imaginaries and transforming our ways of conceiving nature, human-nature as well as human-animal relationships. Applying the concept of cultural ecology, Hubert Zapf elaborates the complex relationship between sustainability and literature. Literature with its experimental, non-systematic and creative potential contributes to a particularly complex and ethically responsive form of cultural sustainability. By analyzing the nineteenth-century German media discourse around the 'German forest' as a symbol of cultural identity and by enquiring into the satirical texts and caricatures this discourse produced, Evi Zemanek challenges the distinction of environmental and cultural sustainability and showcases their entangled history. Examining non-literary texts alongside the novels of Charles Dickens, Ursula Kluwick focuses on debates concerning the utilisation of human excrement as manure in the context of sanitary reform in Victorian Britain in order to show how the concept of sustainability was disseminated culturally. Alexa Weik von Mossner discusses eco-dystopian fiction with their pessimistic images of the future as cultural critiques. By presenting the readers a moral dilemma, novels such as Octavia's Butler's *Parable of the Sower* give them the opportunity to experience disturbing imaginary worlds, thus making them viscerally aware of the environmental and social conditions that have caused the dilemma. Gabriele Rippl discusses a similar function of texts as agents of ecological as well as cultural sustainability. By making use of the potential of literary description and eco-ekphrasis, speculative fiction such as Atwood's novel *The Year of the Flood* helps to sharpen readers' awareness of their own values. By addressing relations, motivations, preferences and emotions in individuals and communities, the genre is hence concerned with the underestimated inner, subjective dimension of sustainability.

The way ahead: teaching cultural sustainability

To date, the cultural aspects of sustainability have not been sufficiently researched. This is bound to change in the future, given our thus far unsuccessful attempts to cope with global environmental and humanitarian challenges. As many contributions in this collection convincingly demonstrate, sustainability is very much related to both narrow and wide notions of culture. With respect to sustainability, imagination, for instance the way we imagine our relation to nature, as well as values, beliefs and world views play a crucial role. These dimensions, which we understand as the cultural dimensions of sustainability, have a huge impact on human behaviour and the ways we comprehend and treat our environment. To engender more sustainable lifestyles these cultural dimensions have to be taken seriously. It is only by recognizing the immense role that ethics, values, beliefs, ideas and world views play and by engaging in debates that take the dynamics of culture and the necessity for the recognition of different world views into account that changes in our ecological behaviour will be reached. Hence the question as to where and how debates about the inner aspects of sustainability or cultural sustainability should be carried out in order to raise consciousness of these aspects is pressing. Tackling such questions would help to enable a reform of our imagination and a renewal of the pool of cultural resources, i.e. the social imaginary. Changing our ideas, our world views and behaviour – these are the necessary steps toward the transformation of our (unsustainable) culture – but they may not necessarily be described as development.

This insight asks for an intensified engagement with two aspects of sustainability research. First, on a theoretical level, it will be important to further investigate the role of culture understood in a broad sense, as suggested in this volume. Subsequently, this entails a close examination of the different discourses of cultural sustainability and the broad range of meanings of the term. Central is the question whether cultural sustainability, in addition to ecological, economic and social sustainability, should be considered as a fourth pillar of sustainability or whether culture, and thus cultural sustainability, should rather be the all-encompassing concept, horizon and agonal arena which determines the way the other three dimensions can be understood – a notion we put forward. Second, on a more practical level, teaching cultural sustainability seems to be an equally important desideratum at this crucial moment in time. Researchers will have to put more focus on the topic negotiated in this volume, namely, how important awareness of the ethical, inner dimensions of sustainability is for reaching sustainability goals. Teaching cultural sustainability at both school and university level seems to be as important as the regulations, taxes and financial incentives suggested by political agents and implemented through environmental policies. Thus, the rest of this conclusion will be dedicated to the following question: what could and should the teaching of cultural sustainability look like? We consider three steps indispensible for teaching cultural sustainability at university level: the first is to look closely at the sustainability of culture, i.e. cultural dynamics in their complexity; second, to take stock of the idea of cultural sustainability and its instrumental uses; and, three, to gain more insight into the concrete complex workings of the interior dimensions of sustainability.

In order to take these steps and implement the teaching of cultural sustainability at universities, interdisciplinary approaches are necessary. Insights attained through both theoretical and methodological approaches to this theme have been emerging in a number of different disciplines – many of which are represented in this volume. Yet it is only when these insights are taken together, i.e. at the intersections of disciplines, that systematic knowledge about the role played by culture in sustainable development and the different aspects of cultural sustainability is gained. It is through interdisciplinary discussions – in particular discussions of norms, values, beliefs – that one learns to identify cultural aspects of sustainable development and to relate these to those other, previously recognised aspects. In order to develop the ability to attribute elements of knowledge to different normative frameworks and social forms, an understanding of the theme (which also includes historical aspects) is indispensible.

Stephen Sterling's ideas on sustainable education as well as more recent concepts such as 'transformative science', developed by Uwe Schneidewind and Mandy Singer-Brodowski, plead for re-visioning learning. More than fifteen years ago, Sterling published his critique of the twentieth-century's restricted, managerial and instrumental view of learning geared towards a utilitarian market philosophy both of which "*sustain[] unsustainability*" (Sterling 2001: 14). He presented ideas on how future education for sustainability should encourage critical thinking, creativity and ecological awareness; one that "values and sustains people *and* nature, that recognizes their profound interdependence" (Sterling 2001: 14). Such an understanding of sustainable education involves a paradigm shift, which "is itself a transformative learning process" and embraces "a new participative epistemology" (Sterling 2001: 19). This paradigm shift requires vision, image, design and action, which will lead to a continuous process of transformation. Sterling clearly sees that "in an age of mass communication, the social-cultural milieu arguably affects people and influences values more than formal education programmes do" (Sterling 2001: 33). Like the contributors to this volume, Sterling, too, argues for taking culture, values and affects into account (cf. Sterling 2001: 58–59, for tables of the contrasting educational paradigms).

In his wake, Schneidewind and Singer-Brodowski plead for transdisciplinary sustainability research which goes beyond interdisciplinary research by involving agents from outside the universities, i.e. policy makers, managers, environmental activists, etc. (2014: 42). They appeal to scientific and technical research institutions to include more scholars from the humanities and social scientists (2014: 49). Only by taking such steps, new transformative research and novel ways of knowledge production will be able to boost fresh visions for sustainable development (2014: 68–71). Schneidewind and Singer-Brodowski argue for three turns that are necessary to achieve progress in sustainable development: the 'normative turn,' the 'experimental turn' and the 'institutional turn' (2014: 72–74). The normative turn asks for a discussion of the (subjective) values that underlie (alleged objective) research but are never explicitly addressed and negotiated. The experimental turn ask researchers to distance themselves from the alleged objectivity of their research and to accept a plurality of truths (2014: 73–74). The institutional turn, finally, asks for a rethink of the institutional prerequisites such as research programmes, institutional organization, career tracks and traditional pathways to qualification to enable "transformative literacy" (Scholz 2011) and to make transformative science and research across the disciplines possible (cf. also Kläy et al. 2016).

With respect to education for sustainability, Roman Bartosch, a scholar of Anglophone literatures and organizer of the conference "Cultivating Sustainability: Education and the Environmental Humanities" (to take place at the University of Cologne in autumn 2018), likewise points out that to understand the cultural dimensions of environmental change, both disciplinary and transdisciplinary approaches to the global crises are necessary to bring about the direly needed 'environmental turn' in education. While "concepts such as 'transformative literacy' have begun to shape the field of pedagogy – the role and full potential of education in literatures, cultures, history and other, traditionally humanities-based, fields is less than clear" (Bartosch 2018). Bartosch asks for more engaging explorations of "the potential of global and transcultural learning, inclusive education, and interdisciplinary takes on transformative education from hermeneutic-systematic, historical as well as quantitatively and qualitatively empirical vantage points" (2018). As our own interdisciplinary MA seminars, PhD colloquia and summer schools have demonstrated, and as the contributors of this volume also signal, the way ahead for sustainable development is a closer interdisciplinary and even transdisciplinary engagement with the concept of culture and the multifarious aspects of cultural sustainability. First of all, it is absolutely necessary to look into the 'machine room' of culture, to analyze mechanisms of operationalization, as well as production and reproduction needs that concern the pool of cultural elements comprising the social imaginary. Second, it is necessary to put a stronger focus on the inner or subjective dimensions of sustainability and to scrutinize the decisive role of values, which underlie any debate on sustainability but are hardly ever addressed. Each field involved in sustainable development – be it governance cultures, scientific/scholarly research cultures, educational programmes or the wide range of lifestyle cultures – should consider a stronger engagement in debates on values and ethics. A focus on cultural dimensions of sustainability will make the different academic and non-academic sustainability agents aware of the fact that religious, literary or filmic narratives, graphic novels, computer games, visual works of art or any other cultural products play an important role in the continuous process of transforming the social imaginary of how we humans relate to nature and to each other in our global world. By inviting their recipients to participate in and to reflect on meaning making processes and by emotionally involving them, works of art may help to create new forms of knowledge, new viewpoints and new imaginative approaches which, in turn, help to envision new ways of relating to nature and to others, and thus supports bonding (Tilbury and Wortmann 2004). Like the contributors to our volume, Verena Holz and Ute Stoltenberg have closely examined "cultural and artistic (learning) contexts and participatory design processes" and convincingly argue for the necessity of "an integrative understanding of sustainable development which regards culture as a fundamental structure of societal action and thus as a dimension of sustainable development itself" (2010: 15–16). Without such debates on new paradigms of teaching that engage people in sustainability, and without negotiations over the impact of the cultural dimensions of sustainability and how values determine our behaviour, much-needed progress in sustainable development will not be accomplished and the transformation of cultures, as proclaimed by the annual report "State of the World 2010" by the Worldwatch Institute, will remain nothing but wishful thinking.

References

Bartosch, R. (2018). "CfP: Cultivating Sustainability: Education and the Environmental Humanities". *avldigital BLOG* (24 February 2018). https://avldigital.wordpress.com/2018/02/24/cfp-cultivating-sustainability-education-and-the-environmental-humanities-koeln-23-03–2018/ [Accessed 11 March 2018].

Dessein, J. and Soini, K. (2016). "Culture-Sustainability Relation: Towards a Conceptual Framework". *Sustainability*, 8(2), 167.

Dessein, J., Soini, K., Fairclough, G. and Horlings, L. G. (eds.) (2015). *Culture in, for and as Sustainable Development: Conclusions from the COST Action IS1007 Investigating Cultural Sustainability*. Jyväskylä: The University of Jyväskylä.

Hawkes, J. (2001). *The Fourth Pillar of Sustainability: Culture's Essential Role in Public Planning*. Champaign, IL: Common Ground Publishing.

Holz, V. and Stoltenberg U. (2011). "Mit dem kulturellen Blick auf den Weg zu einer nachhaltigen Entwicklung". In G. Sorgo (ed.). *Die unsichtbare Dimension: Bildung für nachhaltige Entwicklung im kulturellen Prozess*. Vienna: FORUM Umweltbildung im Umweltdachverband, 15–34.

Horlings, L. G. (2015). "The Inner Dimensions of Sustainability: Personal and Cultural Values". *Current Opinion in Environmental Sustainability*, 14, 163–169.

Kläy, A., Zimmermann, A. and Schneider, F. (2016). "Statt Eingreifen wider Willen – Reflexive Transformative Wissenschaft". *VSH Bulletin: Vereinigung der Schweizerischen Hochschuldozierenden*, 42(3–4), 46–52.

Schneidewind, U. and Singer-Brodowski, M. (2014). *Transformative Wissenschaft: Klimawandel im deutschen Wissenschafts- und Hochschulsystem* [2013]. Marburg: Metropolis-Verlag.

Scholz, R. W. (2011). *Environmental Literary in Science and Society: From Knowledge to Decisions*. New York: Cambridge University Press.

Sterling, S. (2001). *Sustainable Education: Re-visioning Learning and Change*. Foxhole: Green Books.

Tilbury, D. and Wortmann, D. (2004). *Engaging People in Sustainability*. Gland: IUCN – The World Conservation Union. Commission on Education and Communication.

Worldwatch Institute (2010). "State of the World 2010". www.worldwatch.org/files/pdf/Education.pdf [Accessed 11 March 2018].

Index

Page numbers in *italics* indicate figures and page numbers in **bold** indicate tables. Notes are indicated by the page number followed by n.

1979 Moon Agreement 103
1984 (Orwell) 193, 205
1992 Rio UN Summit 127
2030 Agenda for Sustainable Development: adoption of 12; cultural sustainability goals and 14, 18–19, 22; peaceful/inclusive societies in 19

Abdel-Karim, Maamoun 32
aesthetics 68, 133
aesthetics of complexity 133, 136
Affective Ecologies (Weik von Mossner) 196–197
Agenda 21 for Culture 18
Alaimo, S. 142
al-Asaad, Khaled 32
Alkemade, R. 89
alternate reality games 239
Altner, G. 48
American literature: art as transformation in 146–147; wildness and 144–147; *see also* literature
American nature writing 144
American poetry: cultural ecology of 148–149; Whitman and 148; wildness and 148–149
"Anecdote of the Jar" (Stevens) 148–149
Anthropocene: climate change and 109, 113; human-nonhuman interaction in 144; poetic approaches to 150; sustainability paradigms and 109, 119; unsustainability of 119, *120*, 221
anthropocentrism 104–105, 115
anti-Semitic discourse 171, *172*, 173
Ape and Essence (Huxley) 221
Arendt, H. 44
arts: community 128; cultural sustainability and 130–136, 143, 217; dystopian fiction and 222, 227–228; ecological 131; meaning-making and 134; sustainability and 127, 142, 250; transformation and 146–147; *see also* literature
Asian values 43
Assmann, A. 250
astrobiology 106n6
Atwood, Margaret 56, 193, 221, 223–228, 251
authenticity 33
Avatar (2009) 56
Awakening, The (Chopin) 145–147

Baccolini, R. 204
Badeni, Kasimir Felix 167
Banse, G. 132
barbarism 34n5
Bartelmus, P. 56
Barthes, R. 52
Bartosch, R. 254
Bateson, G. 133, 140
Bayer, A. 149
Bazalgette, John 186–187, 192n4
beauty 68–69
Beloved (Morrison) 145, 147, 230n8
Benhabib, S. 44
Bentham, J. 91–92
Better Life Index 91
Bianchini, F. 128
Bieling, C. 89, 95n4
bio-art 227–228
biocentrism 104
biological genocide 39–40

biosphere protection: imaginative activities and 228; planetary sustainability and 99, 101, 104, 132; as a value issue 99
Birkeland, I. 55–56, 130
Bishop, J. E. 149
Bogost, I. 237
Bourdieu, P. 83
Braat, L. 89
Braidotti, R. 142
Brave New World (Huxley) 221
Briggs, R. 210
Brock, W. H. 185
Brown, K. R. 82
Brown, K. W. 96n10
Brundtland Commission 110
Brundtland Report 4, 6, 16, 47, 63, 127, 216
Butler, Octavia 193–194, 197–205, 207n7, 223, 230n5, 251

Cameron, James 56
Canavan, G. 204, 228
Cantu, S. 79
capability approach 47
Capital (Marx) 87
Capova, K. A. 101
Capra, F. 56
caricatures: anti-Semitic discourse 171, *172*, 173; cultural function of 155–156; German forests and 162, 164, 166, *169–170*, *172*, *174*, 175, *176*, 177; German oak and 166–168; nationalism and 171
Carlowitz, H. C. von 4, 63, 72n1, 111, 115, 118, 121n1, 153–154
Cashmore, E. 36
Cassirer, E. 50
Celtes, C. 157
Chadwick, E. 185–186, 189, 192n2
Chapman, D. 225
Chapman, J. 225
character-based narrative persuasion 236
Chopin, Kate 145–147
Christen, M. **111**, 250
Christianity: eschatology and 118–119; horizon of culture and 56; human rights and 44; human sacredness 70; piety and 69; religious discourse of 203; stewardship and 54, 56; sustainability and 116–117
Cialdini, R. B. 80–81
Circular Amsterdam, A 242, *242*
"City as Space of Possibility" 135
civic liberalism 65
Claviez, T. 223
climate change: Anthropocene and 109, 113; conservatism and 66; contingency management and 119; culture and 235–236; distributive justice and 64; literary description and 221, 223, 227; narrative ethics and 214; rhetorical wars over 235; as slow violence 234
climate change communication: barriers to 233–234, **234**; counter-productiveness of 232–233; cultural dimension of 232; ecogames and 233, 235–236, 240–241, 243; global warming impact in 233–234; solutions for 234, **234**, 236; strategies and properties for 234, **234**, 235–239; values and 234, 236
climate change fiction 251
Cobb, J. 56
cognitive cultural studies 196, 223
cognitive narratology 223
Collapse (Diamond) 217
Collapsus-Energy Risk Conspiracy 240
Collins, R. 81–82
Columbus Earth Center 244n6
Committee on the Peaceful Uses of Outer Space (UNCOPUOS) 101
community: conception of aesthetics by 68; conservatism and 65–66; conservatives and 65–66; cultural life of 14, 41; defining 68–69; human agency and 204; intrinsic value of nature in 96n10; reader imagination and 220, 228; responsibility and 67
community arts 128, 131
community cultural development 128
Connor, A. 193
conservational sustainability 64–65, 73n5
conservatism: ambiguous attitude toward the state by 65; community and 65–66; cultural concept of nature 65; environmentalism and 64–71; realist sense and 66–67; religion and 71; skepticism of techno-capitalist solutions 65; social change and 65; sustainability and 63–64; traditional practice and 71–72; value modules 71
conservative moral systems 71, 235
conspicuous consumption 78
consumption: conspicuous 78; defining 77; Fair Trade 75, 82; *see also* overconsumption; sustainable consumption

contingency management: fear of overall system collapse and 117–118; individual function of 119; multi-crisis in 116–119; product design and responsibility 121; religion and 117–118; social function of 119; sustainability and 110, **111**, 115–119
Convention on the Prevention and Punishment of the Crime of Genocide 39–41, 44
Convention on the Protection and Promotion of the Diversity of Cultural Expressions (UNESCO) 16
Convention Relating to the Status of Refugees 44
Conway, Kellyanne 193
corporate greenwashing 142
COST Action 89, 96n5, 129
costly signalling theory 80
counter-cultures 52–53
Crassus *see* Krassberger, S.
creative industries: community development and 6, 19; cultural diversity and 3; job creation and 21
critical dystopia 194
critical eco-dystopia fiction: dilemmas in 195, 223; interdisciplinary approach in 196; liberated embodied simulation 196; sustainability and 194, 221; utopian elements in 204
Cultivating Humanity (Nussbaum) 195
cultural capital 77
cultural codes 51–52
cultural cues 83
cultural diversity: creativity and 3; development and 16; moral universalism and 44; plurality of perspectives and 44–45; protection/promotion of 19; sustainability and 6, 19–20; UN resolutions 17–18; as world heritage 16
cultural ecology: defining 140; literature as 140–150, 155, 195, 251
cultural ecosystems 140
cultural education 129, 131
cultural exceptionalism 43–44
cultural genocide: concept of 36–37, 39; defining 36, 40; human rights and 41; indigenous peoples and 42; Islamic State (IS) and 36–37; legal enforcement 42; physical genocide and 36–37
cultural heritage: acts of purification and 31–32; authenticity and 33; concept of 29–30; destruction and 31–33; globalization and 49; internationalism and 30; Islamic State (IS) assaults on 28, 31–32; reconstruction and 33–34; sustainability and 6; transmission of 28; UNESCO and 15, 27–28; UN resolutions 17–18; vandalism and 30; wartime destruction of 31
cultural imaginary 191
cultural memory: attention to 27–29; survival of cultural heritage in 33
cultural narratives: literature and 141; operationability and 251; situated knowledge and 48
cultural planning 128
cultural relativism 20
cultural rights 14–15, 19
cultural self-reflection 141, 143
cultural separatism 43–44
cultural sustainability: aesthetics of complexity and 133, 136; arts-based approach to 130–136, 143; becoming and 142–143; communitarian discourse of 49; concept of 4–6, 22–23; conservative discourse of 49; constructivist approach to 98–99, 105; cultures of reference and 52–53; defining 209, 217; discourse of 53–55, 127–132, 135–136; dynamics of 250; ecocentrism and 104–105; graphic narratives and 208, 210, 215; human rights and 43–44, 53; instrumental view of 48–49, 55, 249; interdisciplinary approach in 253; literature and 220–221; mediating function of 249; moral discourse of 47, 49–50; multilateral practice 12–22; nature and 153–154; neoliberal discourse of 49; non-state advocacy for 18; normativity in 132, 216–217; operationability and 250–251, 254; prioritization and 132; racial discourse and 170; research on 251–254; significance of 3–4; social creativity and 130; speculative fiction and 194, 197; storylines of 49; sufficiency and 56; Swiss position on 18–20; teaching of 252–255; values in 254–255; verbo-visual discourse on 155, 159, 171; waste management and 184–186, 191
cultural texts 223
culture: agonal area of distinction and 52–53, 56; agonal arena for sustainability discourse and 54–55; ascription and 51–52; cognitive component of 76,

76; consumption and 75–83; defining 5–6, 13, 15; development theories and 13–16; discourse of 130; dual process theory 83; economic development and 21; economy and 76; ecosystem services and 88–90; expressive component of 76, **76**; in four-pillar model of sustainability 101, 128, 183, 249–250, 252; functions of 127; goalability of 20; as a horizon for sustainability discourse 53–54, 56; human rights and 14–15, 19–20; instrumental view of 55; latent knowledge in 51–53; as a means to an end 49–50, 53; moral discourse of 53; network analysis and 83; normative positions of 50; as production of symbolic imagery 209, 216; social change and 127, 135; social imaginaries of 6; sociological dimensions of 51–53; struggle over 37; subcultural 52; sustainable development and 16–22; symbolic structure of 50–53; as symbolic universe 209, 216; toolkit perspective of 76; UNESCO and 15–16; in UN policy documents 15–23; valuative component of 76, **76**
Culture Action Europe 18
culturecide 36–37
culture-of-poverty 77
cultures of reference 52–53
Curie, E. 211
Curie, Marie 208–209, 211–213, 215–216
Curie, Pierre 208, 211–212, 215
Currents of the Universal Being (Slovic, Bishop, and Lyndgaard) 149

Daly, H. 56–57
Day, Doris 26, 34n1
Decade of Education for Sustainable Development 3
Declaration of the Principles of International Cultural Cooperation (UNESCO) 15
Deep Space Industries 100
De Groot, R. 89
Deleuze, G. 142
Derrida, J. 215, 223
Der wahre Jacob 155
descriptive ethics: in ecological fiction 224; non-human nature and 222–223
descriptized narrations 230n6
Detjens, F. 160
development: culture and 13–16, 19; economic growth and 13; human rights and 14; *see also* sustainable development
Dewey, J. 133
Diamond, J. 217
Dickens, Charles: sanitary reform and 186–187, 190–191, 251; water imagery of 188–190
Die Gartenlaube 155
Dijks, S. 89, 95n4
DiMaggio, P. J. 76–77
Disasters of War, The (Goya) 225, 230n7
Dromi, S. 195, 197
dry conservancy sewage removal 183–187, 192n4
DTV Consultants 237
duality 67
dual process theory 83
Durkheim, E. 92
dust heaps 187–189, 192n6, 192n7
Dynamic of Destruction (Kramer) 34n5
dystopian fiction: arts and 222, 227–228; cultural sustainability and 222; eco-dystopias 194–196, 204, 221, 223–228; eco-ekphrasis and 222; embodied simulation in 197; emotional reaction to 205; narrative style in 198–199, 230n5; pocket utopias and 200; reader imagination and 220–221, 228, 230n5; sustainability discourse and 193–194, 198, 203; utopian elements in 204, 228; *see also* critical eco-dystopia fiction; speculative fiction

Earth Charter 114
Ebersberg, Ottokar Franz 171
ecocentric ethics 223
ecocentrism 104
ecocriticism 131, 142
ecocultures 129
eco-ekphrasis 221–222, 225
eco-fiction 251
ecogames: citizenship and 232; climate change and 233, 235–236; culture and 233; frame of play in 238–239; imagination and 251; impact of 243; narratives in 236–237; persuasion in 236–237; policy-making and 241–242; procedural rhetoric and 237–238; social change and 233, 239–243; social standards and 240–241; social use 238; sustainable action and 239–240; sustainable action model *240*
ecological art 131
ecological imaginary 208–216
ecologically responsible behaviour (ERB) 96n10

ecological sustainability 64, 209–210; *see also* sustainability
Ecological Sustainability (Northrop and Connor) 193
Ecology of Law (Capra and Mattei) 56
economic growth 13
economy: contingency management and 118; culture and 76
ecosystem assessment: ecosystem services and 87–88; policy-making and 87; well-being and 87, 95n3
ecosystem services (ES): assessment of 87–88; cultural services and 88–90, 95n2, 95n4; methodology of 88–89; provisioning and 88; regulation and 88; shared values and 90, 96n11; social issues and 88–90
ecotopian dreaming 204
Edelmann, A. 83
education: cultural 129, 131; environmental turn in 254; as human right 19; social game play and 238, 242; sustainable 21, 129, 253–254
Education for Sustainable Development 129
Eichendorff, J. von 159–160, 162
ekphrasis 225–226, 230n9
"Elegy written in a Country Churchyard" (Gray) 149
"Elegy Written in a Country Coal-Bin" (Morley) 149
embodied simulation 196–197, 201–203, 205
emergent cultures 130
Encyclopedia of Race and Ethnic Studies (Cashmore) 36
"Energy: A Villanelle" (Updike) 149
environmental change: group action and 235; social citizenship and 235; social justice and 235; stringent legislation and 235; transdisciplinary approaches to 254
environmentalism: beauty and 68–69; community and 68–69, 72; conservatism and 64–71; cultural sustainability and 49; eco-networks in 238; green 66; moral systems and 235; motivations for 69; piety and 69–70; Protestant tradition and 54, 57; sustainable development and 13
environmental management: environmental crisis and 112–113; environmental reporting 121; intrinsic value of nature in 113–114; protection and 113; sustainability and 110, **111**, 112–114
epistemology 106n2
eschatology 118–119
ethical turn 214–215
ethnic cleansing 39
European Space Agency (ESA) 100
Evans, Ray 34n1

Fair Trade consumption 75, 82
Fall-Out (Pausewang) 210
Finer, S. E. 186
Finke, P. 140, 154, 175
Fliegende Blätter 155, 160, *161*
Foote, S. 142
forest management: Carlowitz and 4, 63, 111, 115, 153–154; German 110, 153–154, 156; Palestine 90; principles for 112, 115
Forest Steward Council (FSC) 112, 115
Forst, R. 44
Frazier, C. 198
From5to4 237–238
Fry, J. 207n7
future: concepts of 25–27; narrative of progress 25; sustainability and 26–27; uncertainty/risk and 26
"Future We Want, The" (UN General Assembly) 17, 21

Gaia hypothesis 104
Gallagher, C. 192n5
Gallese, V. 196, 201
games for change 232–233, 242
Gayá, P. 121
Geertz, C. 50
Genette, G. 224
genocide: biological 39–40; concept of 37–39; defining 36–37, 39–40; ethnic cleansing and 39; human rights and 44; international prosecution of 38–39; physical 39–40; *see also* cultural genocide
Gerber, M. 4, 249
German forests: anti-Semitic discourse 171, *172*, 173; association with 154; caricatures of 155–156, 162, 164, 166–167, *167*, 168, *169*–170, 171, *172*, 173, *174*, *176*, 177; as cultural symbol 156–160, 162–164, 166, 168, 175, 180n38, 251; deforestation 157–158; destruction of 162, 164, 175, 178n2; early descriptions of 157; as ethical and artistic inspiration 159–160; fairytales and 158; as imagined landscape 178n7;

management of 110, 153–154, 156, 162; political connotations of 158; popular culture and 175; racial discourse and 171, 173, *174*, 175
German Museum Munich 150
German oak: caricatures of 166–167, *167*, 168, *169*–170; as cultural symbol 156, 158, 166, 168, 170, 175; destruction of 159; early descriptions of 157; national identity and 154, 157, 166, 175
Germany: national identity and 154, 157, 166, 170171; racial discourse and 170–171171; war reparations 166, 179n22; wind farms in 175, 177
Gerrig, R. 197
Ginsberg, Allen 148
global ecological citizenship 232
Global Indigenous Caucus 42
globalization 49, 214
Global Reporting Initiative (GRI) 121
global warming: in climate change communication 233–234; as doom scenario 234; environmental change and 235; global South and 244n1; procedural rhetoric and 237; sense of identity and 234; visible impact of 233–234, 244n1; *see also* climate change
Goldstein, N. J. 80–81
Goudzwaard, B. 56
Goya, Francisco 225, 230n7
graphic narratives: counter-discourse in 210–211; as cultural products 208; emotions/empathy in 210, 218n4–219n4; immersive power of 210; narrative ethics in 215–216; sustainability and 215, 217
Gray, Thomas 149
Great Deeds against the Dead (Chapman and Chapman) 225
green environmentalism 66
Greenpeace 65, 240–241
Grégoire, Henri 30
Grendel Games 239
Grimm, J. 158
Grimm, W. 158
Griskevicius, V. 79–81
Gross National Happiness Index 91
"*Group of Friends on Culture and Development*" 18
Grunewald Forest 160, 162, 164, 179n21
Gulbransson, O. *170*

Haidara, Abdel Kader 32
Haidt, J. 71–72
Hamlin, C. 185
Handmaid's Tale, The (Atwood) 193
Hanghzou Declaration 4, 16–18
happiness: defining 92; in disciplines 96n9; indexes for 91–93; personal 91–94; psychological 91–92, 94; Self-Determination Theory and 91, 94; shared values and 88, 91–92; short-/long-term 95; social 88, 91–93; sustainability and 93–95; utilitarian principle of 91; well-being and 88, 93
Happy Planet Index 91, 93
Hard Times (Dickens) 187
Hauser, R. 132
Hawkes, J. 128–129
Hawthorne, Nathaniel 145
Hegel, Georg 25
Hein, L. 89
Heine, Heinrich 166
Heine, T. *169*
Hennicke, P. 56
Herman, D. 197, 224
historicism 29
history 31–32
Hitchcock, Alfred 34n1
Hoffman, M. 203
Hogan, P. 202
Höller, Carsten 227
Holz, V. 254
Hörisch, J. 118
Horlings, L. 208
Huber, W. 250
Huizinga, J. 238
human agency 204
Human Development Index (HDI) 13
human ecological sustainability 193–194
human rights: Asian values 43; covenant related to 39; cultural genocide and 41; cultural separatism in 43; cultural sustainability and 43–44, 53; culture and 14–15, 19–20; development and 17; genocide and 44; plurality of perspectives and 45; refugees and 44; universality of 43–44; Western values 43
Hume, D. 96n8
Huxley, Aldous 221

Illouz, E. 195, 197
Illustrirte Zeitung 155
Imagine Dragons 209
Incredible Hulk, The (Lee) 210
Indiana, Robert 227

indigenous peoples: cultural genocide and 42; cultural heritage and 41–42; rights of 42–43
interaction ritual chains 81–83
intercultural dialogue 105
Intergovernmental Conference on Cultural Policies for Development 3
International Covenant on Economic, Social and Cultural Rights of 1966 41
International Federation of Arts Councils and Culture Agencies (IFACCA) 18
International Federation of Coalitions for Cultural Diversity (IFCCD) 18
International Institute of Space Law 102
International Space Station (ISS) 100
Iovino, S. 223
Islamic State (IS): assaults on cultural heritage sites 28, 31–32; attacks on universal values by 31; cultural genocide and 36–37; physical destruction and 36–37
Island (Huxley) 221
Israel 87, 89

Jackson, T. 56–57
Jacobs, R. 236
James, E. 197
Johannesburg Roundtable on Cultural Diversity and Biodiversity (2002) 3
Johnson, A. 163, *163*, 164
Johnston, L. F. 115
Join the Rebellion and Turn VW Away From the Dark Side (Greenpeace) 241, *241*
Jonas, H. 67
Joyce, James 34n2
Jungkeit, R. 56

Kac, Eduardo 227
Kagan, S. 250
Kakutani, M. 193
Kappas, A. 219n4
Kasser, T. 94, 96n10
Keats, John 226
Keller, L. 142
Kempner, Robert M. W. 38
Kikeriki 171, *172*
Kingsley, C. 185
Kladderadatsch 155
Klopstock, F. G. 157
Kluwick, U. 251
Kraftwerk 209
Kramer, A. 34n5
Krassberger, S. 160, *161*
Kuhn, T. 109–110, 119–121
Kurt, H. 48

Lakoff, G. 235
Lange, H. de 56
Lange-Christopher, K. *174*
latent knowledge 51–53
Leaves of Grass (Whitman) 148
Lee, Stan 210
Leist, A. 250
Leistikow, Walter 164
LeMenager, S. 142
Lemkin, Raphael 38–40, 44
Lepsius, Johannes 37–38
Letters on Modern Agriculture (Liebig) 184
Levinas, E. 215, 223
Lewis, D. 195
liberalism 64–65
liberal realism 71–72
liberal utopianism 71
liberated embodied simulation 205
libraries: cultural heritage and 6, 32; destruction of 31; preservation of 32
Liebe, U. 251
Liebig, Justus von 184, 186, 192n4
Lindloff, Hans Maria 164
Linz, M. 56
literary description: defining 223–224; ekphrasis in 225–226; reader imagination and 221, 223, 251; in utopian/dystopian fiction 221–228
literary ecology 230n2
literature: critical eco-dystopia fiction 194–196, 204, 221, 223; cultural ecology of 140–150, 155, 195; cultural narrative in 141, 208–209; cultural self-reflection and 141, 143; cultural sustainability and 220–221, 251; culture-nature relationship in 141; descriptive ethics and 222–224; dystopian fiction 193–200, 203–205, 221–228; ecocentric ethics in 223; ecocritical 131, 142; embodied simulation in 196–197; emotions/empathy in 210, 215, 218n4; environmental critique and 160; ethical turn in 214–215; forest symbolism in 158–159; graphic narratives 208–212; mind/body dualism in 141; narrative ethics of 210, 214–216, 222; narrative style in 207n6; nature writing 144; parody and 160, 163–164; reader imagination and 6, 215, 221, 223; situational empathy in 202; speculative fiction 193–195, 200, 204; sustainability and 141–143; as sustainable text 143; symbolic regeneration in 144–145, 147–148, 150; unsustainability in

194–195, 198; utopian/dystopian fiction 220–221; water imagery in 183, 186–190; wildness and 144; *see also* American literature; American poetry
live action role-playing games (LARP) 232
Livingston, Jay 34n1
living systems 104
Lo, M. 205
Loewenthal, L. 31
LOI Kidzz ThuisTrainer 238
London: sewerage system construction in 186–187; waste management in 185–187, 192n4
Losch, A. 251
low Earth orbit (LEO) 101
Lowen, Irving S. 213
Luckin, B. 185
Lueger, Karl 171, 180n35
Lukács, G. 224
Luxembourg 100, 102
Lyndgaard, K. 149

Madame Curie (Curie, E.) 211
MaddAddam (Atwood) 221, 228
MaddAddam Trilogy (Atwood) 193, 221
Malezer, Les 42
Man Who Knew Too Much, The (1956) 34n1
Maraña, M. 13
Marie Curie (Quinn) 211
Marx, K. 87
Mattei, U. 56
McCarthy, Cormac 197–198, 221
McGonigal, J. 238–239
Meireis, T. 250–251
Melville, Herman 145–146
Mendelssohn-Bartholdy, F. 160
Mexico Declaration 15
Meyer-Abich, K. 48
Millennium Development Goals (MDGs): culture in 13–14, 17; sustainable development and 17
Millennium Ecosystem Assessment (WHO) 87–91, 95n2, 95n3
Milton, John 27
mind/body dualism 141
Mishori, D. 87
Mitchell, Joni 32, 34n7
Moby-Dick (Melville) 145–146
modernity 68
modern morality 71–72
Moraine, S. 194
moral discourse: cultural sustainability and 47, 49–50; sustainable development and 47–48
moral systems 235
moral universalism 43–44
Morin, E. 133
Morley, Christopher 149
Morrison, T. 230n8
Morrison, Toni 145, 147
Morton, T. 149
Mosher, H. F. 224, 230n6
Moylan, T. 194, 197
Müller, M. G. 219n4
multi-planetary society 99–100

"Narrate or Describe" (Lukács) 224
narrative ethics: defining 214; in literature 214–216, 222; social imaginaries and 210, 214–215
narrative of progress 25
narratives: character-based persuasion in 236; creativity and 236; ecogames and 236
narratized descriptions 230n6
NASA: multi-planetary society and 99–100; planetary sustainability and 98–100
National Ecosystem Assessments (NEA) 87, 89, 94
national heritage 29–30; *see also* cultural heritage
Natural History of the People (Riehl) 154
nature: conquest of 155; cultural concept of 65; cultural sustainability and 153–154
nature writing 144
Neitzel, S. 31
neoliberalism 65
network analysis 83
New Economics Foundation (NEF) 93
nightsoil 184–185, 192n2
Nixon, R. 234
Northrop, R. 193
nuclear energy: harmful effects of 209, 213; plant meltdowns 209–210, 213; social movements and 210
nuclear weapons 208–209
nudging 81
Nussbaum, M. 195, 205, 215

O'Brien, C. 94
Occupy Movement 87
"Ode to a Grecian Urn" (Keats) 226
Office for Outer Space Affairs (UNOOSA) 101
Organiq 237
Orwell, George 193, 205
Oryx and Crake (Atwood) 221

Oteros-Rozasc, E. 89, 95n4
Otto, E. 197, 204
Our Creative Diversity (WCCD) 15
Our Mutual Friend (Dickens): dust heaps in 187–189; water imagery in 186–190, 192n8
Outer Space Treaty 102
overconsumption: cultural change and 75; strategies for reducing 79, 84
Overview Effect 244n6

Pacific Edge (Robinson) 200
Paech, N. 132
Palmyra: discovery of 28; Islamic State (IS) assaults on 33; occupation of 32–33; preservation of artifacts in 32; reconstruction of 33
Parable of the Sower (Butler): dystopia in 193, 197, 204–205; ecotopian dreaming 204; egoism and bias in 203; human agency in 204; hyperempathy in 198, 201–202; liberated embodied simulation in 200–203, 205, 251; narrative style in 198–200, 202, 223; religious discourse in 203; unsustainability in 194–195, 198
Parable of the Talents (Butler) 204–205
paradigm shifts 109
Pausewang, G. 210
persuasive games 236–237
Phillips, J. 204
Phillips, M. 121
physical genocide: cultural genocide and 36–37; defining 39–40
piety 69–70
planetary protection 104
Planetary Resources 100
planetary sustainability: biosphere protection and 99, 101, 104; defining 98; regulation and 101–103; space activities in 99–105
play 238–239
Plieninger, T. 89, 95n4
pocket utopias 200, 204
Podbielski, Victor von 162, 179n20, 179n21
Poetry in the Anthropocene (Bayer and Seel) 149
policy-making: community arts and 128; cultural planning and 128; ecogames and 241–242; social happiness and 92–93
Post-2015 UN Development Agenda 19, 23
post-truth era 193
practice theory 81–83
Pratt, H. J. 211
Princen, T. 56–57
Prinz, J. 203
procedural rhetoric 237–238
procedural sustainability 134–135
productive sustainability 64, 66, 73n5
progressive moral systems 235
Protestant tradition: equal worth in 56; human stewardship in 56; sufficiency and 56–57; sustainability and 57
psychological happiness 91–92, 94
purification 31–32

Quinn, S. 211
Quintilian 224

Rachel Carson Institute 149
"Radioactive" (Imagine Dragons) 209
Radioactive (Redniss): ambivalence of radioactivity in 213; cultural sustainability and 217; ecological sustainability in 209–210; imagery in 211–213; narrative ethics of 215–216; reader imagination and 208, 210–212; social imaginaries of 214
radioactivity: ambivalence of 209, 212–213, 215–216; graphic narratives and 208–209, 211–212; in popular culture 209–210, 212, 214; social imaginaries of 214
"Radio-Aktivität/Radio-Activity" (Kraftwerk) 209
Raessens, J. 251
Rathenau, W. 166, 179n22
ratiocentrism 104
Rawls, J. 73n7
realistic utopia 73n7
realist sense 66–67
'Rebuild Palmyra?' exhibition 33–34, 34n8
reconstruction 33–34
Redclift, M. 110
Redniss, Lauren 208–217, 251
refugees 44
religion: contingency and 117–118; piety and 69–70; sufficiency and 56–57; sustainability and 56–57, 115–116
Report on the Sanitary Condition of the Labouring Population of Britain (Chadwick) 185
resource management: corporate activity reporting in 121; economic principles of 112; intergenerational principle of justice 112; resource crisis and 111; sustainability and 110–111, **111**, 112

responsibility 67
Rhodes, R. 208
Richter, A. L. 158
Ricoeur, P. 215
Riehl, Wilhelm Heinrich 154
Rimini Protokoll 232
Rio+20 UN Conference 17
Rippl, G. 251
River Thames: association with crime 187–188; water-borne sewage removal 186–187, 189–190
Road, The (McCarthy) 197–198, 221
Robinson, J. 134–135
Robinson, Kim Stanley 200
Rogers, D. 195
Ronen, R. 224
Ruffin, K. 203

Sachs, W. 56
sacredness 68, 70, 73n9, 73n11
Sanders, J. 205
sanitary reform 186–187, 190–191, 192n5
Sargent, L. T. 194
Sariel, A. 87
satirical magazines: caricatures in 155–156, 164, *165*, 166, 171, 173; German oak in 166, *167*; nature discourse in 155–156, 160, *161*, 162, *163*, 164; verbo-visual discourse in 160, 162, 166, 170
Scarlet Letter, The (Hawthorne) 145
Scherhorn, G. 56
Schneidewind, W. 253
Scruton, Roger: on beauty 68–69; community and 65, 68–69; conservative green position of 63–66, 71–72; duality and 67; on modernity 68; on piety 69–70; realist sense and 66–67; responsibility and 67; on sacredness 68, 70; social change and 66
Seel, D. 149, 250
Seele, P. **111**
Self-Determination Theory (SDT) 91, 94
Sen, A. 13, 47
sentientism 104
serendipity 130
shared values: biosphere protection 99; culture and 89–90; happiness and 88, 91–92; monetization of 89–90; social issues and 89–90; well-being and 90–91
Shay, J. 44
Shelley, Percy Bysshe 27
Simplicissimus 155, 168, *169*, *170*
Singer-Brodowski, M. 253
situated knowledge 48
situational empathy 202
Skidelski, E. 56
Skidelski, R. 56
Slovic, S. 142, 149
Smith, A. 96n8
Smithson, Robert 227
Snyder, Gary 148
social change: conservatism and 65; culture and 127, 135; ecogames and 233, 239–241; environmentalism and 65–66
social citizens 235
social contract 67
social creativity 130
social facilitation 238
social happiness: cultural value of 250; defining 92, 96n8; measurement of 91; policy-making and 92–93; sustainability and 88
Social Happiness (Thin) 92
social imaginaries: defining 213–214, 230n3; impact of 6; memories and future in 131–132; as narrative ethics 210, 214–215; social transformation and 135; sustainability and 8, 54, 134, 217, 220, 228
social impact games 232, 242
socialism: ecological disasters and 65, 73n2; self-directed criticism and 72n2; social justice and 73n2; sustainability and 64; values of 64
social justice 73n2, 235
social management: anthropocentrism and 115; quality of life for all in 114–115; social indicators in 121; sustainability and 110, **111**, 114–115; system crisis and 114
social networks 83
social standards 240–241
social sustainability 77
Soini, K. 55–56, 130
Spaargaren, G. 81–82
S.P.A.C.E. (Spurring Private Aerospace Competitiveness and Entrepreneurship) act 100
Space 2030 101–102
space environment: commercial use of 100–103; exoburials in 103; mining in 102–103; orbital debris in 101; protected zones 103–105; sustainable development and 99, 101–104; values in 104–105
space tourism 100, 106n4

speculative fiction: cultural sustainability and 193–194, 197; environmental 197; liberated embodied simulation in 200–203, 205; narrative ethics of 222; pocket utopias and 200, 204; post-truth era and 193; unsustainability in 195, 198, 205; *see also* critical eco-dystopia fiction; dystopian fiction
Spicer, Sean 193
Spider-Man (Lee) 210
Steps to an Ecology of Mind (Bateson) 140
Sterling, S. 253
Stevens, Wallace 148–149
Stewart, James 34n1
Stoknes, P. E. 232, 235, 238
Stoltenberg, U. 254
Stuttmann, K. 175, *176*
subcultural symbolic codes 52
subjective well-being (SWB) 96n10
sufficiency: agonal arena of culture and 56–57; cultural element of sustainability and 57; horizon of culture and 56
suicide 91–92
sustainability: arts-informed approach to 127, 132–135; concept of 4–5, 63–64, 155; conservational 64–65, 73n5; conservativism and 63; cultural dimension of 3–5, 12–14; cultural patterns of 55; defining 115, 131–132; discourse of 135, 142; ecological 64; future and 6, 26–27; happiness and 93–95; inner dimensions of 208, 220; liberalism and 64; literature and 141–143, 223, 251; procedural 134–135; productive 64, 66, 73n5; religion and 56–57, 115–116; research on 129–135; social happiness and 88; social imaginaries and 217, 220, 228; socialism and 64; subjective dimensions of 254; well-being and 96n10
sustainability concept epochs 110
"Sustainability in an Imaginary World" 135
sustainability paradigms: Anthropocene and 109; conceptions of 109–111, **111**, 112–120, *120*, 121; contingency management 110, **111**, 115–119, 121; environmental management 110, **111**, 112–114, 121; implementation of 121; resource management 110–111, **111**, 112, 121; social management 110, **111**, 114–115, 121
sustainability religion 116
sustainable action model 239, *240*, 244n3
sustainable becoming 142–143
sustainable consumption: activation of status and 80; analytically distinct perspective on 78, *78*, 79, *79*, 80–81, 84; cultural cues and 83; culture and 75–83; impact of 75; interaction ritual chains 81–83; marketing for 79–80; mutually generative perspective on 78, *78*, 79, *79*, 81–82, 84; nudging for 81; practice theory and 81–83; social networks and 83
sustainable development: concept of 13; cultural diversity and 16; cultural services and 88–90; culture and 17–22, 48–49, 98–99, 128–129, 252; defining 16, 47; ecology and 87; economics of 87; ecosystem assessment and 87–88; ethics of 47; four-pillar model of 101, 128, 183, 249–250, 252; goals of 132; holistic approach to 99; moral discourse of 47–49; social issues and 87–90; three-pillar model of 3, 12; *see also* planetary sustainability
Sustainable Development Goals (SDGs): adoption of 12; culture in 4, 14, 20–21; space environment in 101
Sustainable Development Summit 20, 22
sustainable education 21, 129, 253–254
sustainable happiness 94
sustainable tourism 21
sustainable urban development 135
Suvin, D. 197
Swiss position 18–20
Sylvicultura Oeconomica (von Carlowitz) 72n1, 153
symbolism 50–52

Talaat Pascha 37–38
Taylor, C. 213
techno-capitalist solutions 65
Tehlirjan, Soghomon 37–38
Thames River *see* River Thames
Theory of the Leisure Class (Veblen) 78
thick description 50
Thin, N. 92–93, 96n8, 96n9
Thomas, L. 180n33
Thoreau, Henry D. 144, 236–237
traditional farming knowledge 21
traditional morality 71–72
traditional practice 71–72
transformative literacy 254
triangle of sustainability 3

Trump, Donald J. 53, 193
Tybur, J. M. 80

Uexküell, Jakob von 140
UN Development Assistance Frameworks (UNDAF) 23
UNESCO *see* United Nations Educational, Scientific and Cultural Organization (UNESCO)
UNESCO Summit on Culture and Sustainable Development 127
UN General Assembly Resolution of 2013 19
United Cities and Local Governments (UCLG) 128
United Nations Declaration on the Rights of Indigenous Peoples 41–42
United Nations Development Programme (UNDP) 13
United Nations Educational, Scientific and Cultural Organization (UNESCO): cultural sustainability and 3, 27–28, 127; culture and 15–16; planetary sustainability and 98–99
United Nations General Assembly: human rights and 14; resolutions of 17–18, 22
United Nations Sustainable Development Summit 12
"Universal Declaration of Human Rights" 14, 16, 39, 41
Universal Declaration on Cultural Diversity (UNESCO) 16
UN-Rio Conference 110
unsustainability: Anthropocene and 119, *120*, 221; in speculative fiction 195, 198
Updike, John 149
utopian fiction 220–221
utopian sense 66–67

Vaisey, S. 83
value modules 71
vandalism 30, 32–33
Van den Bergh, B. 80
Van Vugt, M. 79
Veblen, T. 78
Victorian Britain: self-sustainability in 185–186; waste management in 183–186, 190
Vienna Conference on Human Rights 42
Vitens 239
von Winterfeld, U. 56
VPRO 240

Wagner, E. 160
Walden (Thoreau) 237
Walden, a Game 236–237, *237*
wars: barbarism and 34n5; humanitarian principles and 30–31; intentional destruction of cultural heritage in 31
waste management: cultural sustainability and 183–186; cyclical exchange in 185–186; dry conservancy sewage removal 183–187, 192n4; literary symbolism of 187–190; nightsoil 184–185; Victorian Britain 183–186, 190; water-borne sewage removal 183–187, 189, 192n4
water-borne sewage removal 183–187, 189, 192n4
water imagery: aesthetics of 190; female transgression and 190; in literature 183, 186–190; River Thames and 187–190; spiritual rebirth and 192n7
WCED *see* World Commission on Environment and Development (WCED)
Wee, Kim Wee 43
Wehrspaun, M. 48
Weigel, S. 53
Weik von Mossner, A. 196–197, 223
well-being: ecosystem assessment and 87, 95n3; happiness and 88, 93; shared values and 90–91; subjective 96n10; sustainability and 96n10
'Wellbeing Manifesto' 93
Wellner, G. P. 250
Western civilization: ancient Greece and 29; assaults on cultural heritage sites 28; cult of antiquities in 31–32; high self-esteem in 96n7; human rights and 43; modern morality in 71; philosophy and 48; political traditions in 64; symbolic warfare against 31
Westling, L. 141
Wheeler, W. 141
When the Wind Blows (Briggs) 210
Whitman, Walt 148–149
Wijk & Water Battle 239–240
wilderness 144, 148
wildness: American literature and 144–147; American poetry and 148–149; as ecosemiotic communication 144; literature and 144
Wilke, G. 56
Willemen, L. 89
Williams, R. 183

Williams, William Carlos 148
Wirth, J. 166
Wojciehowski, H. 196
Wood, Robert 28–29
Woolcock, M. 195
World Climate Conference (2015) 232, *233*
World Commission on Culture and Development (WCCD) 15
World Commission on Environment and Development (WCED) 3–4, 114
World Conference on Cultural Policies 15
World Decade for Cultural Development 15
World Health Organization (WHO) 87–88
world heritage 27
World Without Oil 239

Year of the Flood, The (Atwood): eco-ekphrases in 221, 225–227; literary description in 223–228, 251; non-human nature and 223; sufficiency and 56
Yeast (Kingsley) 185

Zapf, H. 155, 175, 195, 220, 230n2, 251
Zapf, L. **111**, 250
Žižek, S. 91
Zunshine, L. 196